PRAISE FOR

THE BIG PICTURE

"With profound intelligence and lucid, unpretentious language, Sean Carroll beautifully articulates the worldview suggested by contemporary naturalism. Thorny issues like free will, the direction of time, and the source of morality are clarified with elegance and insight. *The Big Picture* shows how the scientific worldview enriches our understanding of the universe and ourselves. A reliable account of our knowledge of the universe, it is also a serene meditation on our need for meaning. This is a book that should be read by everybody."
—CARLO ROVELLI,
AUTHOR OF *SEVEN BRIEF LESSONS ON PHYSICS*

"Language philosophy, quantum mechanics, general relativity—they're all in *The Big Picture*. Sean Carroll is a fantastically erudite and entertaining writer."
—ELIZABETH KOLBERT,
AUTHOR OF *THE SIXTH EXTINCTION*

"From the big bang to the meaning of human existence . . . a magisterial, yet deeply fascinating, grand tour through the issues that really matter. As gripping as it is important, *The Big Picture* can change the way you think about the world."
—NEIL SHUBIN, AUTHOR OF *YOUR INNER FISH*

NEW YORK TIMES BESTSELLER

Praise for *The Big Picture*

"Weaving the threads of astronomy, physics, chemistry, biology, and philosophy into a seamless narrative tapestry, Sean Carroll enthralls us with what we've figured out in the universe and humbles us with what we don't yet understand. Yet in the end, it's the meaning of it all that feeds your soul of curiosity."
—Neil deGrasse Tyson, host of *Cosmos: A Spacetime Odyssey*

"Sean Carroll's lucid *The Big Picture* reveals how the universe works and our place in it. Carroll, a philosophically sophisticated physicist, discusses consciousness without gimmicks, and deftly shows how current physics is so solid that it rules out ESP forever."   —Steven Pinker, author of *The Better Angels of Our Nature*

"Sean Carroll is a leading theoretical cosmologist with the added ability to write about his subject with unusual clarity, flare, and wit."
—Alan Lightman, author of *The Accidental Universe* and *Einstein's Dreams*

"In this timely exploration of the universe and its mysteries—both physical and metaphysical—Sean Carroll illuminates the world around us with clarity, beauty and, ultimately, with much-needed wisdom."
—Deborah Blum, director of the Knight Science Journalism Program at MIT and author of *The Poisoner's Handbook*

"Instead of feeling humbled and insignificant when gazing upward on a clear starry night, Carroll takes us by the hand and shows us how fantastic the inanimate physical universe is and how special each animate human can be. It is lucid, spirited, and penetrating."
—Michael S. Gazzaniga, author of *Who's in Charge?* and *Tales from Both Sides of the Brain*

"A nuanced inquiry into 'how our desire to matter fits in with the nature of reality at its deepest levels,' in which Carroll offers an assuring dose of what he calls 'existential therapy' reconciling the various and often seemingly contradictory dimensions of our experience."   —Maria Popova, *Brain Pickings*

"[*The Big Picture* is] a tour de force that offers a comprehensive snapshot of the human situation in our infinitely strange universe, and it does this with highly accessible language and engaging storytelling."   —*Salon*

"Carroll is the perfect guide on this wondrous journey of discovery. A brilliantly lucid exposition of profound philosophical and scientific issues in a language accessible to lay readers." —*Kirkus Reviews* (starred review)

"Guides us through several centuries' worth of scientific discoveries to show how they have shaped our understanding and indeed how the laws of nature are linked to the most fundamental human questions of life, death, and our place in the cosmos." —*Library Journal* (prepub alert)

"Sean Carroll's holistic vision accommodates the sciences and the humanities and has a high probability of provoking readers into clarifying their own views about the complex relations among science, religion, and morality." —*The Times Literary Supplement*

"Carroll presents a means through which people can better understand themselves, their universe, and their conceptions of a meaningful life." —*Publishers Weekly*

"[Carroll] sets out to show how various phenomena, including thought, choice, consciousness, and value, hang together with the scientific account of reality that has been developed in physics in the past one hundred years. He attempts to do all this without relying on specialized jargon from philosophy and physics and succeeds spectacularly in achieving both aims." —*Science*

"True to the grand scope of its title. . . . Anyone who enjoys asking big questions will find a lot to consider." —*Booklist*

"*The Big Picture* impresses. Carroll is a lively and sympathetic author who writes as well about biology and philosophy as he does about his own field of physics." —*Financial Times*

"Until now you might have gotten away believing modern physics is about things either too small or too far away to care much about. But no more. Sean Carroll's new book reveals how physicists' quest to better understand the fundamental laws of nature has led to astonishing insights into life, the universe, and everything. Above all, a courageous book, and an overdue one." —Sabine Hossenfelder, Frankfurt Institute for Advanced Studies

THE
BIG
PICTURE

ON THE ORIGINS OF LIFE,
MEANING, AND
THE UNIVERSE ITSELF

SEAN CARROLL

DUTTON

DUTTON
An imprint of Penguin Random House LLC
penguinrandomhouse.com

Previously published as a Dutton hardcover, May 2016

First paperback printing, May 2017

LIBRARY OF CONGRESS CATALOGING-IN-PUBLICATION DATA
Names: Carroll, Sean M., 1966– author.
Title: The big picture : on the origins of life, meaning, and the universe itself / Sean Carroll.
Description: New York, New York : Dutton, an imprint of Penguin
Random House LLC [2016] | 2016 | Includes bibliographical references and index.
Identifiers: LCCN 2015050590 | ISBN 9780525954828 (hc) |
| ISBN 9780698409767 (eBook) | ISBN 9781101984253 (pbk)
Subjects: LCSH: Life—Origin. | Meaning (Philosophy) | Cosmology. |
Naturalism. | Discoveries in science. | Evolution—Philosophy. | Physical laws.
Classification: LCC QH325.C36 2016 | DDC 577—dc23
LC record available at http://lccn.loc.gov/2015050590

Printed in the United States of America
9   10   8

Set in Garamond Premier
Designed by Nancy Resnick

To my teachers:
Mrs. Eberhardt, Edwin Kelly, Edward Guinan, Jack Doody,
Colleen Sheehan, Peter Knapp, George Field, Sidney Coleman,
Nick Warner, Eddie Farhi, Alan Guth, and so many others.
Thank you for challenging me.

Contents

Prologue   1

PART ONE: COSMOS

1. The Fundamental Nature of Reality   9
2. Poetic Naturalism   15
3. The World Moves by Itself   23
4. What Determines What Will Happen?   30
5. Reasons Why   38
6. Our Universe   47
7. Time's Arrow   54
8. Memories and Causes   60

PART TWO: UNDERSTANDING

9. Learning about the World   69
10. Updating Our Knowledge   75
11. Is It Okay to Doubt Everything?   84
12. Reality Emerges   93
13. What Exists, and What Is Illusion?   105
14. Planets of Belief   115
15. Accepting Uncertainty   123
16. What Can We Know about the World without Actually Looking at It?   130

17. Who Am I?   139

18. Abducting God   144

PART THREE: ESSENCE

19. How Much We Know   153

20. The Quantum Realm   159

21. Interpreting Quantum Mechanics   166

22. The Core Theory   172

23. The Stuff of Which We Are Made   178

24. The Effective Theory of the Everyday World   186

25. Why Does the Universe Exist?   195

26. Body and Soul   205

27. Death Is the End   215

PART FOUR: COMPLEXITY

28. The Universe in a Cup of Coffee   225

29. Light and Life   237

30. Funneling Energy   244

31. Spontaneous Organization   250

32. The Origin and Purpose of Life   260

33. Evolution's Bootstraps   273

34. Searching through the Landscape   279

35. Emergent Purpose   291

36. Are We the Point?   302

PART FIVE: THINKING

37. Crawling into Consciousness   317

38. The Babbling Brain   327

39. What Thinks?   336

40. The Hard Problem   348

41. Zombies and Stories   355

42. Are Photons Conscious?   363

43. What Acts on What?   372

44. Freedom to Choose   378

PART SIX: CARING

45. Three Billion Heartbeats   387

46. What Is and What Ought to Be   394

47. Rules and Consequences   403

48. Constructing Goodness   412

49. Listening to the World   419

50. Existential Therapy   428

Appendix: The Equation Underlying You and Me   435

References   443

Further Reading   451

Acknowledgments   455

Index   456

Prologue

Only once in my life have I been truly close to dying.

My judgment was a bit off. It was dark, the traffic was heavy. An inattentive driver on the 405 freeway in Los Angeles veered in front of me to avoid an exit ramp, and I swerved to avoid him. The enormous eighteen-wheeler in the lane to my left wasn't as far back as I thought. The very last inch of my back bumper caught the very front corner of the truck's cab. That was enough. I lost all control of my car, which executed a slow and stately counterclockwise turn, ending with my driver's side flush into the front of the truck, still speeding down the freeway. It was slow and stately from my perspective, anyway. I felt as if I were trapped in amber, watching helplessly as my car moved of its own volition, until it nestled against the truck's grill, perpendicular to the direction of traffic, a blinding headlight shining in my face.

I was shaken but unhurt. The car was a bit rumpled, and needed some serious work in the body shop, but it was able to drive me home once all the police reports had been filled out. A few inches here, a change of speed there, a bit more panic on the part of the truck driver—things could have been different.

Many of us come close to dying, long before we do die. We confront the finitude of our lives.

In my professional capacity as a physicist I study the universe as a whole. It's a big universe. Fourteen billion years after the Big Bang, the region of space we can directly see is populated by a few hundred billion galaxies, averaging a hundred billion stars each. We human beings, by contrast, are quite tiny—a recent arrival on an insignificant planet orbiting a nondescript star. Whatever the outcome of my freeway misadventure had been, my lifetime would be measured in decades, not in billions of years.

A person is a diminutive, ephemeral thing, standing smaller in comparison with the universe than a single atom stands in comparison with the Earth. Can any one individual existence really *matter*?

In some sense it obviously can. I live a fortunate life, with family and friends who care about me, and who would be extremely upset were I to die. I myself would be quite unhappy if I somehow knew ahead of time that my life was going to end. But from the perspective of a vast, seemingly indifferent cosmos, does it really matter all that much?

I like to think that our lives do matter, even if the universe would trundle along without us. But we have to respect the question, and work hard to understand how our desire to matter fits in with the nature of reality at its deepest levels.

A friend of mine, a neuroscientist and biologist, can make individual cells young again. Scientists have developed techniques for taking stem cells in the adult human body, which have aged and taken on some more mature characteristics, and reverse-aging them until they are just like newborn stem cells.

There is a long road from cells to complete organisms. So I asked her, half-jokingly, whether we would someday be able to reverse-age human beings, and potentially keep them young forever.

"You and I are going to die someday," she mused. "But if either of us has grandchildren, I wouldn't be so sure."

That's thinking like a biologist. As a physicist, I know it doesn't violate any laws of nature to imagine living beings lasting for millions or even billions of years, so I have no objection there. But eventually all of the stars will have exhausted their nuclear fuel, their cold remnants will fall into black holes, and those black holes will gradually evaporate into a thin gruel of elementary particles in a dark and empty universe. We won't *really* live forever, no matter how clever biologists get to be.

Everybody dies. Life is not a substance, like water or rock; it's a process, like fire or a wave crashing on the shore. It's a process that begins, lasts for a while, and ultimately ends. Long or short, our moments are brief against the expanse of eternity.

We have two goals in front of us. One is to explain the story of our universe and why we think it's true, the big picture as we currently understand it. It's a fantastic conception. We humans are blobs of organized mud, which through the impersonal workings of nature's patterns have developed the capacity to contemplate and cherish and engage with the intimidating complexity of the world around us. To understand ourselves, we have to understand the stuff out of which we are made, which means we have to dig deeply into the realm of particles and forces and quantum phenomena, not to mention the spectacular variety of ways that those microscopic pieces can come together to form organized systems capable of feeling and thought.

The other goal is to offer a bit of existential therapy. I want to argue that, though we are part of a universe that runs according to impersonal underlying laws, we nevertheless *matter*. This isn't a scientific question—there isn't data we can collect by doing experiments that could possibly measure the extent to which a life matters. It's at heart a philosophical problem, one that demands that we discard the way that we've been thinking about our lives and their meaning for thousands of years. By the old way of thinking, human life couldn't possibly be meaningful if we are "just" collections of atoms moving around in accordance with the laws of physics. That's exactly what we are, but it's not the *only* way of thinking about what we are. We are collections of atoms, operating independently of any immaterial spirits or influences, *and* we are thinking and feeling people who bring meaning into existence by the way we live our lives.

We are small; the universe is big. It doesn't come with an instruction manual. We have nevertheless figured out an amazing amount about how things actually work. It's a different kind of challenge to accept the world for what it is, to face reality with a smile, and to make our lives into something valuable.

In the first section of the book, "Cosmos," we examine some important aspects of the wider universe of which we are a small part. There are many ways to talk about the world, which leads us to the framework called *poetic naturalism*. "Naturalism" claims that there is just one world, the natural world; we'll explore some of the indications that point us in that direction, including how the universe moves and evolves. "Poetic" reminds us that there is more than one way of talking about the world. We find it natural to use a vocabulary of "causes" and "reasons why" things happen, but those ideas aren't part of how nature works at its deepest levels. They are emergent phenomena, part of how we describe our everyday world. The difference between the everyday and deeper descriptions arises from the arrow of time, the distinction between past and future that can ultimately be traced to the special state in which our universe began near the Big Bang.

In the second section, "Understanding," we consider how we should go about trying to understand the world. Or, at least, move closer and closer to the truth; we have to be willing to accept uncertainty and incomplete knowledge, and always be ready to update our beliefs as new evidence comes in. We will see how our best approach to describing the universe is not a single, unified story but an interconnected series of models appropriate at different levels. Each model has a domain in which it is applicable, and the ideas that appear as essential parts of each story have every right to be thought of as "real." Our task is to assemble an interlocking set of descriptions, based on some fundamental ideas, that fit together to form a stable planet of belief.

We then turn to "Essence," where we think about the world as it actually is: the fundamental laws of nature. We'll discuss quantum field theory, the basic language in which modern physics is written. We will appreciate the triumph of the Core Theory, the enormously successful model of the particles and forces that make up you, me, the sun, the moon, the stars, and everything you have ever seen, touched, or tasted in all your life. There is much we don't know about how the world works, but we have extremely good reason to think that the Core Theory is the correct description of nature in its domain of applicability. That domain is wide enough to immediately exclude a number of provocative phenomena: from telekinesis and astrology to survival of the soul after death.

With some laws of physics in hand, there is still much work to be done in connecting these deeper principles to the richness of the world around us. In the fourth section, "Complexity," we begin to see how those connections come about. The emergence of complex structures isn't a strange phenomenon in tension with the general tendency of the universe toward greater disorder; it is a natural consequence of that tendency. In the right circumstances, matter self-organizes into intricate configurations, capable of capturing and using information from their environments. The culmination of this process is life itself. The more we learn about the basic workings of life, the more we appreciate how they are in harmony with the fundamental physical principles governing the universe as a whole. Life is a process, not a substance, and it is necessarily temporary. We are not the reason for the existence of the universe, but our ability for self-awareness and reflection makes us special within it.

This brings us to one of the knottier problems faced by naturalism, the puzzle of consciousness. We confront this issue in "Thinking," where we go beyond "naturalism" all the way to "physicalism." Modern neuroscience has made tremendous strides in understanding how thought actually works inside our brains, and there is little question that our personal experiences have definite correlates in physical processes therein. We can even begin to see how this remarkable ability evolved over time, and what kinds of abilities are crucial to achieving consciousness. The most difficult problem is a philosophical one: how is it even possible that inner experience, the uniquely experiential *aboutness* of our lives inside our heads, can be reduced to mere matter in motion? Poetic naturalism suggests that we should think of "inner experiences" as part of a way of talking about what is happening in our brains. But ways of talking can be very real, even when it comes to our ability to make free choices as rational beings.

Finally, in "Caring" we confront the hardest problem of all, that of how to construct meaning and values in a cosmos without transcendent purpose. A common charge against naturalism is that such a task is simply impossible: without something beyond the physical world to guide us, there is no reason to live at all, and certainly no reason to live one way rather than another. Some naturalists respond by agreeing, and getting on with their lives; others react strongly the other way, by arguing that values can be determined scientifically just as much as the age of the universe can be. Poetic naturalism strikes a middle ground, accepting that values are human constructs, but denying that they are therefore illusory or meaningless. All of us have cares and desires, whether given to us by evolution, our upbringing, or our environment. The task before us is to reconcile those cares and desires within ourselves, and amongst one another. The meaning we find in life is not transcendent, but it's no less meaningful for that.

PART ONE

COSMOS

1

The Fundamental Nature of Reality

In the old Road Runner cartoons, Wile E. Coyote would frequently find himself running off the edge of a cliff. But he wouldn't, as our experience with gravity might lead us to expect, start falling to the ground below, at least not right away. Instead, he would hover motionless, in puzzlement; it was only when he realized there was no longer any ground beneath him that he would suddenly crash downward.

We are all Wile E. Coyote. Since human beings began thinking about things, we have contemplated our place in the universe, the reason why we are all here. Many possible answers have been put forward, and partisans of one view or another have occasionally disagreed with each other. But for a long time, there has been a shared view that there is some meaning, out there somewhere, waiting to be discovered and acknowledged. There is a point to all this; things happen for a reason. This conviction has served as the ground beneath our feet, as the foundation on which we've constructed all the principles by which we live our lives.

Gradually, our confidence in this view has begun to erode. As we understand the world better, the idea that it has a transcendent purpose seems increasingly untenable. The old picture has been replaced by a wondrous new one—one that is breathtaking and exhilarating in many ways, challenging and vexing in others. It is a view in which the world stubbornly refuses to give us any direct answers about the bigger questions of purpose and meaning.

The problem is that we haven't quite admitted to ourselves that this transition has taken place, nor fully accepted its far-reaching implications. The issues are well-known. Over the course of the last two centuries, Darwin has upended our view of life, Nietzsche's madman bemoaned the death of God, existentialists have searched for authenticity in the face of absurdity, and modern atheists have been granted a seat at society's table. And yet, many continue on as if nothing has changed; others revel in the new order, but placidly believe that adjusting our perspective is just a matter of replacing a few old homilies with a few new ones.

The truth is that the ground has disappeared beneath us, and we are just beginning to work up the courage to look down. Fortunately, not everything in the air immediately plummets to its death. Wile E. Coyote would have been fine if he had been equipped with one of those ACME-brand jet packs, so that he could fly around under his own volition. It's time to get to work building our conceptual jet packs.

What is the fundamental nature of reality? Philosophers call this the question of *ontology*—the study of the basic structure of the world, the ingredients and relationships of which the universe is ultimately composed. It can be contrasted with *epistemology*, which is how we obtain knowledge about the world. Ontology is the branch of philosophy concerned with the nature of reality; we also talk about "an" ontology, referring to a specific idea about what that nature actually is.

The number of approaches to ontology alive in the world today is somewhat overwhelming. There is the basic question of whether reality exists at all. A *realist* says, "Of course it does"; but there are also *idealists*, who think that capital-*M* Mind is all that truly exists, and the so-called real world is just a series of thoughts inside that Mind. Among realists, we have *monists*, who think that the world is a single thing, and *dualists*, who believe in two distinct realms (such as "matter" and "spirit"). Even people who agree that there is only one type of thing might disagree about whether there are fundamentally different kinds of properties (such as mental properties and physical properties) that those things can have. And even people who agree that there is only one kind of thing, and that the world is purely physical, might diverge when it comes to asking which aspects of that world are "real" versus "illusory." (Are colors real? Is consciousness? Is morality?)

Whether or not you believe in God—whether you are a *theist* or an *atheist*—is part of your ontology, but far from the whole story. "Religion" is a completely different kind of thing. It is associated with certain beliefs, often including belief in God, although the definition of "God" can differ substantially within religion's broad scope. Religion can also be a cultural force, a set of institutions, a way of life, a historical legacy, a collection of practices and principles. It's much more, and much messier, than a checklist of doctrines. A counterpart to religion would be *humanism*, a collection of beliefs and practices that is as varied and malleable as religion is.

The broader ontology typically associated with atheism is *naturalism*—there is only one world, the natural world, exhibiting patterns we call the "laws of nature," and which is discoverable by the methods of science and empirical investigation. There is no separate realm of the supernatural, spiritual, or divine; nor is there any cosmic teleology or transcendent purpose inherent in the nature of the universe or in human life. "Life" and "consciousness" do not denote essences distinct from matter; they are ways of talking about phenomena that emerge from the interplay of extraordinarily complex systems. Purpose and meaning in life arise through fundamentally human acts of creation, rather than being derived from anything outside ourselves. Naturalism is a philosophy of unity and patterns, describing all of reality as a seamless web.

Naturalism has a long and distinguished pedigree. We find traces of it in Buddhism, in the atomists of ancient Greece and Rome, and in Confucianism. Hundreds of years after the death of Confucius, a Chinese thinker named Wang Chong was a vocal naturalist, campaigning against the belief in ghosts and spirits that had become popular in his day. But it is really only in the last few centuries that the evidence in favor of naturalism has become hard to resist.

All of these isms can feel a bit overwhelming. Fortunately we don't need to be rigorous or comprehensive about listing the possibilities. But we do need to think hard about ontology. It's at the heart of our Wile E. Coyote problem.

The last five hundred or so years of human intellectual progress have completely upended how we think about the world at a fundamental level. Our everyday experience suggests that there are large numbers of truly different *kinds of stuff* out there. People, spiders, rocks, oceans, tables, fire, air, stars—these all seem dramatically different from one another, deserving of independent entries in our list of basic ingredients of reality. Our "folk ontology" is pluralistic, full of myriad distinct categories. And that's not even counting notions that seem more abstract but are arguably equally "real," from numbers to our goals and dreams to our principles of right and wrong.

As our knowledge grows, we have moved by fits and starts in the direction of a simpler, more unified ontology. It's an ancient impulse. In the sixth century BCE, the Greek philosopher Thales of Miletus suggested that *water* is a primary principle from which all else is derived, while across the world, Hindu philosophers put forward *Brahman* as the single ultimate reality. The development of science has accelerated and codified the trend.

Galileo observed that Jupiter has moons, implying that it is a gravitating body just like the Earth. Isaac Newton showed that the force of gravity is universal, underlying both the motion of the planets and the way that apples fall from trees. John Dalton demonstrated how different chemical compounds could be thought of as combinations of basic building blocks called atoms. Charles Darwin established the unity of life from common ancestors. James Clerk Maxwell and other physicists brought together such disparate phenomena as lightning, radiation, and magnets under the single rubric of "electromagnetism." Close analysis of starlight revealed that stars are made of the same kinds of atoms as we find here on Earth, with Cecilia Payne-Gaposchkin eventually proving that they are mostly hydrogen and helium. Albert Einstein unified space and time, joining together matter and energy along the way. Particle physics has taught us that every atom in the periodic table of the elements is an arrangement of just three basic particles: protons, neutrons, and electrons. Every object you have ever seen or bumped into in your life is made of just those three particles.

We're left with a very different view of reality from where we started. At a fundamental level, there aren't separate "living things" and "nonliving things," "things here on Earth" and "things up in the sky," "matter" and "spirit." There is just the basic stuff of reality, appearing to us in many different forms.

How far will this process of unification and simplification go? It's impossible to say for sure. But we have a reasonable guess, based on our progress thus far: it will go all the way. We will ultimately understand the world as a single, unified reality, not caused or sustained or influenced by anything outside itself. That's a big deal.

Naturalism presents a hugely grandiose claim, and we have every right to be skeptical. When we look into the eyes of another person, it doesn't seem like what we're seeing is simply a collection of atoms, some sort of immensely complicated chemical reaction. We often feel connected to the universe in some way that transcends the merely physical, whether it's a sense of awe when we contemplate the sea or sky, a trancelike reverie during meditation or prayer, or the feeling of love when we're close to someone we care about. The difference between a living being and an inanimate object seems much more profound than the way certain molecules are arranged. Just looking around, the idea that everything we see and feel can somehow be explained by impersonal laws governing the motion of matter and energy seems preposterous.

It's a bit of a leap, in the face of all of our commonsense experience, to think that life can simply start up out of non-life, or that our experience of consciousness needs no more ingredients than atoms obeying the laws of physics. Of equal importance, appeals to transcendent purpose or a higher power seem to provide answers to questions to some of the pressing "Why?" questions we humans like to ask: Why this universe? Why am I here? Why anything at all? Naturalism, by contrast, simply says: those aren't the right questions to ask. It's a lot to swallow, and not a view that anyone should accept unquestioningly.

Naturalism isn't an obvious, default way to think about the world. The case in its favor has built up gradually over the years, a consequence of our relentless quest to improve our understanding of how things work at a deep level, but there is still work to be done. We don't know how the universe began, or if it's the only universe. We don't know the ultimate, complete laws of physics. We don't know how life began, or how consciousness arose. And we certainly haven't agreed on the best way to live in the world as good human beings.

The naturalist needs to make the case that, even without actually having these answers yet, their worldview is still by far the most likely framework in which we will eventually find them. That's what we're here to do.

The pressing, human questions we have about our lives depend directly on our attitudes toward the universe at a deeper level. For many people, those attitudes are adopted rather informally from the surrounding culture, rather than arising out of rigorous personal reflection. Each new generation of people doesn't invent the rules of living from scratch; we inherit ideas and values that have evolved over vast stretches of time. At the moment, the dominant image of the world remains one in which human life is cosmically special and significant, something more than mere matter in motion. We need to do better at reconciling how we talk about life's meaning with what we know about the scientific image of our universe.

Among people who acknowledge the scientific basis of reality, there is often a conviction—usually left implicit—that all of that philosophical stuff like freedom, morality, and purpose should ultimately be pretty easy to figure out. We're collections of atoms, and we should be nice to one another. How hard can it really be?

It can be really hard. Being nice to one another is a good start, but it doesn't get us very far. What happens when different people have incompatible conceptions of niceness? Giving peace a chance sounds like a swell idea, but in the real world, there are different actors with different interests, and conflicts will inevitably arise. The absence of a supernatural guiding force doesn't mean we can't meaningfully talk about right and wrong, but it doesn't mean we instantly know one from the other, either.

Meaning in life can't be reduced to simplistic mottos. In some number of years I will be dead; some memory of my time here on Earth may linger, but I won't be around to savor it. With that in mind, what kind of life is worth living? How should we balance family and career, fortune and pleasure, action and contemplation? The universe is large, and I am a tiny part of it, constructed of the same particles and forces as everything else: by itself, that tells us precisely nothing about how to answer such questions. We're going to have to be both smart and courageous as we work to get this right.

2

Poetic Naturalism

One thing *Star Trek* never really got clear on was how transporter machines are supposed to work. Do they disassemble you one atom at a time, zip those atoms elsewhere, and then reassemble them? Or do they send only a blueprint of you, the information contained in your arrangement of atoms, and then reconstruct you from existing matter in the environment to which you are traveling? Most often the ship's crew talks as if your actual atoms travel through space, but then how do we explain "The Enemy Within"? That's the episode, you'll remember, in which a transporter malfunction causes two copies of Captain Kirk to be beamed aboard the *Enterprise*. It's hard to see how two copies of a person could be made out of one person-sized collection of atoms.

Fortunately for viewers of the show, the two copies of Kirk weren't precisely identical. One copy was the normal (good) Kirk, and the other was evil. Even better, the evil one quickly got scratched on the face by Yeoman Rand, so it wasn't hard to tell the two apart.

But what if they had been identical? We would then be faced with a puzzle about the nature of personal identity, popularized by philosopher Derek Parfit. Imagine a transporter machine that could disassemble a single individual and reconstruct multiple exact copies of them out of different atoms. Which one, if any, would be the "real" one? If there were just a single copy, most of us would have no trouble accepting them as the original person. (Using different atoms doesn't really matter; in actual human bodies, our atoms are lost and replaced all the time.) Or what if one copy were made of new atoms, while the original you remained intact—but the original suffered a tragic death a few seconds after the duplicate was made. Would the duplicate count as the same person?

All good philosophical fun and games of course, but without much relevance to the real world, at least not at our current level of technology. Or maybe not. There's an older thought experiment called the Ship of Theseus that raises some of the same issues. Theseus, the legendary founder of Athens, had an impressive ship in which he had fought numerous battles. To honor him, the citizens of Athens preserved his ship in their port. Occasionally a plank or part of the mast would decay beyond repair, and at some point that piece would have to be replaced to keep the ship in good order. Once again we have a question of identity: is it the same ship after we've replaced one of the planks? If you think it is, what about after we've replaced *all* of the planks, one by one? And (as Thomas Hobbes went on to ask), what if we then took all the old planks and built a ship out of them? Would that one then suddenly become the Ship of Theseus?

Narrowly speaking, these are all questions about identity. When is one thing "the same thing" as some other thing? But more broadly, they're questions about ontology, our basic view of what exists in the world. What kinds of things are there at all?

When we ask about the identity of the "real" Captain Kirk or Ship of Theseus, a whole bundle of unstated assumptions come along for the ride. We are assuming that there are things called "persons," and things called "ships," and that these things have some persistence over time. And everything goes swimmingly, until we come up against a puzzle, such as these duplication scenarios, that puts a strain on how we define these kinds of objects.

All this matters, not because we're on the verge of building a working transporter, but because our attempts to make sense of the big picture inevitably involve different kinds of overlapping ways of talking about the world. We have atoms, and we have biological cells, and we have human beings. Is the notion of "this particular human being" an important one to how we think about the world? Should categories like "persons" and "ships" be part of our fundamental ontology at all? We can't decide whether an individual human life actually matters if we don't know what we mean by "human being."

As knowledge generally, and science in particular, have progressed over the centuries, our corresponding ontologies have evolved from quite rich to relatively sparse. To the ancients, it was reasonable to believe that there were all kinds of fundamentally different things in the world; in modern thought, we try to do more with less.

We would now say that Theseus's ship is made of atoms, all of which are made of protons, neutrons, and electrons—exactly the same kinds of particles that make up every other ship, or for that matter make up you and me. There isn't some primordial "shipness" of which Theseus's is one particular example; there are simply arrangements of atoms, gradually changing over time.

That doesn't mean we can't talk about ships just because we understand that they are collections of atoms. It would be horrendously inconvenient if, anytime someone asked us a question about something happening in the world, we limited our allowable responses to a listing of a huge set of atoms and how they were arranged. If you listed about one atom per second, it would take more than a trillion times the current age of the universe to describe a ship like Theseus's. Not really practical.

It just means that the notion of a ship is a derived category in our ontology, not a fundamental one. It is a useful *way of talking* about certain subsets of the basic stuff of the universe. We invent the concept of a ship because it is useful to us, not because it's already there at the deepest level of reality. Is it the same ship after we've gradually replaced every plank? I don't know. It's up to us to decide. The very notion of "ship" is something we created for our own convenience.

That's okay. The deepest level of reality is very important; but all the different ways we have of talking about that level are important too.

What we're seeing is the difference between a rich ontology and a sparse one. A rich ontology comes with a large number of different fundamental categories, where by "fundamental" we mean "playing an essential role in our deepest, most comprehensive picture of reality."

In a sparse ontology, there are a small number of fundamental categories (maybe only one) describing the world. But there will be very many ways of talking about the world. The notion of a "way of talking" isn't mere decoration—it's an absolutely crucial part of how we apprehend reality.

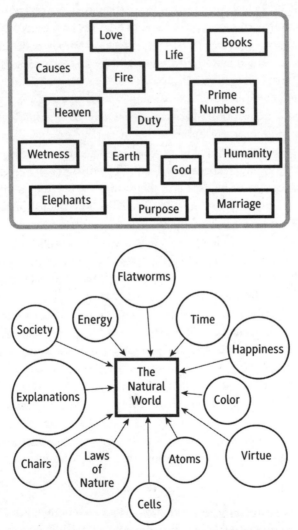

Two different kinds of ontologies, rich and sparse. Boxes are fundamental concepts, while circles are derived or emergent concepts—ways of talking about the world.

One benefit of a rich ontology is that it's easy to say what is "real"—every category describes something real. In a sparse ontology, that's not so clear. Should we count only the underlying stuff of the world as real, and all the different ways we have of dividing it up and talking about it as merely illusions? That's the most hard-core attitude we could take to reality, sometimes called *eliminativism*, since its adherents like nothing better than to go around eliminating this or that concept from our list of what is real. For an eliminativist, the question "Which Captain Kirk is the real one?" gets answered by "Who cares? People are illusions. They're just fictitious stories we tell about the one true real world."

I'm going to argue for a different view: our fundamental ontology, the best way we have of talking about the world at the deepest level, is extremely sparse. But many concepts that are part of non-fundamental ways we have of talking about the world—useful ideas describing higher-level, macroscopic reality—deserve to be called "real."

The key word there is "useful." There are certainly non-useful ways of talking about the world. In scientific contexts, we refer to such non-useful ways as "wrong" or "false." A way of talking isn't just a list of concepts; it will generally include a set of rules for using them, and relationships among them. Every scientific theory is a way of talking about the world, according to which we can say things like "There are things called planets, and something called the sun, all of which move through something called space, and planets do something called orbiting the sun, and those orbits describe a particular shape in space called an ellipse." That's basically Johannes Kepler's theory of planetary motion, developed after Copernicus argued for the sun being at the center of the solar system but before Isaac Newton explained it all in terms of the force of gravity. Today, we would say that Kepler's theory is fairly useful in certain circumstances, but it's not as useful as Newton's, which in turn isn't as broadly useful as Einstein's general theory of relativity.

※

The strategy I'm advocating here can be called *poetic naturalism*. The poet Muriel Rukeyser once wrote, "The universe is made of stories, not of atoms." The world is what exists and what happens, but we gain enormous insight by talking about it—telling its story—in different ways.

Naturalism comes down to three things:

1. There is only one world, the natural world.
2. The world evolves according to unbroken patterns, the laws of nature.
3. The only reliable way of learning about the world is by observing it.

Essentially, naturalism is the idea that the world revealed to us by scientific investigation is the one true world. The poetic aspect comes to the fore when we start talking about that world. It can also be summarized in three points:

1. There are many ways of talking about the world.
2. All good ways of talking must be consistent with one another and with the world.
3. Our purposes in the moment determine the best way of talking.

A poetic naturalist will agree that both Captain Kirk and the Ship of Theseus are simply ways of talking about certain collections of atoms stretching through space and time. The difference is that an eliminativist will say "and therefore they are just illusions," while the poetic naturalist says "but they are no less real for all of that."

Philosopher Wilfrid Sellars coined the term *manifest image* to refer to the folk ontology suggested by our everyday experience, and *scientific image* for the new, unified view of the world established by science. The manifest image and the scientific image use different concepts and vocabularies, but ultimately they should fit together as compatible ways of talking about the world. Poetic naturalism accepts the usefulness of each way of talking in its appropriate circumstances, and works to show how they can be reconciled with one another.

Within poetic naturalism we can distinguish among three different kinds of stories we can tell about the world. There is the deepest, most fundamental description we can imagine—the whole universe, exactly described in every microscopic detail. Modern science doesn't know what that description actually is right now, but we presume that there at least *is* such an underlying reality. Then there are "emergent" or "effective" descriptions, valid within some limited domain. That's where we talk about ships and people, macroscopic collections of stuff that we group into individual entities as part of this higher-level vocabulary. Finally, there are values: concepts of right and wrong, purpose and duty, or beauty and ugliness. Unlike higher-level scientific descriptions, these are not determined by the scientific goal of fitting the data. We have other goals: we want to be good people, get along with others, and find meaning in our lives. Figuring out the best way to talk about the world is an important part of working toward those goals.

Poetic naturalism is a philosophy of freedom and responsibility. The raw materials of life are given to us by the natural world, and we must work to understand them and accept the consequences. The move from description to prescription, from saying what happens to passing judgment on what should happen, is a creative one, a fundamentally human act. The world is just the world, unfolding according to the patterns of nature, free of any judgmental attributes. The world exists; beauty and goodness are things that we bring to it.

Poetic naturalism may seem like an appealing idea—or it may seem like an absurd bunch of hooey—but it certainly leaves us with a lot of questions. Most obviously, what is the unified natural world that underlies everything? We've been bandying about words like "atoms" and "particles," but we know from discussions of quantum mechanics that the truth is a bit more slippery than that. And we certainly don't claim to know the ultimate final Theory of Everything—so how much do we actually know? And what makes us think that it's enough to justify the dreams of naturalism?

There are equally many, if not more, questions about connecting that underlying physical world to our everyday reality. There are "Why?" questions: Why this particular universe, with these particular laws of nature? Why does the universe exist at all? There are also "Are you sure?" questions: Are we sure that a unified physical reality could naturally give rise to life as we know it? Are we sure it is sufficient to describe consciousness, perhaps the most perplexing aspect of our manifest world? And then there are the

"How?" questions: How do we decide what ways of talking are the best? How do we agree on judgmental questions about right and wrong? How do we find meaning and purpose in a world that is purely natural? Above all, how do we *know* any of this?

Our task is to put together a rich, nuanced picture that reconciles all the different aspects of our experience. To put ourselves in the right frame of mind, in the next few chapters we'll survey some of the ideas that helped set humanity on the road to naturalism.

3

The World Moves by Itself

I n 1971, viewers watching live TV got to see *Apollo 15* astronaut David Scott perform a fun demonstration. Near the end of an extravehicular moon walk, Scott held up a hammer and a feather, then proceeded to let go of them simultaneously. Both objects, under the gentle pull of the moon's gravity, fell to the ground, landing at precisely the same time.

That's not what would have happened here on Earth, unless you were practicing your spacesuit drills in one of NASA's giant vacuum chambers. Under ordinary circumstances, air resistance would greatly slow the fall of the feather, while the hammer would be largely unaffected. But in the vacuum on the moon's surface, their trajectories were indistinguishable.

Scott had confirmed an important insight put forward by Galileo Galilei back in the late sixteenth century: the natural motion of all objects is to fall in the same way under the influence of gravity, and it is only friction caused by air that makes heavier objects seem to fall faster than lighter ones in our everyday experience. And a good thing too. As mission controller Joe Allen put it, this experimental result was "predicted by well-established theory, but a result nonetheless reassuring considering both the number of viewers that witnessed the experiment, and the fact that the homeward journey was based critically on the validity of the particular theory being tested."

The story is told that Galileo performed a version of the experiment himself, dropping balls of different weights (but comparable air resistance) from the top of the Leaning Tower of Pisa. Galileo doesn't seem to have claimed that he did this, but it was later asserted by his pupil Vincenzo Viviani in a biography of his master.

The Leaning Tower of Pisa.
(Courtesy of W. Lloyd MacKenzie)

The experiment we know Galileo actually performed was an easier one to construct and control: he rolled balls of different masses down inclined planes. He was able to show that the balls accelerated in a uniform fashion, by an amount that depended on the angle of the plane but not on the masses of the balls. He then suggested that if we could trust this result all the way to planes that were inclined absolutely perpendicular to the floor, that would be exactly like dropping objects straight down, without a plane there at all. Therefore, he concluded, all masses would fall in a uniform way under the force of gravity, if it weren't for the influence of air resistance.

More important than this specific finding is the underlying message it conveys: we can learn about the natural motion of objects by imagining we can get rid of various nuisance effects, such as friction and air resistance, and then perhaps recovering more realistic kinds of motion by putting those effects back in later.

That is no small insight. It is arguably the biggest idea in the history of physics.

Physics is, by far, the simplest science. It doesn't seem that way, because we know so much about it, and the required knowledge often seems esoteric and technical. But it is blessed by this amazing feature: we can very often make ludicrous simplifications—frictionless surfaces, perfectly spherical bodies—ignoring all manner of ancillary effects, and nevertheless get results that are unreasonably good. For most interesting problems in other sciences, from biology to psychology to economics, if you modeled one tiny aspect of a system while pretending all the others didn't exist, you would just end up getting nonsense. (Which doesn't stop people from trying.)

This enormous, paradigm-shifting idea—in idealized situations where friction and dissipation can be ignored, physics becomes simple—was in large part responsible for helping to establish an equally influential, arguably more world-shattering concept: *conservation of momentum*. It might not sound like a principle of such dramatic import, but momentum is at the very heart of a shift in how we view the world, from an ancient cosmos of causes and purposes to a modern one of patterns and laws.

Before Galileo and others revolutionized the study of motion in the sixteenth and seventeenth centuries, Aristotle had long reigned as the leading thinker on the subject. Aristotle's view of physics was resolutely teleological: he thought of objects as having a natural state of being, and processes as being directed toward a goal. Famously, he suggested that we could distinguish between four different kinds of "causes," although "kinds of explanation" might be a better translation of what he had in mind. The four kinds were *material cause*, the stuff of which an object is made; *formal cause*, the essential property that makes an object what it is; *efficient cause*, the thing that brings the object about (closest to our informal notion of "cause"); and *final cause*, the purpose for which an object exists. Understanding why things change and move and behave the way they do comes down to putting them in the context of these causes.

For Aristotle, the nature of an object determines how it moves. Of the four classical elements, earth and water tend to fall to lower elevations, whereas air and fire tend to rise. An object can be in its natural state of rest or motion, where it will tend to remain until a "violent motion" causes it to change, after which it will return.

Consider a coffee cup sitting at rest on a table. It is in its natural state, in this case at rest. (Unless we were to pull the table out from beneath it, in which case it would naturally fall, but let's not do that.) Now imagine we exert a violent motion, pushing the cup across the table. As we push it, it moves; when we stop, it returns to its natural state of rest. In order to keep it moving, we would have to keep pushing on it. As Aristotle says, "Everything that is in motion must be moved by something."

This is manifestly how coffee cups do behave in the real world. The difference between Galileo and Aristotle wasn't that one was saying true things and the other was saying false things; it's that the things Galileo chose to focus on turned out to be a useful basis for a more rigorous and complete understanding of phenomena beyond the original set of examples, in a way that Aristotle's did not.

In the sixth century, John Philoponus, a philosopher and theologian living in Egypt, began the journey from Aristotle to our present understanding of motion. He suggested that we should think of a motive power or "impetus," which was imparted to a body by the initial act of pushing, and kept the body in motion until all of the impetus had dissipated. It was a small step forward, but one that opened up a new vista on how to think about the nature of motion. Rather than talking about causes, the focus shifted to quantities and properties of matter itself.

Ibn Sina (Avicenna), Persian philosopher and polymath, d. 1037.

Another crucial contribution was made by the Persian thinker Ibn Sina (sometimes Romanized as Avicenna), one of the leading lights of the Islamic Golden Age, around the year 1000. He elaborated on Philoponus's idea of impetus, calling it "inclination" (*mayl*). It was Ibn Sina who proposed that inclination didn't disperse on its own, but only due to air resistance or other external influences. And in a vacuum, he points out, there is no such resistance: an undisturbed projectile would keep moving at a constant rate, forever.

This brings us remarkably close to the modern idea of *inertia*—the concept that bodies will move uniformly unless acted upon. In the fourteenth century, Jean Buridan, a French cleric who was probably influenced by Ibn Sina, came up with a quantitative formula equating the impetus with the weight of an object times its velocity. At the time, however, the distinction between mass and weight was not understood. Galileo, influenced in turn by Buridan, coined the term "momentum" and said it would remain constant in a body that was not being acted on by any forces, but he didn't clearly differentiate between momentum and velocity. It was René Descartes who equated momentum with mass times speed, but even he (despite being the inventor of analytic geometry) didn't appreciate that momentum has a direction as well as a magnitude; that was left to Dutch scientist Christiaan Huygens in the seventeenth century. Then, it was Isaac Newton who put the notion to brilliant use in his systematic reinvention of the science of motion, which we still teach in high schools and colleges today.

Why is conservation of momentum such a big deal? We're not here to study Newtonian mechanics, as rewarding as that would be. There will be no exercises involving pulleys or inclined planes. We're here to think about the fundamental nature of reality.

For Aristotle, physics was a story of natures and causes. Whenever there was motion of any sort, there had to be a mover: an efficient cause that led to that motion. Aristotle had a more expansive definition of "motion" than we use today, one that is really closer to "transformation." It would include, for example, an object changing its color, or possibilities becoming actualities. But the same principles apply; Aristotle's conviction was that all of these transformations implied the existence of a transforming cause. There's nothing absurd about such an idea. In our everyday experience, things don't "just happen"—something works to cause them, to bring them about. Aristotle, without any of the benefit of modern scientific knowledge, was trying to codify what he knew about the way the world works into some kind of systematic framework.

So Aristotle observes a world populated by countless changing things, and infers a cause in each case. A is caused to move by B, which in turn is caused to move by C, and so on. It's reasonable to ask: What started it all? To what can we trace back this chain of motions and causes? He quickly rejects the possibilities that any motions are self-caused, or that the chain of causes goes back infinitely far. It needs to terminate somewhere, in something that causes motion but does not itself move: an unmoved mover.

Aristotle's theory of motion was largely set forth in his book *Physics*, but the details of the unmoved mover were left to a later one, *Metaphysics*. There, despite being nominally a pagan, he identifies the unmoved mover with God: not just an abstract principle but a being, immortal and benevolent. It's not a bad argument for God's existence, although it's easy to poke holes in it by denying the underlying assumptions. Maybe some motions do cause themselves, or maybe infinite regresses are perfectly okay. But this "cosmological argument" was extremely influential, picked up and elaborated on by Thomas Aquinas and others.

Most important for our purposes, the whole structure of Aristotle's argument for an unmoved mover rests on his idea that motions require causes. Once we know about conservation of momentum, that idea loses its steam. We can quibble over the details—I have no doubt Aristotle would have been able to come up with an ingenious way of accounting for objects on frictionless surfaces moving at constant velocity. What matters is that the new physics of Galileo and his friends implied an entirely new ontology, a deep shift in how we thought about the nature of reality. "Causes" didn't have the central role that they once did. The universe doesn't need a push; it can just keep going.

It's hard to overemphasize the importance of this shift. Of course, even today, we talk about causes and effects all the time. But if you open the contemporary equivalent of Aristotle's *Physics*—a textbook on quantum field theory, for example—words like that are nowhere to be found. We still, with good reason, talk about causes in everyday speech, but they're no longer part of our best fundamental ontology.

What we're seeing is a manifestation of the layered nature of our descriptions of reality. At the deepest level we currently know about, the basic notions are things like "spacetime," "quantum fields," "equations of motion," and "interactions." No causes, whether material, formal, efficient, or final. But there are levels on top of that, where the vocabulary changes. Indeed, it's possible to recover pieces of Aristotle's physics quantitatively, as limits of Newtonian mechanics in an appropriate regime, where dissipation and friction are central. (Coffee cups do come to a stop, after all.) In the same way, it's possible to understand why it's so useful to refer to causes and effects in our everyday experience, even if they're not present in the underlying equations. There are many different useful stories we have to tell about reality to get along in the world.

4

What Determines What Will Happen?

Isaac Newton, the most influential scientist of all time, was a very religious man. His views were undoubtedly heterodox by the standards of his childhood Anglican faith; he rejected the Trinity, and wrote numerous works on prophesy and biblical interpretation, with chapter titles such as "Of the power of the eleventh horn of Daniel's fourth Beast, to change times and laws." He couldn't rely on an argument for God's existence along the lines of Aristotle's unmoved mover. His own work seemed to depict a universe moving perfectly well under its own power, but as he pointed out in the "General Scholium" (an essay appended to later editions of his masterwork, *Principia Mathematica*), someone had to set it all up:

> This most excellently contrived System of the Sun, and Planets, and Comets, could not have its Origin from any other than from the wise Conduct and Dominion of an intelligent and powerful Being.

Elsewhere, Newton seemed to imply that the mutual perturbations of the planets on one another would gradually cause the system to get out of whack, at which point God would intervene to set things back in order.

Pierre-Simon Laplace, a French physicist and mathematician born a century after Newton, thought differently. Scholars debate over his true religious views, which seem to have vacillated between deism (God created the world, but did not subsequently intervene in its operation) and outright atheism. Laplace is the one who, when asked by Emperor Napoleón why God didn't appear in his book on celestial mechanics, purportedly replied, "I had no need of that hypothesis." Whatever his ultimate beliefs, it seems that Laplace held steadfastly against the idea of a Creator who would ever directly interfere in the motions of the world.

Pierre-Simon Marquis de Laplace, 1749–1827.

Laplace was one of the first thinkers to truly understand classical (Newtonian) mechanics, deep in his bones—better than Newton himself. Someone was bound to do it. Science progresses, and we learn more and more about our best theories; there are many physicists today who understand relativity better than Einstein, or quantum mechanics better than Schrödinger or Heisenberg. Laplace tackled problems from the stability of the solar system to the foundations of probability, routinely inventing the required new mathematics along the way. He suggested that Newtonian gravity could be thought of as a *field theory*, positing a "gravitational potential field" that filled all of space, thereby resolving Newton's puzzlement about actions at a distance between faraway bodies.

Perhaps Laplace's greatest contribution to our understanding of mechanics was not a technical or mathematical advance, but a philosophical one. He realized that there was a simple answer to the question "What determines what will happen next?" And the answer is "The state of the universe right now."

There's a worry that this result threatens the existence of human agency, our ability to make choices about what to do next. As we'll see, that's not really an issue of physics, but one of description: What is the best way we have to talk about human beings? When we talk about simple Newtonian systems, like the planets moving through the solar system, determinism is part of the picture. When we talk about enormously more complex things like people, there's no way for us to have enough information to make ironclad predictions. Our best theories of people, presented on their own terms and without reference to underlying particles and forces, leave plenty of room for human choice.

The world, according to classical physics, is not fundamentally teleological. What happens next is not influenced by any future goals or final causes toward which it might be working. Nor is it fundamentally historical; to know the future—in principle—requires only precise knowledge of the present moment, not any additional knowledge of the past. Indeed, the entirety of both the past and future history are utterly determined by the present. The universe is resolutely focused on the current moment; it marches forward, instant to instant, under the grip of unbreakable physical laws, with no heed paid to its glorious accomplishments or to its hopeful prospects. Much later, the biologist Ernst Haeckel would dub this viewpoint *dysteleology*, though the term is so ungainly that it never really caught on.

In modern parlance, Laplace was pointing out that the universe is something like a computer. You enter an input (the state of the universe right now), it does a calculation (the laws of physics) and gives you an output (the state of the universe one moment later). Similar ideas had previously been suggested by Gottfried Wilhelm Leibniz and Roger Boscovich, and were prefigured over two millennia earlier by Ajivika, a heterodox school of ancient Indian philosophy. Since computers hadn't been invented yet, Laplace imagined a "vast intellect" that knew the positions and velocities of all the particles in the universe, and understood all the forces they were subject to, and had sufficient computational power to apply Newton's laws of motion. In that case, as he put it, "for such an intellect nothing would be uncertain, and the future just like the past would be present before its eyes." His contemporaries immediately judged "vast intellect" to be too boring, and renamed it *Laplace's Demon*.

It's convenient to say "one moment later," but for Newton and Laplace, and to the best of our current understanding in theoretical physics, the flow of time is continuous rather than discrete. That's no problem at all; this is a job for calculus, which Newton and Leibniz invented for just this reason. By the "state" of the universe, or any subsystem thereof, we mean the position and the velocity of every particle within it. The velocity is just the rate of change (the derivative) of the position as time passes; the laws of physics provide us with the acceleration, which is the rate of change of the velocity. Together, you give me the state of the universe at one time, and I can use the laws of physics to integrate forward (or backward) and get the state of the universe at any other time.

We're using the language of classical mechanics—particles, forces—but the idea is much more powerful and general. Laplace introduced the idea of "fields" as a centrally important concept in physics, and the notion became entrenched with the work of Michael Faraday and James Clerk Maxwell on electricity and magnetism in the nineteenth century. Unlike a particle, which has a position in space, a field has a value at every single point in space—that's just what a field is. But we can treat that field value like a "position," and its rate of change as a "velocity," and the whole Laplacian thought experiment goes through undisturbed. The same is true for Einstein's general theory of relativity, or Schrödinger's equation in quantum mechanics, or modern speculations such as superstring theory. Since the days of Laplace, every serious attempt at understanding the behavior of the universe at a deep level has included the feature that the past and future are determined by the present state of the system. (One possible exception is the collapse of the wave function in quantum mechanics, which we'll discuss at greater length in chapter 20.)

This principle goes by a simple, if potentially misleading, name: *conservation of information*. Just as conservation of momentum implies that the universe can just keep on moving, without any unmoved mover behind the scenes, conservation of information implies that each moment contains precisely the right amount of information to determine every other moment.

The term "information" here requires caution, because scientists use the same word to mean different things in different contexts. Sometimes "information" refers to the knowledge you actually have about a state of affairs. Other times, it means the information that is readily accessible, embodied in what the system macroscopically looks like (whether you are looking at it and have the information or not). We are using a third possible definition, what we might call the "microscopic" information: the complete specification of the state of the system, everything you could possibly know about it. When speaking of information being conserved, we mean literally all of it.

These two conservation laws, of momentum and information, imply a sea change in our best fundamental ontology. The old Aristotelian view was comfortable and, in a sense, personal. When things moved, there were movers; when things happened, there were causes. The Laplacian view—one that continues to hold in science to this day—is based on patterns, not on natures and purposes. If this certain thing happens, we know this other thing will necessarily follow thereafter, with the sequence described by the laws of physics. Why is it that way? Because that's the pattern we observe.

Laplace's Demon is a thought experiment, not one we're going to reproduce in the lab. Realistically, there never will be and never can be an intelligence vast and knowledgeable enough to predict the future of the universe from its present state. If you sit down and think about what such a computer would have to be like, you eventually realize it would essentially need to be as big and powerful as the universe itself. To simulate the entire universe with good accuracy, you basically have to *be* the universe. So our concern here isn't one of practical engineering; it's not going to happen.

Our interest is a matter of principle: the fact that the current state of the universe determines its future, not that we can imagine taking advantage of that fact to make predictions. This feature, *determinism*, rubs some people the wrong way. It's worth taking a careful look at its limitations and prospects.

Classical mechanics, the system of equations studied by Newton and Laplace, isn't perfectly deterministic. There are examples of cases where a unique outcome cannot be predicted from the current state of the system. This doesn't bother most people, since cases like this are extremely rare— they are essentially infinitely unlikely among the set of all possible things a system could be doing. They are artificial and fun to think about, but not of great import to what happens in the messy world around us.

A more popular objection to determinism is the phenomenon of *chaos*. The ominous name obscures its simple nature: in many kinds of systems, very tiny amounts of imprecision in our knowledge of the initial state of that system can lead to very large variations in where it eventually ends up. As far as determinism is concerned, however, the existence of chaos could not possibly be more irrelevant. Laplace's point was always that perfect information leads to perfect prediction. Chaos theory says that slightly imperfect information leads to very imperfect prediction. True, and it doesn't change the picture the slightest bit. Nobody in their right mind was ever under the impression that we would be able to use Laplace's reasoning to build a useful prediction-making device; the thought experiment was always a matter of principle, not one of practice.

The real issue with classical mechanics is that it's not how the world works. These days we know better: quantum mechanics, which came along in the early twentieth century, is an entirely different ontology. There are no "positions" and "velocities" in quantum mechanics; there is only "the quantum state," also known as "the wave function," which we can use to calculate the outcomes of experiments that observe the system.

Quantum mechanics has supplanted classical mechanics as the best way we know to talk about the universe at a deep level. Unfortunately, and to the chagrin of physicists everywhere, we don't fully understand what the theory actually *is*. We know that the quantum state of a system, left alone, evolves in a perfectly deterministic fashion, free even of the rare but annoying examples of non-determinism that we can find in classical mechanics. But when we *observe* a system, it seems to behave randomly, rather than deterministically. The wave function "collapses," and we can state with very high precision the relative probability of observing different outcomes, but never know precisely which one it will be.

There are several competing approaches as to how to best understand the measurement problem in quantum mechanics. Some involve true randomness, while others (such as my favorite, the Everett or Many-Worlds formulation) retain complete determinism. We'll talk about the alternatives in chapter 21. All of the popular versions of quantum mechanics, however, maintain the underlying philosophy of Laplace's analysis, even if they do away with perfect predictability: what matters, in predicting what will happen next, is the current state of the universe. Not a goal in the future, nor any memory of where the system has been. As far as our best current physics is concerned, each moment in the progression of time follows from the previous moment according to clear, impersonal, quantitative rules.

There is a bit of a mismatch between Laplace's notion of determinism and what most people think of when they hear "the future is determined." The latter phrase conjures up images of *destiny* or *fate*—the idea that what will eventually happen has "already been decided," with the implication that it's been decided by someone, or something.

The physical notion of determinism is different from destiny or fate in a subtle but crucial way: because Laplace's Demon doesn't actually exist, the future may be determined by the present, but literally nobody knows what it will be. When we think of destiny, we think of something like the Three Fates of Greek mythology or the Weird Sisters of Shakespeare's *Macbeth*, wizened oracles who will use riddles to indicate our future path, which we will try to escape from and fail. The real universe is nothing like that. It's more like an annoying child who likes to approach people and say, "I know what's going to happen to you next!" Then, when you ask what will happen, the child says, "I can't tell you." And after it happens, they say, "See? I knew that was going to happen!" That's the universe for you.

The momentary or Laplacian nature of physical evolution doesn't have much relevance for the choices we face in our everyday lives. For poetic naturalism, the situation is clear. There is one way of talking about the universe that describes it as elementary particles or quantum states, in which Laplace holds sway and what happens next depends only on the state of the system right now. There is also another way of talking about it, where we zoom out a bit and introduce categories like "people" and "choices." Unlike our best theory of planets or pendulums, our best theories of human behavior are not deterministic. We don't know any way to predict what a person will do based on what we can readily observe about their current state. Whether we think of human behavior as determined depends on what we know.

5

Reasons Why

In November 2003, Dutch pediatric nurse Lucia de Berk was sentenced to life imprisonment without parole, for the murder of four children under her care and the attempted murder of three others. Her case became a media sensation for an unusual reason: it involved the misuse of statistical reasoning.

Some direct evidence was brought against de Berk, but it was flimsy. In one case, for example, the victim ("baby Amber") was alleged to have been poisoned by the drug digoxin, but doctors pointed out that similar chemical signals could have arisen naturally. The crucial part of the case against de Berk wasn't any incontrovertible evidence of individual murders, but rather the supposed statistical unlikelihood of so many deaths occurring while a single nurse was on duty. One expert testified that there was less than 1 chance in 342 million of such a coincidence. The prosecution argued, successfully, that the improbability implied by this calculation meant that a lower burden of proof should be used when evaluating the deaths as a group than would be appropriate when investigating only a single incident.

The problem was that the calculation was entirely bogus. It was plagued by elementary mistakes, from multiplying probabilities that weren't independent to "fishing" for seeming coincidences in large numbers of events. After the conviction, other experts put forward alternative calculations, ranging from 1 in 1 million to 1 in 25, depending on precisely how the questions were asked. Further investigation showed that the infant mortality rate at the hospital had been higher in the years before de Berk had been hired than it became once she started working there, not really the effect one would expect the presence of a serial killer to have. Ultimately, doubts about both the statistical arguments and the direct evidence led to a retrial. In 2010, de Berk was fully acquitted of all charges.

But math mistakes alone are not sufficient to account for Lucia de Berk's wrongful conviction. What started the ball rolling was a psychological conviction: the idea that something as horrible as these infant deaths couldn't just be random; someone must be to blame. There must be a *reason why* it happened. As horrible as the death of a child necessarily is, it becomes more sensible to us if it can somehow be explained as the result of someone's actions, rather than simply random chance.

Looking for causes and reasons is a deeply ingrained human impulse. We are pattern-recognizing creatures, quick to see faces in craters on Mars or connections between the location of Venus in the sky and the state of our love life. Not only do we seek order and causation, but we favor fairness as well. In the 1960s, psychologist Melvin Lerner proposed the "Just World Fallacy" after noticing people's tendency to blame victims of misfortune when something went wrong. To test his idea, he and his collaborator Carolyn Simmons conducted experiments in which subjects were shown other people apparently suffering the effects of electrical shocks. Afterward, many of the subjects—who knew nothing about the people supposedly being shocked—passed harsh judgments against them, berating their character. The more violent the shocks appeared to be, the harder the subjects were on the victims.

Searching for reasons why things happen is by no means an irrational pursuit. In many familiar contexts, things don't "just happen." If you are sitting in your living room and a baseball suddenly crashes through your window, it makes sense to look outside and expect to see some kids at play. Giant whales do not spontaneously come into existence several miles in the air. Our familiar intuitions concerning cause and effect have developed over evolutionary time because they provide useful guides for understanding how the world really works.

The mistake is to elevate this expectation to an unbreakable principle.

We see things happen, and we attribute reasons to them. Not only with events at home and people's personal fates but all the way down to the basics of ontology. If the world consists of certain things and behaves in certain ways, we think, there must be a reason why it is so.

This mistake has a name: the *Principle of Sufficient Reason*. The term was coined by German philosopher and mathematician Gottfried Leibniz, but the essential idea had been anticipated by many earlier thinkers, most notably by Baruch Spinoza in the seventeenth century. One way of stating it would be:

> **Principle of Sufficient Reason:**
> For any true fact, there is a reason why it is so, and why something else is not so instead.

Leibniz once formulated it simply as "Nothing happens without a reason," which is remarkably close to the maxim "Everything happens for a reason," which you can buy on T-shirts and bumper stickers today. (Alternatively, designer and cancer survivor Emily McDowell sells empathy cards reading "Please let me be the first to punch the next person who tells you everything happens for a reason.") Leibniz did grant that sometimes the reasons would be knowable only by God.

Why would anybody believe not only that we can usually attribute reasons to things that happen but that every single fact about the universe is associated with a particular reason? There is an obvious alternative, after all: that some facts have reasons behind them, but that there are also "brute" facts—things that are simply true, with no further explanation possible. How are we to judge whether brute facts are part of the basic ontology of the world?

Whenever we are confronted with questions about belief, we can employ the technique called *abduction*, or "inference to the best explanation." Abduction is a type of reasoning that can be contrasted with *deduction* and *induction*. With deduction, we start with some axioms whose truth we do not question, and derive rigorously necessary conclusions from them. With induction, we start with some examples we know about, and generalize to a wider context—rigorously, if we have some reason for believing that such a generalization is always correct, but often we don't quite have that guarantee. With abduction, by contrast, we take all of our background knowledge about how the world works, and perhaps some preference for simple explanations over complex ones (Occam's razor), and decide what possible explanation provides the best account of all the facts we have. In chapters 9 and 10 we will explore this method of inference more fully under the topic of *Bayesian reasoning*.

In the case of the Principle of Sufficient Reason (PSR), for simplicity let's divide the possibilities into two competing claims: that every fact has a reason that explains it (the PSR is true), or that some facts do not (the PSR is false). To each claim we assign some prior credence—the degree of belief we start out with. Then we gather evidence, by looking at how the world works, and update our credences appropriately.

The usual strategy of defenders of the Principle of Sufficient Reason is not to gather evidence but to proclaim that what we have is a "bedrock metaphysical principle." That is to say, it's the kind of thing we can't even imagine not being true. Accordingly, they assign a prior credence of unity to every fact having a reason, and a prior credence of zero to the existence of brute facts. Given that choice, no evidence is going to have any effect on your credences thereafter; you will always believe that every fact is associated with a sufficient reason.

Our standards for promoting a commonsensical observation to a "metaphysical principle" should be very high indeed. As Scottish philosopher David Hume—who, if anyone, deserves to be called the father of poetic naturalism, perhaps with his Roman predecessor Lucretius as the grandfather—pointed out, the Principle of Sufficient Reason doesn't seem to rise to that level. Hume noted that conceiving of effects without causes might seem unusual, but it does not lead to any inherent contradiction or logical impossibility.

When pressed as to why we can't live without the Principle of Sufficient Reason, its defenders generally fall back on one of two angles. They may try to defend it by appealing to some other bedrock metaphysical principle. Leibniz, for example, had something he called the Principle of the Best, according to which God always acts in the best possible way, including in the creation of the world. This is only a persuasive argument if we accept the new principle as truly inescapable, which is rarely the case for people who were skeptical of the Principle of Sufficient Reason in the first place.

The other possible angle is to claim that something like the Principle of Sufficient Reason is inherent in the very act of logical thinking itself, that rationality is implicitly committed to it. Imagine, for example, that you went to take a shower one day, only to find that there was an accordion sitting in your bathtub. It would be hard for you not to think that there must be some reason why the accordion was there. It probably didn't just happen. Similarly, so this line of thought goes, for every fact we notice about the universe: as soon as we apprehend it, we think there must be a reason behind it.

This isn't an argument that the Principle of Sufficient Reason is logically incontrovertible; it only implies that we often act as if something like it were true. If we're honest, it's an empirical, evidence-based argument, not an a priori one. We're not used to seeing accordions appear without good reason, as a matter of empirical fact; but we could certainly imagine a world in which they did so.

Metaphysical principles are tempting shortcuts but not reliable guides. There are good reasons why things often seem to happen for reasons—and also reasons why that's not a bedrock principle.

It may seem strange to suggest, on the one hand, that we live in a Laplacian universe where one moment follows directly from the next in accordance with unbreakable laws of physics, and on the other hand that there are facts that don't have any reasons to explain them. Can't we always give a reason for what happens, namely "the laws of physics and the prior configuration of the universe"?

That depends on what we mean by a "reason." It's important to first distinguish between two kinds of "facts" we might want to explain. There are *things that happen*—that is, states of the universe (or parts thereof) at specific moments in time. And then there are *features of the universe*, such as the laws of physics themselves. The kinds of reasons that would suffice to explain one have a different character from the other.

When it comes to "things that happen," what we mean by a "reason" is essentially the same as what we mean when we refer to the "cause" of an event. And yes, we are free to say that events are explained or caused by "the laws of physics and the prior configuration of the universe." That's true even in quantum mechanics, which is itself sometimes erroneously offered up as an example of things (like the decay of an atomic nucleus) happening without reasons. If that's what one is looking for in a reason, the laws of physics do indeed provide it. Not as some metaphysical principle but as an observed pattern in our universe.

However, that isn't really what people have in mind when they're searching for reasons. If someone asks "Why did that tragic shooting occur?" or "Why is the average temperature of the Earth's atmosphere rising so rapidly?" answering with "Because of the laws of physics and the prior configuration of the universe" isn't going to be satisfying. What we are really after is some identifiable aspect of the configuration of the universe without which the event in question would not have occurred.

The laws themselves, as we've discussed, make no reference to "reasons" or "causes." They are simply patterns that connect what happens at different places and times. Nevertheless, the concept of a "reason why" something is true is a very useful one in our daily lives. Any sensible poetic naturalist would judge it to be a helpful part of an accurate way of talking about a certain part of the universe. Indeed, we talked that way in the very first paragraph of this chapter.

What we might want to ask is: "What is the reason why it makes sense to talk about 'reasons why'?" And there's a good answer, namely: because of the *arrow of time*.

The observable universe around us isn't just an arbitrary collection of stuff obeying the laws of physics—it's stuff that starts out in a very particular kind of arrangement, and obeys the laws of physics thereafter. By "starts out" we are referring to conditions near the Big Bang, a moment about 14 billion years ago. We don't know whether the Big Bang was the actual beginning of time, but it was a moment in time beyond which we can't see any further into the past, so it's the beginning of our observable part of the cosmos. The particular kind of arrangement the universe was in at that time is one with a very low *entropy*—the scientific way of measuring disorderliness or randomness of a system. Entropy used to be very low, and has been growing ever since—which is to say our observable universe used to be in a specific, orderly arrangement, and has been becoming more disorderly for 14 billion years.

It's that tendency for entropy to increase that is responsible for the existence of time's arrow. It's easy to break eggs, and hard to unbreak them; cream and coffee mix together, but don't unmix; we were all born young, and gradually grow older; we remember what happened yesterday, but we don't remember what will happen tomorrow. Most of all, what causes an event must precede the event, not come afterward.

Just as there is no reference to "causes" in the fundamental laws of physics, there isn't an arrow of time, either. The laws treat the past and future on an equal footing. But the usefulness of our everyday language of explanation and causation is intimately tied to time's arrow. Without it, those terms wouldn't be a useful way of talking about the universe at all.

We'll see how our convictions that things happen for reasons, and effects follow causes, are not bedrock principles. They arise because of a contingent feature of how matter is evolving in our local universe. There is a close connection between *cosmology*, on the one hand, and *knowledge*, on the other. Understanding our universe helps us perceive why we are so convinced that things happen for reasons.

The "reasons" and "causes" why things happen, in other words, aren't fundamental; they are emergent. We need to dig in to the actual history of the universe to see why these concepts have emerged.

An obvious place where it's tempting to look for reasons why is the question of why various features of the universe take the form that they do. Why was the entropy low near the Big Bang? Why are there three dimensions of space? Why is the proton almost 2,000 times heavier than the electron? Why does the universe exist at all?

These are very different questions from "Why is there an accordion in my bathtub?" We're no longer asking about occurrences, so "Because of the laws of physics and the prior configuration of the universe" isn't a good answer. Now we're trying to figure out why the fundamental fabric of reality is one way rather than some other way.

The secret here is to accept that such questions *may or may not have answers.* We have every right to ask them, but we have no right at all to demand an answer that will satisfy us. We have to be open to the possibility that they are brute facts, and that's just how things are.

These kinds of "Why?" questions don't exist in a vacuum. They make sense in some particular kind of context. If we ask "Why is there an accordion in my bathtub?" and someone answers "Because space is three-dimensional," we aren't going to be happy—even if it's arguably true that the accordion wouldn't have been in there if space were only two-dimensional. We ask the question in the context of a world where there are things called accordions, which tend to appear in some places and not others, and that there is something called your bathtub, in which certain things regularly appear and others do not. Part of that context might be that you have a roommate who had some friends over last night, and they had too much to drink, and one of them brought along an accordion, and she wouldn't stop playing it, and ultimately the decision was made to hide it from her. It's only within that kind of context that we can hope for answers to such "Why?" questions.

But the universe, and the laws of physics, aren't embedded in any bigger context, as far as we know. They might be—we should be open-minded about the possibility of something outside our physical universe, whether it's a nonphysical reality or something more mundane, like an ensemble of universes that make up a multiverse. In that context we could start asking questions about what kinds of universes are "natural" or easy to create, and possibly discover an explanation for the particular features we observe. Alternatively, we could discover reasons why the laws of physics themselves necessitate that something we thought was arbitrary (like the masses of the proton and the electron) can actually be derived from a deeper principle. Then, in a different way, we would be able to pat ourselves on the back for having explained something.

What we can't do is demand that the universe scratch our explanatory itches. Curiosity is a virtue, and it's good to look for answers to "Why?" questions whenever we might be able to find them, or when we think that asking such questions might help us to understand things better. But we should be at peace with the possibility that, for some questions, the answer doesn't go any deeper than "That's what it is." We're not used to that—our intuition assures us that every event can be explained in terms of some reason why. To understand why we have that impression, we need to dig more deeply into how our actual universe has evolved.

6

Our Universe

Nothing puts human existence into context quite like contemplating the cosmos. What you might not guess, sitting comfortably in your living room with a glass of wine and a good book, is that what's happening in your immediate neighborhood is dramatically affected by the evolution of the whole universe. Many of the most important features of our lives here on Earth—our notion of the passage of time, the existence of causes and effects, our memories of the past, and freedom to make choices about the future—are ultimately consequences of conditions near the Big Bang. To get ahold of the big picture, we need to put ourselves in cosmological context.

It's hard not to be moved when looking at the night sky. In true darkness, far away from the all-pervasive lights of human civilization, the inky-black background comes alive with thousands of stars, a handful of planets, and the majestic sweep of the Milky Way galaxy stretching from one horizon to the other. It's also hard to grasp the true extent of the universe on the basis of what we see when we look at the sky. There is no sense of scale, no familiar landmarks by which to judge size and separation. The stars bear a close resemblance to the planets, even though we now know they are quite different; they look nothing like the sun, although we now know they are very similar.

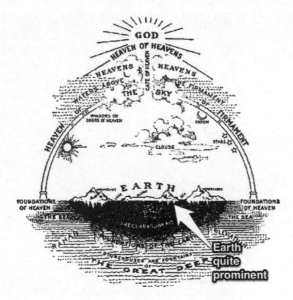

An ancient Hebrew cosmology.
(Illustration by George L. Robinson)

It's not surprising that ancient cosmologists, when theorizing about the universe, took as its fulcrum the thing they understood the best: themselves. Cultures scattered throughout history have devised a number of imaginative cosmological scenarios, and they tended to share a common conviction that our home, the Earth, was somehow special. Sometimes Earth was at the center of it all, sometimes it was at the bottom, very often it held particular significance for whatever force or god was responsible for creation. One way or another, there was a shared belief that we *mattered* in the greater scheme of things.

It wasn't until Giordano Bruno, a sixteenth-century Italian philosopher and mystic, that anyone suggested that the sun was just one star among many, and the Earth one of many planets that orbited stars. Bruno was burned at the stake for heresy in Rome in 1600, his tongue pierced by an iron spike and his jaw wired shut. His cosmological speculations were probably not the part of his heresy that the Church found most objectionable, but they didn't help any.

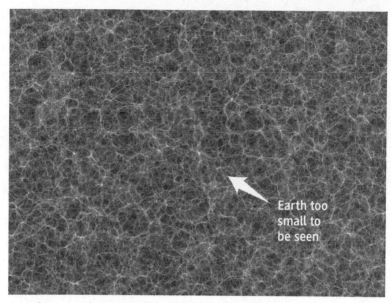

Earth too
small to
be seen

A modern cosmology: a simulation of the universe on very large scales, featuring billions of galaxies, each with many billions of stars, many featuring solar systems like our own. (Courtesy of the Millennium Simulation Project)

Today we understand a great deal about the scale of the universe. Bruno was on the right track: cosmically speaking, there's no indication that we matter at all.

Our modern picture of our cosmos was painstakingly pieced together through data collected by astronomers, who frequently brought back results that defied conventional theoretical wisdom of the time. A century ago, in 1915, Albert Einstein put the finishing touches on his general theory of relativity, which conceives of spacetime itself as a dynamic object whose curvature gives rise to the force we know as gravity. Before that point, it's safe to say that we knew next to nothing about what the universe was really like on large scales. Spacetime was thought to be absolute and eternal, in accordance with Newtonian mechanics, and astronomers were divided on whether the Milky Way was the only galaxy in the universe, or merely one of countless many.

Now the basics have been well established. The Milky Way we see stretching across the dark night sky is a galaxy—a collection of stars orbiting under their mutual gravitational attraction. It's hard to count precisely how many, but there are over 100 billion stars in the Milky Way. It's not alone; scattered throughout observable space we find at least 100 billion galaxies, typically with sizes roughly comparable to that of our own. (By coincidence, the number 100 billion is also a very rough count of the number of neurons in a human brain.) Recent studies of relatively nearby stars suggest that most of them have planets of some sort, and perhaps one in six stars has an "Earth-like" planet orbiting around it.

Perhaps the most notable feature of the distribution of galaxies through space is that, the farther out we look, the more uniform things become. On the very largest scales, the universe is extremely smooth and featureless. There is no center, no top or bottom, no edges, no preferred location at all.

Scatter all that material throughout space, and general relativity says that it's not just going to sit there. Galaxies are going to pull on one another, so the universe must be either expanding from a more dense state, or contracting from a less dense one. In the 1920s, Edwin Hubble discovered that our universe is indeed expanding. Given that discovery, we can use our theoretical understanding to extrapolate backward in time. According to general relativity, if we keep running the movie of the early universe backward, we come to a singularity at which the density and expansion rate approach infinity.

That scenario, developed by Belgian priest Georges Lemaître under the name "the Primeval Atom" but eventually dubbed "the Big Bang model," predicts that the early universe was not only denser but also hotter. So hot and dense that it would have been glowing like the interior of a star, and all of that radiation should still suffuse space today, ready for detection in our telescopes. That's just what happened in the fateful spring of 1964, when astronomers Arno Penzias and Robert Wilson at Bell Laboratories detected the cosmic microwave background radiation, leftover light from the early universe that has cooled off as space expanded. Today it is just a bit less than 3 degrees above absolute zero; it's a cold universe out there.

When we talk about the "Big Bang model," we have to be careful to distinguish that from "the Big Bang" itself. The former is an extraordinarily successful theory of the evolution of the observable universe; the latter is a hypothetical moment that we know almost nothing about.

The Big Bang model is simply the idea that approximately 14 billion years ago the matter in the universe was extremely hot, densely packed, and spread almost uniformly through space, which was expanding very rapidly. As space expanded, matter diluted and cooled, and stars and galaxies condensed out of the smooth plasma under the relentless pull of gravity. Unfortunately, the plasma was so hot and dense at early times that it was essentially opaque. The cosmic microwave background reveals what the universe looked like when it first became transparent, but before that, we cannot directly see.

The Big Bang itself, as predicted by general relativity, is a moment in time, not a location in space. It would not be an explosion of matter into an empty, preexisting void; it would be the beginning of the entire universe, with matter smoothly distributed all throughout space, all at once. It would be the moment prior to which there were no moments: no space, no time.

It's also, most likely, not real. The Big Bang is a prediction of general relativity, but singularities where the density is infinitely big are exactly where we expect general relativity to break down—they are outside the theory's domain of applicability. At the very least, quantum mechanics should become crucially important under such conditions, and general relativity is a purely classical theory.

So the Big Bang doesn't actually mark the beginning of our universe; it marks the end of our theoretical understanding. We have a very good idea, on the basis of observational data, what happened soon after the Bang. The microwave background radiation tells us to a very high degree of precision what things were like a few hundred thousand years afterward, and the abundance of light elements tells us what the universe was doing when it was a nuclear fusion reactor, just a few minutes afterward. But the Bang itself is a mystery. We shouldn't think of it as "the singularity at the beginning of time"; it's a label for a moment in time that we currently don't understand.

Ever since the expansion of the universe was discovered, the question of the future fate of the universe has preoccupied the minds of cosmologists.

Would it keep expanding forever, or eventually reverse course, contracting down to an ultimate "Big Crunch"?

A major clue was uncovered just as the twentieth century was ending, when in 1998 two teams of astronomers announced that the universe wasn't only expanding; it was accelerating. If you focused on a particular faraway galaxy and measured its velocity, then came back a few million or billion years later and measured it again, you would find that it's now moving away from you even faster. (That's not what the astronomers did, of course; they compared the velocities of galaxies at different distances.) If this behavior continues forever—which seems quite plausible—the universe will continue to expand and dilute in perpetuity.

Normally we'd expect the expansion of the universe to slow down as the gravitational forces between the galaxies worked to pull them together. The observed acceleration must be due to something other than matter as we know it. There is a very obvious, robust candidate for what the culprit might be: *vacuum energy*, which Einstein invented and called the *cosmological constant*. Vacuum energy is a kind of energy that is inherent in space itself, remaining at a constant density (amount of energy per cubic centimeter) even as space expands. Due to the interplay of energy and spacetime in general relativity, vacuum energy never runs out or fades away; it can keep pushing forever.

We don't know for sure whether it will keep pushing forever, of course; we can only extrapolate our theoretical understanding into the future. But it's possible, and in some sense would be simplest, for the accelerated expansion to simply continue without end.

That leads to a somewhat lonely future for our universe. Right now the night sky is alive with brightly shining stars and galaxies. That can't last forever; stars use up their fuel, and will eventually fade to black. Astronomers estimate that the last dim star will wink out around 1 quadrillion (10^{15}) years from now. By then other galaxies will have moved far away, and our local group of galaxies will be populated by planets, dead stars, and black holes. One by one, those planets and stars will fall into the black holes, which in turn will join into one supermassive black hole. Ultimately, as Stephen Hawking taught us, even those black holes will evaporate. After about 1 googol (10^{100}) years, all of the black holes in our observable universe will have evaporated into a thin mist of particles, which will grow more and more dilute as space continues to expand. The end result of this, our most likely scenario for the future of our universe, is nothing but cold, empty space, which will last literally forever.

We are small, and the universe is large. It's hard, upon contemplating the scale of the cosmos, to think that our existence here on Earth plays an important role in the purpose or destiny of it all.

That's just what we see, of course. For all we know, the universe could be infinitely big; or it could be just a bit larger than what we observe. The uniformity that characterizes our observable region of space could extend on indefinitely, or other regions could be extremely different from our own. We should be modest when making pronouncements about the universe beyond what we can measure.

One of the most striking features of the universe is the contrast between its uniformity in space and its dramatic evolution over time. We seem to live in a universe with a pronounced temporal imbalance: about 14 billion years between the Big Bang and now, and perhaps an infinite number of years between now and the eventual future. To the best of our knowledge, there's a legitimate sense in which we find ourselves living in a young and vibrant period in the universe's history—a history that will mostly be cold, dark, and empty.

Why is that? Maybe there's a deeper explanation, or maybe that's just how it is. The best a modern cosmologist can do is to take these observed features of the universe as clues to its ultimate nature, and keep trying to put it all into a more comprehensive picture. A crucial question along the way is, why did the matter in the universe evolve over billions of years in such a way as to create us?

7

Time's Arrow

Every human being goes through a process of aging over the course of their life, from a young child to an older adult. The universe, too, changes as it ages—from the hot, dense Big Bang to its cold, empty future. These are two different manifestations of time's arrow, the directionality of time that distinguishes past from future. What is far from obvious, but nonetheless true, is that these two processes are intimately related. The reason why we are all born young and die older; the reason why we can make choices about what to do next but not about things we've already done; the reason why we remember the past and not the future—all of these can ultimately be traced to the evolution of the wider universe, and in particular to conditions near its very beginning, 14 billion years ago at the Big Bang.

Traditionally, people have thought the opposite. It's been popular to imagine that the world is *teleological*—directed toward some future goal. But it's better to think of it as *ekinological*, from the Greek "εκκίνηση," meaning "start" or "departure." Everything interesting and complex about the current state of our universe can be traced directly to conditions near its beginning, the consequences of which we are living out every day.

This fact about the universe is absolutely crucial to our understanding of the big picture. We look at the world around us and describe it in terms of causes and effects, reasons why, purposes and goals. None of those concepts exists as part of the fundamental furniture of reality at its deepest. They emerge as we zoom out from the microscopic level to the level of the everyday. To appreciate why we seem to live in a world of causes and purposes, while nature deep down is a story of impersonal Laplacian patterns, we need to understand the arrow of time.

To understand time, it helps to start with space. Here on the surface of the Earth, you would be forgiven for thinking that there is an intrinsic difference between the directions "up" and "down," something deeply embedded into the fabric of nature. In reality, as far as the laws of physics are concerned, all directions in space are created equal. If you were an astronaut, floating in your spacesuit while you performed an extravehicular activity, you wouldn't notice any difference between one direction in space and any other. The reason why there's a noticeable distinction between up and down for us isn't because of the nature of space; it's because we live in the vicinity of an extremely influential object: the Earth.

Time works the same way. In our everyday world, time's arrow is unmistakable, and you would be forgiven for thinking that there is an intrinsic difference between past and future. In reality, both directions of time are created equal. The reason why there's a noticeable distinction between past and future isn't because of the nature of time; it's because we live in the aftermath of an extremely influential event: the Big Bang.

Remember Galileo and conservation of momentum: physics becomes simple when we ignore friction and other bothersome influences, and consider isolated systems. So let's think of a pendulum rocking back and forth, and for convenience let's imagine that our pendulum is in a sealed vacuum chamber, free of air resistance. Now someone records a movie of the pendulum rocking, and shows it to you. You are not very impressed; you've seen pendulums before. Then they reveal the surprise: they were actually playing the movie backward. You hadn't noticed because a pendulum rocking backward in time looks exactly like one rocking forward in time.

That's a simple example of a very general principle. For every way that a system can evolve forward in time in accordance with the laws of physics, there is another allowed evolution that is just "running the system backward in time." There is nothing in the underlying laws that says things can evolve in one direction in time but not the other. Physical motions, to the best of our understanding, are *reversible*. Both directions of time are on an equal footing.

That seems reasonable enough for simple systems: pendulums, planets moving around the sun, hockey pucks gliding on frictionless surfaces. But when we think about complicated macroscopic systems, everything in our experience tells us that certain things happen as time moves from past to future, but not in the other direction. Eggs break and get scrambled but don't unscramble and unbreak; perfume disperses into a room but never retreats back into its bottle; cream mixes into coffee but never spontaneously unmixes. If there is a purported symmetry between past and future, why do so many everyday processes occur only forward and never backward?

Even for these complicated processes, it turns out, there is a time-reversed process that is perfectly compatible with the laws of physics. Eggs could unbreak, perfume could go back into its bottle, cream and coffee could unmix. All we have to do is to imagine reversing the trajectory of every single particle of which our system (and anything it was interacting with) is made. None of these processes violates the laws of physics—it's just that they are extraordinarily unlikely. The real question is not why we never see eggs unbreaking toward the future; it's why we see them unbroken in the past.

Our basic understanding of these issues was first put together in the nineteenth century by a group of scientists who invented a new field called *statistical mechanics*. One of their leaders was the Austrian physicist Ludwig Boltzmann. It was he who took the concept of entropy, which was recognized as a central idea in the study of thermodynamics and irreversibility, and reconciled it with the microscopic world of atoms.

Ludwig Boltzmann, master of entropy and probability, 1844–1906. (Courtesy of Goethe University of Frankfurt)

Before Boltzmann came along, entropy was understood in terms of the inefficiency of things like steam engines, which were all the rage at the time. Anytime you try to burn fuel to do useful work such as pulling a locomotive, there is always some waste generated in the form of heat. Entropy can be thought of as a way of measuring that inefficiency; the more waste heat emitted, the more entropy you've created. And no matter what you do, the total entropy generated is always a positive number: you can make a refrigerator and cool things down, but only at the cost of expelling even more heat out the back. This understanding was codified in the *second law of thermodynamics*: the total entropy of a closed system never decreases, staying constant or increasing as time passes.

Boltzmann and his colleagues argued that we could understand entropy as a feature of how atoms are arranged inside different systems. Rather than thinking of heat and entropy as distinct kinds of things, obeying their own laws of nature, we can think of them as properties of systems made of atoms, and derive those rules from the Newtonian mechanics that applies to everything in the universe. Heat and entropy, in other words, are convenient *ways of talking* about atoms.

Boltzmann's key insight was that, when we look at an egg or a cup of coffee with cream, we don't actually see the individual atoms and molecules of which it is made. What we see are some observable macroscopic features. There are many possible arrangements of the atoms that give us exactly the same macroscopic appearance. The observable features provide a coarse-graining of the precise state of the system.

Given that, Boltzmann suggested that we could identify the entropy of a system with the number of different states that would be macroscopically indistinguishable from the state it is actually in. (Technically, it's the logarithm of the number of indistinguishable states, but that mathematical detail won't concern us.) A low-entropy configuration is one where relatively few states would look that way, while a high-entropy one corresponds to many possible states. There are many ways to arrange molecules of cream and coffee so that they look all mixed together; there are far fewer arrangements where all of the cream is on the top and all of the coffee on the bottom.

With Boltzmann's definition in hand, it makes perfect sense that entropy tends to increase over time. The reason is simple: there are far more states with high entropy than states with low entropy. If you start in a low-entropy configuration and simply evolve in almost any direction, your entropy is extraordinarily likely to increase. When the entropy of a system is as high as it can get, we say that the system is in *equilibrium*. In equilibrium, time has no arrow.

What Boltzmann successfully explained is why, given the entropy of the universe today, it's very likely to be higher-entropy tomorrow. The problem is that, because the underlying rules of Newtonian mechanics don't distinguish between past and future, precisely the same analysis should predict that the entropy was higher *yesterday*, as well. Nobody thinks the entropy actually was higher in the past, so we have to add something to our picture.

The thing we need to add is an assumption about the initial condition of the observable universe, namely, that it was in a very low-entropy state. Philosopher David Albert has dubbed this assumption the *Past Hypothesis*. With that assumption, and an additional (much weaker) assumption that the initial conditions weren't finely tuned to make the entropy decrease even further with time, everything falls into place. The reason why the entropy was lower yesterday than it is today is simple: because it was even lower the day before yesterday. And that's true because it was even lower the day before that. This reasoning proceeds stepwise all the way back 14 billion years into the past, right to the Big Bang. That may or may not have been the absolute beginning of space and time, but it's certainly the beginning of the part of the universe we can observe. The origin of time's arrow, therefore, is ekinological: it arises from a special condition in the far past.

Nobody knows exactly why the early universe had such a low entropy. It's one of those features of our world that may have a deeper explanation we haven't yet found, or may just be a true fact we need to learn to accept.

What we know is that this initially low entropy is responsible for the "thermodynamic" arrow of time, the one that says entropy was lower toward the past and higher toward the future. Amazingly, it seems that this property of entropy is responsible for *all* of the differences between past and future that we know about. Memory, aging, cause and effect—all can be traced to the second law of thermodynamics and in particular to the fact that entropy used to be low in the past.

8

Memories and Causes

Every person's life is caught in the relentless grip of time. We are born young, grow older, and die. We experience moments of surprise and delight, as well as periods of profound sadness. Our memories are cherished records of the past, and our aspirations help us map our future plans. If we want to situate our everyday lives as human beings in a natural world governed by physical laws, one of our first goals must be to understand how the flow of time relates to our individual lives.

You may be willing to believe that something straightforward and mechanical, such as increasing entropy, can be responsible for something equally straightforward and mechanical, such as how cream mixes into coffee. It seems harder to establish that entropy is responsible for all of our experience of the flow of time. For one thing, the past and future seem not only like different directions but also like completely different kinds of things. The past is fixed, our intuition assures us; it has already happened, while the future is still unformed and up for grabs. The present moment, the *now*, is what actually exists.

And then along came Laplace to tell us differently. Information about the precise state of the universe is conserved over time; there is no fundamental difference between the past and the future. Nowhere in the laws of physics are there labels on different moments of time to indicate "has happened yet" and "has not happened yet." Those laws refer equally well to any moment in time, and they tie all of the moments together in a unique order.

We can highlight three ways that the past and future seem radically different to us:

- We remember the past, but not the future.
- Causes precede their effects.
- We can make choices that affect the future, but not the past.

All of these features of how time works can ultimately be reconciled with the fact that the universe runs according to time-symmetric laws by the additional fact that the past had a lower entropy than the future. Let's look at the first two now, postponing for the moment the contentious issues of choice and free will. We will get there (I predict).

There are few more important manifestations of time's arrow than the phenomenon of memory. We have impressions in our minds—not always perfectly accurate, but often quite good—of events that have happened in the past. We do not, most of us agree, possess analogous impressions of the future. The future may be predicted, but it cannot be remembered. This imbalance accords quite well with our intuitive feeling that the past and the future have very different ontological statuses; one has happened, the other hasn't.

From the Laplacian point of view, where information is present in each moment and conserved through time, a memory isn't some kind of direct access to events in the past. It must be a feature of the present state, since the present state is all we presently have. And yet there is an epistemic asymmetry, an imbalance of knowledge, between past and future. That asymmetry is a consequence of the low entropy of the early universe.

Think of walking down the street and noticing a broken egg lying on the sidewalk. Ask yourself what the future of that egg might have in store, in comparison with its recent past. In the future, the egg might wash away in a storm, or a dog might come by and lap it up, or it might just fester for a few more days. Many possibilities are open. In the past, however, the basic picture is much more constrained: it seems exceedingly likely that the egg used to be unbroken, and was dropped or thrown to this location.

We don't actually have any direct access to the past of the egg, any more than we do its future. But we think we know more about where it came from than where it might be going. Ultimately, even if we don't realize it, the source of our confidence is the fact that entropy was lower in the past. We are very used to unbroken eggs breaking; that's the natural way of things. In principle, the set of things that could befall the egg in the future is precisely the same size as the set of ways it could have arrived in its present condition, as a consequence of conservation of information. But we use the Past Hypothesis to rule out most of those possibilities about the past.

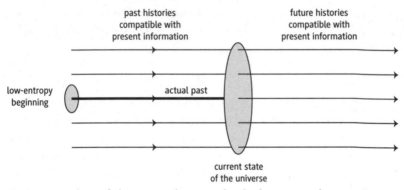

The Past Hypothesis of a low-entropy beginning breaks the symmetry between the past, on the left, and future, on the right.

The story of the egg is a paradigm for every kind of "memory" we might have. It's not just literal memories in our brain; any records that we may have of past events, from photographs to history books, work on the same principle. All of these records, including the state of certain neuronal connections in our brain that we classify as a memory, are features of the current state of the universe. The current state, by itself, constrains the past and future equally. But the current state plus the hypothesis of a low-entropy past gives us enormous leverage over the actual history of the universe. It's that leverage that lets us believe (often correctly) that our memories are reliable guides to what actually happened.

Back in chapter 4 we highlighted how Laplace's conservation of information undermines the central role that Aristotle placed on causality.

Concepts like "cause" appear nowhere in Newton's equations, nor in our more modern formulations of the laws of nature. But we can't deny that the idea of one event being caused by another is very natural, and seemingly a good fit to how we experience the world. This apparent mismatch can be traced back to entropy and the arrow of time.

It might seem strange to describe the world as operating according to unbreakable physical laws, and then turn around and deny causality a central role. After all, if the laws of physics predict what will happen at the next moment from what the situation is now, doesn't that count as "cause and effect"? And if we don't think that every effect has a cause, aren't we unleashing chaos on the world, and saying that basically anything can happen?

The strangeness evaporates once we appreciate the substantial difference between the kind of relationship of the past to the future that we get from the laws of physics, and the kind we usually think of as cause and effect. The laws of physics take the form of rigid patterns: if the ball is at a certain position and has a certain velocity at a certain time, the laws will tell you what the position and velocity will be a moment later, and what they were a moment before.

When we think about cause and effect, by contrast, we single out certain events as uniquely *responsible* for events that come afterward, as "making them happen." That's not quite how the laws of physics work; events simply are arranged in a certain order, with no special responsibility attributed to one over any of the others. We can't pick out one moment, or a particular aspect of any one moment, and identify it as "the cause." Different moments in time in the history of the universe follow each other, according to some pattern, but no one moment causes any other.

Understanding this feature of how nature works has led some philosophers to advocate that we eliminate cause and effect entirely. As Bertrand Russell once memorably put it:

> The law of causality, I believe, like much that passes muster among philosophers, is a relic of a bygone age, surviving, like the monarchy, only because it is erroneously supposed to do no harm.

It's an understandable reaction, but perhaps a bit too extreme. After all, it would be hard to get through the day without appealing to causes at all. Certainly when we speak of the actions taken by human beings, we like to assign credit or blame to them; that won't work if we can't even say that their actions caused any particular outcome. Causality provides a very useful way of talking in our everyday lives.

As with memory, the emergence of everyday causality from the underlying rigid pattern of the laws of physics can be traced to the arrow of time. Think of an example very much like that of the broken egg: a glass of wine spilled on the carpet. There are many future and past histories of the atoms that make up the wine and the glass that are compatible with what we can see about its current state. Now let's add a "mini Past Hypothesis": that five minutes ago the glass of wine was sitting on the table, not moving.

That hypothesis breaks the symmetry between past and future, and constrains the possible histories of the wineglass over the course of the last five minutes. But notice a crucial feature about this constraint: we know that the evolution of the glass of wine was not what it would have been had it simply been left alone, undisturbed. In that case, with overwhelming probability, the glass would simply have stayed there. Glasses of wine don't hop right off the table and onto the floor of their own accord.

Therefore, we can say with confidence that something must have disturbed the glass of wine—a stray elbow, or someone trying to fit a cheese plate onto an already-crowded table. With the information we have we can't say precisely what it was, but we know that something intervened to alter how the wineglass would have behaved had it been left untouched. That something, whatever it was, we justifiably label the "cause" of the glass falling.

All of which sounds innocent enough, but what is really going on here? There's certainly a sense in which the current state of the wineglass can be attributed to "the prior state of the entire universe, plus the laws of physics." Anything that happens can be explained in that way. But we also have access to a more *useful* way of characterizing the situation, which relies crucially on the context in which we are speaking. In this case, it relies on the fact that we know something about wineglasses and their environments, and this particular situation specifically. Left to their own devices, glasses of wine that are sitting peacefully on tables tend to continue doing so. If our glass of wine had been floating in zero gravity on the International Space Station, our analysis would have been quite different.

Understanding context becomes important because our invocation of causality relies on comparing what *actually* happened to what *could* have happened, in a different hypothetical world. Philosophers refer to this as *modal reasoning*—thinking not only about what does happen but about what could happen in possible worlds.

One master of modal reasoning was David Lewis, one of the most influential twentieth-century philosophers whom non-philosophers have never heard of. Lewis suggested that we could make sense of statements like "A causes B" by thinking of different possible worlds: in particular, worlds that were essentially the same except for whether the event A actually occurred. Then, if we see that B occurs in all the worlds where A occurred, and B does not occur when A does not occur, it's safe to say "A causes B." If the wineglass falls and breaks when Sally swings her elbow around, but stays on the table in a closely related world in which she does not, then Sally's elbow swinging caused the glass to fall.

There is one worry about this kind of account. Why can we say that A causes B, rather than B causes A? Why don't we think that the reason why Sally swung her elbow is because the glass was going to be knocked off the table?

The answer has to do with the leverage that different events have on one another. When we're thinking about memories or records, the idea is that the later event (say, a photograph of you at your senior prom) absolutely implies the existence of the former event (you at your senior prom). But not vice versa; we could imagine you going to the prom and avoiding having your photograph taken. Causes are the other way around. Given the wineglass on the ground, we can imagine things other than a stray elbow that could have knocked it down, but given the location of the glass to start, the swinging elbow absolutely implies that the glass will topple. When a later event has great leverage over an earlier one, we call the latter a "record" of the former; when the earlier event has great leverage over a later one, we call the former a "cause" of the latter.

"Memories" and "causes" aren't pieces of our fundamental ontology describing our world that we discover through careful research. They are concepts that we invent in order to provide useful descriptions of the macroscopic world. The arrow of time plays a crucial role in how those contexts relate to the underlying time-symmetric laws of physics. And the origin of that arrow is that we know something specific and informative about the past (it had a low entropy), but there is no corresponding statement we can make about the future. Our progress through time is pushed from behind, not pulled from ahead.

PART TWO

UNDERSTANDING

9

Learning about the World

Not much is known about Rev. Thomas Bayes, who lived during the eighteenth century. Serving mostly as clergyman to his local parish, he published two works in his lifetime. One defended Newton's theory of calculus, back when it still needed defending, and the other argued that God's foremost aim is the happiness of his creatures.

In his later years, however, Bayes became interested in the theory of probability. His notes on the subject were published posthumously, and have subsequently become enormously influential—a Google search on the word "Bayesian" returns more than 11 million hits. Among other people, he inspired Pierre-Simon Laplace, who developed a more complete formulation of the rules of probability. Bayes was an English Nonconformist Presbyterian minister, and Laplace was a French atheist mathematician, providing evidence that intellectual fascination crosses many boundaries.

The question being addressed by Bayes and his subsequent followers is simple to state, yet forbidding in its scope: How well do we know what we think we know? If we want to tackle big-picture questions about the ultimate nature of reality and our place within it, it will be helpful to think about the best way of moving toward reliability in our understanding.

Even to ask such a question is to admit that our knowledge, at least in part, is not perfectly reliable. This admission is the first step on the road to wisdom. The second step on that road is to understand that, while nothing is perfectly reliable, our beliefs aren't all equally unreliable either. Some are more solid than others. A nice way of keeping track of our various degrees of belief, and updating them when new information comes our way, was the contribution for which Bayes is remembered today.

Among the small but passionate community of probability-theory aficionados, fierce debates rage over What Probability Really Is. In one camp are the *frequentists*, who think that "probability" is just shorthand for "how frequently something would happen in an infinite number of trials." If you say that a flipped coin has a 50 percent chance of coming up heads, a frequentist will explain that what you really mean is that an infinite number of coin flips will give equal numbers of head and tails.

In another camp are the *Bayesians*, for whom probabilities are simply expressions of your states of belief in cases of ignorance or uncertainty. For a Bayesian, saying there is a 50 percent chance of the coin coming up heads is merely to state that you have zero reason to favor one outcome over another. If you were offered to bet on the outcome of the coin flip, you would be indifferent to choosing heads or tails. The Bayesian will then helpfully explain that this is the only thing you could *possibly mean* by such a statement, since we never observe infinite numbers of trials, and we often speak about probabilities for things that happen only once, like elections or sporting events. The frequentist would then object that the Bayesian is introducing an unnecessary element of subjectivity and personal ignorance into what should be an objective conversation about how the world behaves, and they would be off.

Our job here isn't to decide anything profound about the nature of probability. We're interested in beliefs: things that people think are true, or at least likely to be true. The word "belief" is sometimes used as a synonym for "thinking something is true without sufficient evidence," a concept that drives nonreligious people crazy and causes them to reject the word entirely. We're going to use the word to mean anything we think is true regardless of whether we have a good reason for it; it's perfectly okay to say "I believe that two plus two equals four."

Often—in fact all the time, if we're being careful—we don't hold our beliefs with 100 percent conviction. I believe the sun will rise in the east tomorrow, but I'm not absolutely certain of it. The Earth could be hit by a speeding black hole and completely destroyed. What we actually have are *degrees of belief,* which professional statisticians refer to as *credences.* If you think there's a 1 in 4 chance it will rain tomorrow, your credence that it will rain is 25 percent. Every single belief we have has some credence attached to it, even if we don't articulate it explicitly. Sometimes credences are just like probabilities, as when we say we have a credence of 50 percent that a fair coin will end up heads. Other times they simply reflect a lack of complete knowledge on our part. If a friend tells you that they really tried to call on your birthday but they were stuck somewhere with no phone service, there's really no probability involved; it's true or it isn't. But you don't know which is the case, so the best you can do is assign some credence to each possibility.

Bayes's main idea, now known simply as *Bayes's Theorem*, is a way to think about credences. It allows us to answer the following question. Imagine that we have certain credences assigned to different beliefs. Then we gather some information, and learn something new. How does that new information change the credences we have assigned? That's the question we need to be asking ourselves over and over, as we learn new things about the world.

Say you're playing poker with a friend. The game is five-card draw, so you each start with five cards, then choose to discard and replace a certain number of them. You can't see their cards, so to begin, you have no idea what they have, other than knowing they don't have any of the specific cards in your own hand. You're not completely ignorant, however; you have some idea that some hands are more likely than others. A starting hand of one pair, or no pairs at all, is relatively likely; getting dealt a flush (five cards of the same suit) right off the bat is quite rare. Running the numbers, a random five-card hand will be "nothing" about 50 percent of the time, one pair about 42 percent of the time, and a flush less than 0.2 percent of the time, not to mention the other possibilities. These starting chances are known as your *prior* credences. They are the credences you have in mind to start, prior to learning anything new.

But then something happens: your friend discards a certain number of cards, and draws an equal number of replacements. That's new information, and you can use it to update your credences. Let's say they choose to draw just one card. What does that tell us about their hand?

It's unlikely that they have one pair; if they had, they probably would have drawn three cards, maximizing the chance that they would improve to three or four of a kind. Likewise, if they had three of a kind to start, they probably would have drawn two cards. But drawing one card fits very well with the idea that they have two pair or four of a kind, in which case they would want to hold on to all four of the relevant cards. It's also somewhat consistent with them having either four cards of the same suit (hoping to draw to a flush) or four cards in a row (hoping to complete a straight). These likely behaviors, sensibly enough, are called the *likelihoods* of the problem. By combining the prior credences with the likelihoods, we arrive at updated credences for what their starting hand was. (Figuring out what their hand probably is after the drawing is complete requires a bit more work, but nothing a good poker player can't handle.) Those updated chances are naturally known as the *posterior* credences.

Bayes's Theorem can be thought of as a quantitative version of the method of inference we previously called "abduction." (Abduction places emphasis on finding the "best explanation," rather than just fitting the data, but methodologically the ideas are quite similar.) It's the basis of all science and other forms of empirical reasoning. It suggests a universal scheme for thinking about our degrees of belief: start with some prior credences, then update them when new information comes in, based on the likelihood of that information being compatible with each original possibility.

The interesting thing about Bayesian reasoning is the emphasis on those prior credences. In the case of poker hands it's not such a challenging idea; the priors come directly from the chances of being dealt different cards. But the concept enjoys a much wider range of applicability.

You're having coffee with a friend one afternoon, and they make one of the following three statements:

- "I saw a man bicycling by my house this morning."
- "I saw a man riding a horse by my house this morning."
- "I saw a headless man riding a horse by my house this morning."

In each of these three cases, you're given essentially the same kind of evidence: a statement uttered by your friend in a matter-of-fact tone. But the credence, or degree of belief, you would subsequently assign to each possibility is utterly different in the three cases. If you live in a city or the suburbs, you are much more likely to believe that your friend saw a bicyclist than a man on horseback—unless, perhaps, police officers in your neighborhood frequently ride horses, or there is a traveling rodeo in town. Whereas if you live out in the country where horses are frequent and the roads aren't paved, it might be easier to accept the horse than the bicycle. In either case, you're going to be much more skeptical that anyone was riding anything while lacking a head.

What's happening is simply that you have priors. Depending on where you live, the prior credence you would assign to seeing bicyclists or horseback riders will be different, and no matter what, your prior for riders having heads is much higher than your prior for riders lacking them. And that's perfectly okay. In fact, any Bayesian will tell you, there's no way around it. Every time we reason about the probable truth of different claims, our answers are a combination of the prior credence we assign to that claim and the likelihood of various bits of new information coming to us if that claim were true.

Scientists are often in the position of judging dramatic-sounding claims. In 2012, physicists at the Large Hadron Collider announced the discovery of a new particle, most likely the long-sought-after Higgs boson. Scientists around the world were immediately ready to accept the claim, in part because they had good theoretical reasons for expecting the Higgs to be found exactly where it was; their prior was relatively high. In contrast, in 2011 a group of physicists announced that they had measured neutrinos that were apparently moving faster than the speed of light. The reaction in that case was one of universal skepticism. This was not a judgment against the abilities of the experimenters; it simply reflected the fact that the prior credence assigned by most physicists to any particle moving faster than light was extremely low. And, indeed, a few months later the original team announced that their measurement had been in error.

There is an old joke about an experimental result being "confirmed by theory," in contrast to the conventional view that theories are confirmed or ruled out by experiments. There is a kernel of Bayesian truth to the witticism: a startling claim is more likely to be believed if there is a compelling theoretical explanation ready to hand. The existence of such an explanation increases the prior credence we would assign to the claim in the first place.

10

Updating Our Knowledge

O nce we admit that we all start out with a rich set of prior credences, the crucial step is to update those credences when new information comes in. To do that, we need to describe Bayes's Theorem in more precise terms.

Let's return to our friendly poker game. We know what cards we have, but we don't know our opponent's cards. This puts us in a situation where there are various different "propositions" (assertions that something is true), and we have a comprehensive list of all the possible propositions. In this case, the propositions correspond to all the various cards our opponent could start with in a poker hand (nothing; a pair; something better than a pair). In other cases they could be the possible interpretations of an outlandish claim a friend makes (they're correct; they're sincere but misguided; they're lying), or a set of competing ontologies (naturalism; supernaturalism; something more exotic).

To every proposition we consider, we assign a prior credence. To help visualize things, we can represent our credences by dividing some grains of sand among a collection of jars. Each jar stands for a different proposition, and the number of grains of sand in each jar is proportional to the credence assigned to that proposition. The credence for proposition X is just the fraction, out of the grains in all the jars, that are in the jar labeled X:

$$\text{Credence in } X = \frac{\text{Grains in jar } X}{\text{Grains in all jars}}$$

Call this the grains-of-sand rule.

Bayes's Theorem tells us how to update those credences when we get some new information. Let's say we get information in the form of some new data, such as the number of cards our opponent draws. Then for each jar, we *remove* a fraction of the sand corresponding to the likelihood that we would *not* have obtained that data if the corresponding proposition were correct. If we think our opponent would draw precisely one card only 10 percent of the time if they had a pair, we remove nine-tenths of the grains of sand from the jar labeled "pair" when we see them draw a single card. Then we do the analogous thing for all the other jars. At the end, our grains-of-sand rule is once again true: the credence of proposition X is the number of grains of sand in jar X divided by the total number in all the jars.

What this procedure does is to re-weight the prior credences by the likelihoods, in order to obtain posterior credences. We might start with a situation where several jars have approximately the same amount of sand, corresponding to equal credences. But then we obtain some new information, which would be likely under some propositions and unlikely under some other ones. We remove just a little sand from the jars where the information was likely, and a lot of sand from those where the information was unlikely. We're left with a relatively greater amount of sand in the more-likely jars, corresponding to greater posterior credence for those propositions. Of course, if our prior credence in one proposition was incredibly large compared to that for its competitors, we would have to remove a very large amount of sand (collect data that was very unlikely under that proposition) for that credence to become small. When priors are very large or very small, the data has to be very surprising in order to shift our credences.

Consider a different scenario: you're a high school student, you have a crush on someone, and you want to ask them to the prom. The question is, will they say yes, or no? So there are two different propositions: "Yes" (they will go to the prom with you) and "No" (they won't), and for each we have a prior credence. Let's be optimistic and assign credence 0.6 to Yes, and 0.4 to No. (Clearly the total credences must always add up to 1.) We set up two jars of sand, in which we place 60 grains in the Yes jar and 40 grains in the No jar. The total number of grains doesn't matter, only the relative proportion.

Our next step is to collect new information and update our priors by using likelihoods. You're standing at your locker, and you see your crush walking down the hall. Will they say hi, or just walk right by you? That depends on how they think about you—they're more likely to stop and say hi if they're also inclined to go with you to the prom than if they're not so inclined. Using your keen knowledge of human interaction, under proposition Yes they will stop and say hi 75 percent of the time, and walk right by 25 percent (maybe they were just distracted). But under proposition No, the odds aren't as good: 30 percent of the time they'll say hi, and 70 percent they'll walk right by. Those are your likelihoods for various information to be gathered under the different propositions. Time to collect some data and update your credences!

Let's say that your crush does, to your delight, stop and say hi. How does that affect the chances that they would accept an invitation to the prom? Reverend Bayes tells us to remove 25 percent of the sand from the Yes jar, and 70 percent of the sand from the No jar (corresponding in each case to

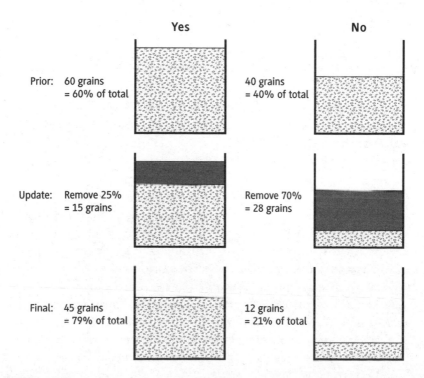

	Yes		No
Prior:	60 grains = 60% of total		40 grains = 40% of total
Update:	Remove 25% = 15 grains		Remove 70% = 28 grains
Final:	45 grains = 79% of total		12 grains = 21% of total

the fraction of the time the observed outcome would not have happened). We're left with 60 x 0.75 = 45 grains in the Yes jar, and 40 x 0.30 = 12 grains in the No jar. According to the grains-of-sand rule stated earlier, the updated credence of Yes is the number of grains in the Yes jar (45) divided by the total number in both jars (45 + 12 = 57). That gives 0.79.

Not bad! The credence that they will say yes if we ask them to the prom has risen from 60 percent, our prior, all the way up to a posterior credence of 79 percent, just because they stopped by to say hi. I think it's time to start shopping for formalwear.

Don't let the crunch of numerical detail obscure the main message. In the Bayesian philosophy, to every proposition that may or may not be true about the world, we assign a prior credence. Each such proposition also comes with a collection of likelihoods: the chances that various other things would be true if that proposition were true. Every time we observe new information, we update our degrees of belief by multiplying our original credences by the relevant likelihood of making that observation under each of the propositions. Symbolically,

$$\left(\begin{array}{c} \text{Credence in proposition } X \\ \text{given observation } D \end{array} \right) \propto \left(\begin{array}{c} \text{Likelihood of observation } D \\ \text{given proposition } X \end{array} \right) \times \left(\begin{array}{c} \text{Prior credence} \\ \text{in proposition } X \end{array} \right)$$

That's Bayes's Theorem in a nutshell. The symbol "$\propto$" means "is proportional to." It's just a reminder that we should make sure all of our credences add up to 1 at the end of the day.

It feels natural to assign numerical credences in certain cases, like poker hands or flips of a coin, where we can simply count all the possibilities. We're also familiar with using probability-talk when referring to future events: "There is less than a 1 percent chance that the oncoming asteroid will impact the Earth and cause a mass extinction."

The Bayesian approach is much more general than this, however. It reminds us that we assign prior credences, and update them appropriately, to *every factual proposition that may or may not be true about the world*. Does God exist? Can our inner conscious experiences be explained in purely physical terms? Are there objective standards of right and wrong? All of the possible answers to such questions are propositions for which each of us has a prior credence (whether we admit it or not), and which we update when relevant new information comes in (whether we do so correctly or not).

Bayes's Theorem allows us to be quantitative about our degrees of belief, but it also helps us keep in mind how belief works at all. Thinking about credences in this way provides a number of useful lessons.

Prior beliefs matter. When we're trying to understand what is true about the world, everyone enters the game with some initial feeling about what propositions are plausible, and what ones seem relatively unlikely. This isn't an annoying mistake that we should work to correct; it's an absolutely necessary part of reasoning in conditions of incomplete information. And when it comes to understanding the fundamental architecture of reality, none of us has complete information.

Prior credences are a starting point for further analysis, and it's hard to say that any particular priors are "correct" or "incorrect." There are, needless to say, some useful rules of thumb. Perhaps the most obvious is that simple theories should be given larger priors than complicated ones. That doesn't mean that simpler theories are always correct; but if a simple theory is wrong, we will learn that by collecting data. As Albert Einstein put it: "The supreme goal of all theory is to make the irreducible basic elements as simple and as few as possible without having to surrender the adequate representation of a single datum of experience."

Simplicity is sometimes easy to gauge; sometimes it is less so. Consider three competing theories. One says that the motion of planets and moons in the solar system is governed, at least to a pretty good approximation, by Isaac Newton's theories of gravity and motion. Another says that Newtonian physics doesn't apply at all, and that instead every celestial body has an angel assigned to it, and these angels guide the planets and moons in their motions through space, along paths that just coincidentally match those that Newton would have predicted.

Most of us would probably think that the first theory is simpler than the second—you get the same predictions out, without needing to invoke vaguely defined angelic entities. But the third theory is that Newtonian gravity is responsible for the motions of everything in the solar system *except* for the moon, which is guided by an angel, and that angel simply chooses to follow the trajectory that would have been predicted by Newton. It is fairly uncontroversial to say that, whatever your opinion about the first two theories, the third theory is certainly less simple than either of them. It involves all of the machinery of both, without any discernible difference in empirical predictions. We are therefore justified in assigning it a very low prior credence. (This example seems frivolous, but analogous moves become common when we start talking about the progress of biological evolution or the nature of consciousness.)

Some people don't like the Bayesian emphasis on priors, because they seem subjective rather than objective. And that's right—they are. It can't be helped; we have to start somewhere. On the other hand, ideally the likelihoods of making certain observations can be objectively determined. If you have a certain theory about the world, and that theory is precise and well-defined, you can say with confidence what the chances are of observing various bits of data under the assumption that your theory is correct. In realistic circumstances, of course, we are often stuck trying to evaluate theories that aren't so rigorously defined in the first place. ("Consciousness transcends the physical" is a legitimate proposition, but it's not sufficiently precise to make quantitative predictions.) Nevertheless, it's our job to try to make our propositions as well-defined as possible, to the point where we can use them to objectively establish the likelihoods of different observations.

Everyone's entitled to their own priors, but not to their own likelihoods.

Evidence should move us toward consensus. You might worry that having subjective priors could make it hard for some people to ever reach agreement. If I assign a prior credence of 0.000001 to an idea like "God created the universe," and you assign a prior credence of 0.999999 to the same proposition, it would require some serious updating on the basis of observations before one of us changed our view.

In practice, that's a real problem. People have certain views that they're just never going to change, which in Bayesian language corresponds to priors set to 0 or 1. That's too bad, and something we need to learn to deal with in the real world.

But in principle, if we are all trying to be fair and open-minded and willing to change our beliefs in the face of new information, evidence will win out in the end. You can assign a very high prior credence to some idea, but if that idea predicts that certain outcomes happen only 1 percent of the time, and those outcomes keep happening, an honest Bayesian updating will eventually lead you to assign a very low posterior credence. You might assign a high prior credence to "Drinking coffee will give me the ability to accurately predict the future." Then you drink some coffee, make predictions, find that your predictions didn't come true, and update appropriately. If you do that enough, the data will wipe out your original prior. That's called "changing your mind," and it's a good thing. Furthermore, since the likelihoods are meant to be objective, gathering more and more data nudges everyone in the direction of the same set of ultimate beliefs about the world.

That's how it's supposed to work anyway. It's up to each of us to honestly carry out the process in good faith.

Evidence that favors one alternative automatically disfavors others. Imagine we are comparing two propositions, X and Y, and we observe an outcome that has a 90 percent chance of happening under X and a 99 percent chance of happening under Y. According to Bayes's Theorem, after collecting that information, the credence we assign to X will go *down*.

That can seem counterintuitive. After all, if X were true, we would have a 90 percent chance of obtaining that outcome—how can observing it count as evidence against this theory? The answer is just that it's even more likely under the other theory. The shift in credences might not be large, but it will always be there. As a result, the fact that you can come up with an explanation for some event within some theory doesn't mean that event doesn't lower the credence you have for the theory. The converse is also true: if some observation would have favored one theory, but we obtained the opposite of that observation, that result necessarily decreases our credence for the theory.

Consider two theories: theism (God exists) and atheism (God doesn't exist). And imagine we lived in a world where the religious texts from different societies across the globe and throughout history were all perfectly compatible with one another—they all told essentially the same stories and promulgated consistent doctrine, even though there was no way for the authors of those texts to have ever communicated.

Everyone would, sensibly, count that as evidence in favor of theism. You could cook up some convoluted explanation for the widespread consistency even under atheism: maybe there is a universal drive toward telling certain kinds of stories, implanted in us by our evolutionary history. But we can't deny that theism provides a more straightforward explanation: God spread his word to many different sets of people.

If that's true, it follows as a matter of inescapable logic that the absence of consistency across sacred texts counts as evidence *against* theism. If data D would increase our credence in theory X, then not-D necessarily decreases it. It might not be hard to explain such inconsistency, even if theism is true: maybe God plays favorites, or not everyone was listening very carefully. That is part of estimating our likelihoods, but it doesn't change the qualitative result. In an honest accounting, the credence we assign to a theory should go down every time we make observations that are more probable in competing theories. The shift might be small, but it is there.

All evidence matters. It's not hard to pretend we're being good Bayesians while we're actually cooking the books by looking at some evidence but not all of it.

Let's say a friend tells you that they believe in the Loch Ness Monster. There are pictures, they say, and they provide good evidence. Surely, you must admit, the likelihood of such pictures being taken is larger under the theory that Nessie is real than under the theory that she isn't.

True, but that's far from the whole story. First, your prior for a monster living in a remote Scottish lake should be pretty small. Even then, if the evidence were sufficiently compelling, you should change your mind. But a few grainy pictures aren't all the evidence. We should also take into account all of the searches in the loch that tried to find a monster and came up empty. Not to mention the evidence that the original famous photograph of Nessie was eventually admitted to be a hoax. We can't pick and choose which evidence we want to consider; everything relevant should be brought to bear.

Bayes's Theorem is one of those insights that can change the way we go through life. Each of us comes equipped with a rich variety of beliefs, for or against all sorts of propositions. Bayes teaches us (1) never to assign perfect certainty to any such belief; (2) always to be prepared to update our credences when new evidence comes along; and (3) how exactly such evidence alters the credences we assign. It's a road map for coming closer and closer to the truth.

11

Is It Okay to Doubt Everything?

Ludwig Wittgenstein, one of the greatest philosophers of the twentieth century, began his doctoral studies at Cambridge as a student of Bertrand Russell, a massively influential thinker in his own right. Russell liked to tell the story of how a young Wittgenstein would deny that anything empirical—an assertion about the real world, rather than a logical provable statement—was truly knowable. In his relatively small quarters at Cambridge, Russell challenged Wittgenstein to admit that there was not a rhinoceros in the room. Wittgenstein refused. "My German engineer, I think, is a fool," Russell wrote in a letter, though he later changed his mind. (Wittgenstein was Austrian, not German, and certainly no fool.)

It's an old parlor game among philosophers, seeing who can be the best at doubting seemingly obvious truths about the world. Skepticism, in the sense of doubting anything, was a popular school of thought in ancient Greece. The champions were the Pyrrhonists, followers of Pyrrho of Elis, who insisted that we couldn't even be sure about the fact that we can never be sure about anything.

A more recent contestant in the game was the seventeenth-century thinker René Descartes. He was not only a philosopher but also a mathematician and scientist, laying the foundations for analytic geometry and contributing to early work in mechanics and optics. If you have ever drawn x and y axes on a piece of graph paper, your life has been affected by René Descartes; he invented that little trick, which we now call "Cartesian coordinates." In his philosophizing, Descartes was very influenced by the practice of mathematics. In particular, he was enchanted by the fact that in math we can prove statements beyond any doubt—at least, once we accept the relevant postulates.

René Descartes, philosopher, mathematician, and doubter of many things other than his own existence, 1596–1650. (Painting after Frans Hals)

In 1641, Descartes published his celebrated *Meditations on First Philosophy*. To this day it is one of the books most likely to be assigned to college students taking their first philosophy course. In *Meditations*, Descartes attempts to be as skeptical as possible about our knowledge of the world. You might think, for example, that you are sitting on a chair, and that the existence of that chair is beyond dispute. But is it really? After all, you've undoubtedly been quite sure about this or that belief in the past, and turned out to be wrong. When we are dreaming or hallucinating, there's no question that we are "experiencing" things that aren't actually happening. It's possible, Descartes suggests, that we are dreaming even now, or that our senses are being tricked by an evil demon, one who (for whatever inscrutable demonic reason) wants us to believe in a chair that doesn't really exist.

But not to lose hope. Descartes concludes that there is one belief about which skepticism is impossible: his own existence. Sure, he reasons, we can doubt the existence of the sky and the Earth—our senses could be fooled. But he can't be skeptical about himself; if he didn't exist, who was it who was being skeptical? Descartes summarized this view in his famous *cogito ergo sum*: I think, therefore I am. (He first wrote that Latin phrase in the later work *Principles of Philosophy*, but the French formulation *je pense, donc je suis* appears in the earlier *Discourse on Method*, aimed at a broader audience.)

It would be an unsatisfying, solipsistic existence if each person could be convinced only that they themselves existed, and had to reserve judgment about everyone else. Descartes wants to build a foundation for justified belief about the whole world, not just himself. But he's not allowed to appeal to anything he sees or experiences—after all, even if he himself exists, that evil demon could still be tricking him when it comes to the evidence of his senses.

So as Descartes's meditations continue, he realizes that he can salvage the reality of the world without ever leaving the comfort of his armchair. Not only do I think, he says to himself, but I can hold in my mind an idea of perfection—a clear and distinct idea, as a matter of fact. This idea, as well as my own existence, must have some cause, and the only possible cause is God. Indeed, God is himself perfect, and the property of "existing" is a necessary aspect of perfection—it is more perfect to exist than to not exist. Therefore, God exists.

And then we are off to the races. If we are confident not only in our own existence but also in God's, then we can be confident in much more than that. After all, God is perfect, and a perfect being wouldn't allow me to be utterly deceived in everything I see and hear. God can overrule any tricky demons that might be trying to mislead me. So the evidence of my senses, and the objective reality of the world, can largely be trusted. Now we can start doing science, secure in the knowledge that we are discovering truths about the universe.

Descartes was a Catholic, and thought of himself as defending his religious beliefs against the nagging doubts of skepticism. Not everyone else saw it that way. His proofs for the existence of God were perceived as bloodless and philosophical, divorced from the intense spiritual experience of lived faith. He was accused of atheism, which for most of recorded history was a way of saying "You don't believe in God the way you are supposed to." (Atheism was one of the crimes for which Socrates was sentenced to death, even though he talked about gods all the time. Meletus, one of his adversaries, ended up accusing him both of atheism and of belief in demigods.) Eventually, in 1663, Pope Alexander VII would place all of Descartes's works on the Church's *Index Librorum Prohibitorum*, the list of officially prohibited writings, where it joined books by Copernicus, Kepler, Bruno, and Galileo, among others.

One of my college professors once told me that nobody could get a PhD in philosophy without writing a refutation of Descartes. It remained unclear which part of Descartes was supposed to be refuted—his initial skepticism and ability to doubt everything, or his laying foundations for secure belief through his conviction that both he and God certainly existed?

Opinions on the existence of God, and in particular on Descartes's purported proofs, vary widely. But before even getting to that part of the argument, most people feel a visceral reaction against "Cartesian doubt." It strikes us as ridiculous and irritating to imagine that we can't be sure of anything at all, not even the existence of the chair on which we are sitting.

But in that part of his method, Descartes was completely correct. We may be quite convinced that the world around us is real, but we can't be *absolutely* certain, beyond any conceivable doubt. We can even come up with a number of scenarios under which we could be fooled, beyond Descartes's suggestions that we might be dreaming or being fooled by an evil demon. We could be a brain in a vat, receiving false impulses from wires hooked directly into our neurons rather than the real outside world. We could be living in a computer simulation like in *The Matrix*, and the true external reality could be something very different than we suppose. Finally, as his critics have pointed out, Descartes shouldn't only worry that he is dreaming; he should also worry that he is being dreamed. (In the Hindu Vedanta tradition, all the world is a dream of Brahma.)

In 1857, naturalist Philip Henry Gosse published a book, *Omphalos*, in which he attempted to reconcile the age of the Earth as inferred from geological evidence (very old) with that inferred from the evidence of the Bible (very young). His idea was simple: God had created the world a few thousand years ago, but with all the signs of being much older, including mountain ranges that would take millions of years to form, and fossils of apparently great antiquity. Gosse's title came from the Greek word for "navel," since part of his inspiration was that the first human, Adam, must have been a complete person, and therefore had a navel, even though no woman had given birth to him. Versions of his idea are promoted to this day by some Christian and Jewish creationists, who use it to account for cosmological evidence of light that left distant galaxies billions of years ago.

It's easy to see how the Omphalos hypothesis leads to yet another skeptical scenario, which has waggishly been labeled "Last Thursdayism"—the idea that the entire universe was created intact just last Thursday, complete with all of the records and artifacts that seem to point to the existence of an extended past. Bertrand Russell once pointed out that there's no way of being completely sure that the world didn't spring into existence five minutes ago. You might think that this can't be true, since you have clear memories of last Wednesday. But a memory—just like a picture, or a diary—exists *now*. We take memories and records as (somewhat) reliable guides to the past, since that seems to have worked for us thus far. It's logically possible, however, that all of those purported memories, as well as our impressions that they are reliable, were created along with everything else.

Without really meaning to, physicists have been led to consider cosmological models that veer uncomfortably close to the Omphalos hypothesis. In the nineteenth century, Ludwig Boltzmann contemplated a universe that has lasted forever but has almost everywhere and almost always been in a state of uniform, uninteresting disorder. The individual atoms in such a universe would be in constant motion, randomly shuffling and bumping into one another. But eventually, if we wait long enough, the motions of the atoms will bring them just by chance into a highly ordered state—for example, much like the Milky Way galaxy, which astronomers of the time thought was the entire universe. (The ancient Roman poet Lucretius suggested a very similar picture; like Boltzmann, he was an atomist, trying to account for the origin of order in the world.) This configuration would evolve as normal, eventually dissipating back into the surrounding chaos

as the universe reaches its ultimate heat death. At least until the next fluctuation.

There is one quite significant problem with Boltzmann's idea. Fluctuations from disorder to order are rare, and larger fluctuations are much more
rare than smaller ones. So if Boltzmann had been right, there's no need to
wait for something as impressive and grand as the Milky Way, with hundreds of billions of stars, to shuffle its way into existence. It's far easier for
something smaller, like the sun and its planets, to emerge out of the chaos.
And when you think about it, the vast majority of conscious, thinking creatures in this kind of universe will be single individuals who have fluctuated
into existence all by themselves—just long enough to think, "Hm, I seem
to be all alone in this universe," and then die. Indeed, why even bother with
an entire body? Most of these lonely souls will be the minimal possible
amount of matter that could qualify as a thinking being: a disembodied
brain, floating in space.

For obvious reasons, this has become known as the "Boltzmann Brain"
scenario. To be clear, nobody thinks the universe is actually like that. The
problem is that it seems like it *should* be true, if the universe is infinitely old
and randomly fluctuating. In that case, the appearance of Boltzmann
Brains seems inevitable. And since the overwhelming majority of observers
in such a universe are disembodied brains, why am I not one?

There is a way out of the Boltzmann Brain problem that is simple, but
wrong. It's to say "Maybe most observers in the universe are random fluctuations, but I'm not one, so I don't really care." How do you know you're
not a random fluctuation? You can't say that you have memories of a long
and fascinating life, since those memories could have fluctuated into exis
tence. You might point to your surroundings—there's a room, and a window, and outside seems to be an elaborate environment, all of which is
much more than would be predicted by this crazy fluctuation scenario.

And that's true; *most* people in this crazy fluctuation scenario shouldn't
find themselves surrounded by rooms and neighborhoods and all the stuff
we are pretty sure constitutes our local environments. But *some* of them
will. If the universe is truly infinitely old, there will be an infinite number
of such environments. And the overwhelming majority of them will have
randomly fluctuated into existence directly from the surrounding chaos.

You may think, for example, that you are reading a book by a person named Sean Carroll, who probably exists (or once did, depending on when you are reading). But given an infinite universe, it's much easier for this book, with my name on the cover and picture on the flap, to randomly fluctuate into existence by itself than for this book *and* my actual person to fluctuate into existence. Even if we grant you the reality of what you seem to experience in your local environment, in Boltzmann's cosmology you have no reason whatsoever to actually trust in the existence of anything else—including anything beyond your immediate perception, or anything you might think you are remembering about the past. All of your memories and impressions, with probability close to 1, just fluctuated into existence themselves. It's the ultimate skeptical scenario.

Are you sure you're not a Boltzmann Brain? Or at least, do you know your local environment didn't recently fluctuate into existence? How do you know you're not a brain in a vat, or a character in some more advanced being's video game?

You don't. You can't. If by "know" we mean "know with absolute, metaphysical certainty, without any conceivable possibility of being wrong," then we cannot ever know that none of these scenarios is correct.

Later in life, Wittgenstein himself contemplated a way out of this conundrum. In *On Certainty*, one of the first things he writes is "From its *seeming* to me—or to everyone—to be so, it doesn't follow that it *is so*." But he immediately follows this with "What we can ask is whether it can make sense to doubt it." Put conversely, something might conceivably be true, but there might not be any point in assigning much credence to it.

Consider the most dramatic kinds of skeptical scenarios, like Descartes's worry that all of his knowledge of the external world is unreliable because he is being fooled by an evil demon. We would like to prove that this is wrong, or at least collect some strong evidence against it. But we can't. A sufficiently powerful and clever demon would be able to influence all of our appeals to logic and evidence. "I think, therefore I am"; "Existence is an attribute of perfection, therefore God exists"—these might very well seem logically sound to you (or at least to Descartes). But that's just what the evil demon would want you to think! How can we be sure that the demon isn't tricking us into logical fallacies?

Any of the various skeptical scenarios about the existence of external reality, and our knowledge thereof, might very well be true. But at the same time, that doesn't mean we should attach high credence to them. The problem is that it is never *useful* to believe them. That's what Wittgenstein means by "making sense."

Let's compare two possibilities: first, that our impression of the reality around us is basically correct, and second, that reality as we know it doesn't exist and we are being fooled by an evil demon. Our inclination is to collect as much information as possible, calculate the likelihood of that information under each scenario, and update our credences accordingly. But in the second scenario, the evil demon could be feeding us the same information we would expect under the first scenario. There is no way to distinguish between the scenarios by collecting new data.

What we're left with is our choice of prior credences. We're allowed to pick priors however we want—and every possibility should get some non-zero number. But it's okay to set our prior credence in radically skeptical scenarios at very low values, and attach higher prior credence to the straightforwardly realistic possibilities.

Radical skepticism is less useful to us; it gives us no way to go through life. All of our purported knowledge, and all of our goals and aspirations, might very well be tricks being played on us. But what then? We cannot actually act on such a belief, since any act we might think is reasonable would have been suggested to us by that annoying demon. Whereas, if we take the world roughly at face value, we have a way of moving forward. There are things we want to do, questions we want to answer, and strategies for making them happen. We have every right to give high credence to views of the world that are productive and fruitful, in preference to those that would leave us paralyzed with ennui.

Some skeptical scenarios aren't merely fanciful concoctions like Descartes's demon—they are situations that we worry could actually be true. A world dominated by Boltzmann Brains is what we would expect if the universe were infinitely old and constantly fluctuating. *The Matrix* was a science-fiction conceit, but philosopher Nick Bostrom has argued that it's more likely we are living in a simulation than directly in the "real world." (The idea is essentially that it's easy for a technologically advanced civilization to run powerful computer simulations, including simulated people, so most "people" in the universe are most likely part of such simulations.)

Is it possible that you and your surrounding environment, including all of your purported knowledge of the past and the outside world, randomly fluctuated into existence out of a chaotic soup of particles? Sure, it's possible. But you should never attach very high credence to the possibility. Such a scenario is *cognitively unstable*, in the words of David Albert. You use your hard-won scientific knowledge to put together a picture of the world, and you realize that in that picture, it is overwhelmingly likely that you have just randomly fluctuated into existence. But in that case, your hard-won scientific knowledge just randomly fluctuated into existence as well; you have no reason to actually think that it represents an accurate view of reality. It is impossible for a scenario like this to be true and at the same time for us to have good reasons to believe in it. The best response is to assign it a very low credence and move on with our lives.

The simulation argument is a little different. Is it possible that you, and everything you've ever experienced, are simply a simulation being conducted by a higher level of intelligent being? Sure, it's possible. It's not even, strictly speaking, a skeptical hypothesis: there is still a real world, presumably structured according to laws of nature. It's just one to which we don't have direct access. If our concern is to understand the rules of the world we do experience, the right attitude is: so what? Even if our world has been constructed by higher-level beings rather than constituting the entirety of reality, by hypothesis it's all we have access to, and it's an appropriate subject of study and attempted understanding.

It makes sense, as Wittgenstein would say, to apportion the overwhelming majority of our credence to the possibility that the world we see is real, and functions pretty much as we see it. Naturally, we are always willing to update our beliefs in the face of new evidence. If there comes a clear night, when the stars in the sky rearrange themselves to say, "I AM YOUR PROGRAMMER. HOW DO YOU LIKE YOUR SIMULATION SO FAR?" we can shift our credences appropriately.

12

Reality Emerges

With our Bayesian knowledge-building tool kit in hand, we can return to fleshing out some of the ideas behind poetic naturalism. In particular, the innocuous-seeming but secretly profound idea that there are many ways of talking about the world, each of which captures a different aspect of the underlying whole.

The progress of human knowledge has bequeathed to us a couple of insights that, taken together, suggest a world that is profoundly different from the picture we construct from our everyday experience. There is conservation of momentum: the universe doesn't need a mover; constant motion is natural and expected. It is tempting to hypothesize—cautiously, always with the prospect of changing our minds if it doesn't work—that the universe doesn't need to be created, caused, or even sustained. It can simply *be*. Then there is conservation of information. The universe evolves by marching from one moment to the next in a way that depends only on its present state. It neither aims toward future goals nor relies on its previous history.

These discoveries indicate that the world operates by itself, free of any external guidance. Together they have dramatically increased our credence in naturalism: there is only one world, the natural world, operating according to the laws of physics. But they also highlight a looming question: Why does the world of our everyday experience *seem* so different from the world of fundamental physics? Why aren't the basic workings of reality perfectly obvious at first glance? Why is the vocabulary we use to describe the everyday world—causes, purposes, reasons why—so different from that of the microscopic world—constant motion, Laplacian patterns?

This brings us to the "poetic" part of poetic naturalism. While there is one world, there are many ways of talking about it. We refer to these ways as "models" or "theories" or "vocabularies" or "stories"; it doesn't matter. Aristotle and his contemporaries weren't just making things up; they told a reasonable story about the world they actually observed. Science has discovered another set of stories, harder to perceive but of greater precision and wider applicability. It's not good enough that the stories succeed individually; they have to fit together.

One pivotal word enables that reconciliation between all the different stories: *emergence*. Like many magical words, it's extremely powerful but also tricky and liable to be misused in the wrong hands. A property of a system is "emergent" if it is not part of a detailed "fundamental" description of the system, but it becomes useful or even inevitable when we look at the system more broadly. A naturalist believes that human behavior emerges from the complex interplay of the atoms and forces that make up individual human beings.

The Starry Night. (Painting by Vincent van Gogh)

Emergence is ubiquitous. Consider a painting, such as van Gogh's *The Starry Night*. The canvas and paint constitute a physical artifact; on one level, it is just a collection of certain atoms in certain locations. There is nothing to the painting other than those atoms. Van Gogh didn't infuse it with any form of spiritual energy; he put the paint onto the canvas. If the atoms making up the paint had been put in different locations, it would have been a different painting.

But it's obvious that specifying an arrangement of atoms isn't the only way of talking about this physical artifact, and it's not even the best way for most purposes. When we talk about *The Starry Night*, we refer to the color palette, the mood it evokes, the swirling of the moon and stars in the sky, and perhaps to van Gogh's period in the asylum at Saint-Paul de Mausole. All of these higher-level concepts are something in addition to a dry (but accurate) list of all the atoms that make up the paint. They are emergent properties.

The classic example of emergence, one you should constantly return to whenever these things get confusing, involves the air in the room around you. That air is a gas, and we can speak of it as having various properties: a temperature, a density, a humidity, a velocity, and so on. We think of the air as a continuous fluid, and all of those properties take on numerical values at every point in the room. (Remember that gases, like liquids, are fluids.) But we know that the air isn't "really" a fluid. If we look at it very closely, down at a microscopic level, we see that it's composed of individual atoms and molecules—mostly nitrogen and oxygen, with trace bits of other elements and compounds. One way of talking about the air would simply be to list every one of those molecules—perhaps 10^{28} of them—and specify their positions, velocities, orientations in space, and so on. This is sometimes called *kinetic theory*, and it's a perfectly legitimate way of talking.

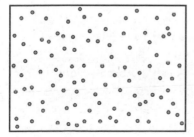

Two ways of thinking about air: as a collection of discrete molecules, or as a smooth fluid.

Specifying the state of each molecule at every moment in time is a consistent and self-contained description of the system; if you were as smart as Laplace's Demon, that would be enough to determine the state at any other time. In practice it's incredibly cumbersome, and nobody ever talks that way.

Describing the air in terms of its macroscopic fluid properties such as temperature and density is also a perfectly legitimate way of talking. Just as there are equations that can tell us how the individual molecules bump into one another and move over time, there are separate equations that tell us how the fluid parameters evolve over time. And the good news is, you don't need to be nearly as smart as Laplace's Demon to actually find the solution; real computers are completely up to the task. Atmospheric scientists and aeronautical engineers solve such equations every day.

So the fluid description and the molecular description are two different ways of talking about the air, both of which—at least in certain circumstances—tell very precise and useful stories about how air behaves. This example illustrates a number of features that commonly appear in discussions of emergence:

- The different stories or theories use utterly different vocabularies; they are different ontologies, despite describing the same underlying reality. In one we talk about the density, pressure, and viscosity of the fluid; in the other we talk about the position and velocity of all the individual molecules. Each story comes with an elaborate set of ingredients— objects, properties, processes, relations—and those ingredients can be wildly different from one story to another, even if they are all "true."

- Each theory has a particular *domain of applicability*. The fluid description wouldn't be legitimate if the number of molecules in a region were so small that the effects of particular molecules were important individually, rather than only in aggregate. The molecular description is effective under wider circumstances, but still not always; we could imagine packing enough molecules into a small enough region of space that they collapsed to make a black hole, and the molecular vocabulary would no longer be appropriate.

- Within their respective domains of applicability, each theory is *autonomous*—complete and self-contained, neither relying on the other. If we're speaking the fluid language, we describe the air using density and pressure and so on. Specifying those quantities is enough to answer whatever questions we have about the air, according to that theory. In particular, we don't need to ever refer to any ideas about molecules and their properties. Historically, we talked about air pressure and velocity long before we knew it was made of molecules. Likewise, when we are talking about molecules, we don't ever have to use words like "pressure" or "viscosity"—those concepts simply don't apply.

The important takeaway here is that stories can invoke utterly different ideas, and yet accurately describe the same underlying stuff. This will be crucially important down the line. Organisms can be alive even if their constituent atoms are not. Animals can be conscious even if their cells are not. People can make choices even if the very concept of "choice" doesn't apply to the pieces of which they are made.

If we have two different theories that both accurately describe the same underlying reality, they must be related to each other and mutually consistent. Sometimes that relationship is simple and transparent; other times we just have to trust that it's there.

The case of fluid dynamics emerging from molecules is as simple as it gets. One theory can directly be obtained from the other by a process known as *coarse-graining*. There is an explicit map from one theory (molecules) to the other (fluid). A particular state in the first theory—a list of all the molecules, their positions, and velocities—corresponds to some particular state in the second one—a density and pressure and velocity of the fluid at every point.

Moreover, many different states in the molecular theory get mapped to the *same* state in the fluid one. When this is the case, we often call the first theory the "microscopic" or "fine-grained" or "fundamental" one, and the second the "macroscopic" or "coarse-grained" or "emergent" or "effective" one. These labels aren't absolute. To a biologist working with an emergent theory of cells and tissue, the theory of atoms and their interactions might be a microscopic description; to a string theorist working on the quantum theory of gravity, superstrings might be the microscopic entities, and atoms are emergent. One person's microscopic is another person's macroscopic.

We want our theories to give physical predictions that are consistent with each other. Imagine that a state x in the microscopic theory evolves into some state y. And imagine that the "emergence" map sends x and y to states X and Y in the emergent fluid theory. Then it had better be the case that X evolves to Y under the rules of the emergent theory, at least with very high probability. Starting with a microscopic state, the process "evolve forward in time, and see what that corresponds to in the emergent theory" should give the same answer as "see what it corresponds to in the emergent theory, then evolve forward in time."

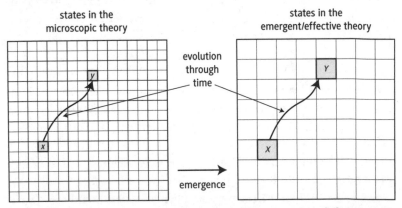

Emergence of one theory from another. Boxes in each image represent different possible states the entire system could be in, as described by each theory. Time evolution and emergence should be compatible: microstates that map to the same emergent state should evolve into microstates that also map to the same emergent state. Several microstates map to each emergent state.

Coarse-graining goes one way—from microscopic to macroscopic—but not the other way. You can't discover the properties of the microscopic theory just from knowing the macroscopic theory. Indeed, emergent theories can be *multiply realizable*: there can, in principle, be many distinct microscopic theories that are incompatible with one another but compatible with the same emergent description. You can understand the air as a fluid without knowing anything about its molecular composition, or even if there is a description in terms of particles at all.

The reason why emergence is so helpful is that different theories are not created equal. Within its domain of applicability, the emergent fluid theory is enormously more computationally efficient than the microscopic molecular theory. It's easier to write down a few fluid variables than the states of all those molecules. Typically—though not necessarily—the theory that has a wider domain of applicability will also be the one that is more computationally cumbersome. There tends to be a trade-off between comprehensiveness of a theory and its practicality.

Our ability to construct two different theories about the air in your room, once as a fluid and another time as a collection of molecules, is an especially concrete and vivid example of emergence, and more generally of the poetic-naturalist idea of telling multiple stories about the same underlying reality. There are, as you might guess, some subtleties worth exploring.

One of the features of the molecules/fluid example is that we can *derive* the macroscopic fluid theory from the microscopic molecular theory. That is, we can start with the molecules, assume that there is a high density of molecules at every point in space, and then "smooth out" the distribution to obtain explicit formulas for fluid properties such as pressure and temperature in terms of what the molecules are doing. This is what is meant by "coarse-graining" above.

Sneakily, however, we have taken advantage of a very special feature of kinetic theory, one that doesn't readily extend to other situations we might be interested in. At heart, the molecules in the air are simple objects, mindlessly bumping into one another when they pass through the same point in space. All we're really doing to derive the fluid description is calculating the average properties of all the molecules. The average number of molecules gives us the density, the average energy gives us the temperature, the average momentum moving in different directions gives us the pressure, and so on.

We can't take such features for granted. Quantum mechanics, in particular, features the phenomenon of *entanglement*. It's not possible to specify the state of a system by listing the state of all of its subsystems individually; we have to look at the system as a whole, because different parts of it can be entangled with one another. To dig a bit deeper, when we combine quantum mechanics with gravity, it is widely believed (although not known for certain, since we know almost nothing for certain about quantum gravity) that space itself is emergent rather than fundamental. Then it doesn't even make sense to talk about "a location in space" as a fundamental concept.

We needn't ascend to esoteric realms of quantum gravity to find situations in which a straightforward smoothing-out process isn't enough to take us from a microscopic theory to an emergent one. Perhaps we want to have a theory of the human brain that emerges out of the behavior of many neurons. Or a theory of a single neuron that emerges out of the interactions of the molecules of which it is made. The problem is that both neurons and the complicated organic molecules in each neuron are pretty complex in their own right; their behavior depends in subtle ways on the specific inputs they are receiving from their environments. Simply averaging over all of them in some region isn't going to capture all of that subtlety. That's not to say that there can't be a useful emergent theory, with a many-to-one map from neuron states to brain states, or molecular states to neuron states; it's just that obtaining it is going to be a bit more indirect than it was for the air in our room.

The molecular and fluid descriptions of air in a room provide an innocent, uncontroversial example of emergence. Everyone agrees on what is happening and how to talk about it. But its simplicity can be misleading. Seeing how relatively easy it is to derive fluid mechanics from molecules, one can get the idea that deriving one theory from another is what emergence is all about. It's not—emergence is about different theories speaking different languages, but offering compatible descriptions of the same underlying phenomena in their respective domains of applicability. If a macroscopic theory has a domain of applicability that is a subset of the domain of applicability of some microscopic theory, and both theories are consistent, then the microscopic theory can be said to *entail* the macroscopic one; but that's often something we take for granted, not something that can explicitly be demonstrated. The ability to actually go through the steps to derive one theory from another is great when it happens, but not at all crucial to the idea.

As systems evolve through time, perhaps in response to changes in their external environment, they can pass from the domain of applicability of one kind of emergent description to a different one—what's known as a *phase transition*. Water is the most familiar example. Depending on the temperature and pressure, water can find itself in the form of solid ice, liquid water, or gaseous water vapor. The underlying microscopic description remains the same—molecules of H_2O—but the macroscopic properties shift from one "phase" to another. Because of the different conditions, the way that we talk about the water changes: the density, hardness, speed of sound through the medium, and other characteristics of the water can be completely altered, and our vocabulary changes along with them. (You wouldn't talk about *pouring* a block of ice, or *chipping* a cup of liquid water.)

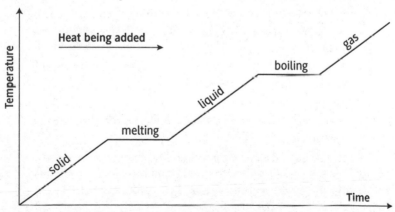

How water changes phase from solid to liquid to gas, as heat is added to it and the temperature rises. The melting and boiling points exhibit plateaus; here the internal structure of the molecules is being rearranged, even though the temperature remains fixed.

The way that phase transitions actually occur is a subject of endless fascination to scientists. Some transitions are rapid, some are slow; some change the substance utterly, others represent a more gradual evolution. The figure illustrates one interesting feature of phase transitions: not all changes are visible on the surface. As we add heat to water, it goes from ice to liquid to vapor, and the temperature rises along the way. At the precise transition point, there is a period where the temperature remains constant while the molecular structure of the water is being rearranged. Entirely new physical properties can come into existence as we change phases, such as solidity or transparency or electrical conductivity. Or life, or consciousness.

When we're talking about simple molecular systems, it's often possible to pinpoint precisely what kind of theoretical vocabulary is appropriate, as well as where we transition from one phase to another. The boundary lines become fuzzier when we start discussing biology or human interactions, but the same basic ideas apply. We've all witnessed phase transitions in the mood of a roomful of people, when someone says the right (or wrong) thing, or when a new person enters the dynamic. Here is a partial list of important phase transitions in the history of the cosmos:

- The formation of protons and neutrons out of quarks and gluons in the early universe.
- Electrons combining with atomic nuclei to make atoms, several hundred thousand years after the Big Bang.
- The formation of the first stars, filling the universe with new light.
- The origin of life: a self-sustaining complex chemical reaction.
- Multicellularity, when different living organisms merged to become one.
- Consciousness: the awareness of self and the ability to form mental representations of the universe.
- The origin of language and the ability to construct and share abstract thoughts.
- The invention of machines and technology.

There are phase transitions in the realm of ideas as well as that of materials. Philosopher of science Thomas Kuhn popularized the idea of a "paradigm shift" to describe how new theories could induce scientists to conceptualize the world in starkly different ways. Even an individual person changing their mind about something can be thought of as a phase transition: our best way of talking about that person is now different. People, like water, can exhibit plateaus in their thinking, where outwardly they hold the same beliefs but inwardly their mental gears are gradually turning.

The fact that each theory or way of talking works only within a specified domain of applicability is absolutely crucial. Again, the example of air is a simple one, but perhaps so simple that it lulls us into a false sense of complacency.

Even though we think of the air in the room as "really" being made of various molecules, that theory's domain of applicability fails to include some situations, such as when the density becomes so high that the air would collapse into a black hole. (Not to worry, that's far removed from the physical situation in most rooms you will find yourself in.) But the fluid description also fails in those cases. In fact, the domain of applicability of the emergent fluid theory is a strict subset of the domain of applicability of the molecular theory.

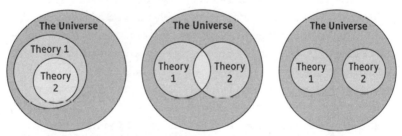

How domains of applicability of different theories could relate to each other.

That situation—two ways of talking, one of whose domain of applicability fits inside that of the other—is by no means necessary. In the diagram, we have shown various ways that domains of applicability might fit together. One might be a subset of the other; or the two might be distinct but overlapping; or they could just be completely different, not sharing any situations in common. For example, in string theory, a leading candidate for a quantum theory of gravity, there are "duality relations" between theories that leave us in the middle situation, where we have two theories with overlapping domains of applicability.

Another example—controversially—might be human consciousness. People are made of particles, and we have a successful picture of how individual particles behave, the Core Theory we'll discuss more in chapter 22. You might think that we could fully describe a person if only we knew the complete state of all of their particles. We have every reason to believe that the domain of applicability of particle physics includes the particles that make up human beings. But it's possible, however unlikely, that there is one set of rules obeyed by particles when there are only a handful of them interacting with one another, as studied by particle physicists, and a slightly different set of rules that they obey when they come together to make a person. This is called *strong emergence*, which we'll discuss in the next chapter. There's no direct evidence that this is true for human beings, but it might help you avoid the ramifications of having all of human behavior described in principle by the known rules of particle physics, if those are the kinds of ramifications you find unpleasant.

These non-hierarchical domains of applicability are not the situation we most often encounter in discussions of emergence. It is far more common to find situations like the leftmost one in the diagram, where one theory is appropriate in a subset of the domain of another theory, perhaps in a nested chain of multiple theories. Indeed, this is closest to the notion of a "hierarchy of sciences," introduced by French philosopher Auguste Comte in the nineteenth century. In this view, we start with physics at the most microscopic and comprehensive level; out of that emerges chemistry, and then biology, and then psychology, and finally sociology.

It is this hierarchical picture that leads people to talk about "levels" when they discuss emergence. Lower levels are more microscopic, fine-grained descriptions, while higher levels are more macroscopic and coarse-grained. That can be convenient when it happens, but what matters is not the existence of a hierarchy but the existence of different ways of talking that describe the same underlying world, and are compatible with each other when their domains of applicability overlap.

13

What Exists, and What Is Illusion?

Auguste Comte helped coin the term "sociology," and put it at the top of his pyramid of science; he thought of the study of societies as the "crowning edifice" of this hierarchy. Subsequently, the dazzling success of physics at describing the microscopic world has flipped things around in some people's minds; they prefer to focus on the deepest, most fundamental way of talking about reality. Ernest Rutherford, a New Zealand–born experimental physicist who was as responsible as anyone for discovering the structure of the atom, once remarked that "all of science is either physics or stamp collecting." It should come as no surprise that scientists who are not physicists—the very large majority of scientists, in other words—would beg to differ.

From the point of view of emergence, the question becomes: how *new and different* are emergent phenomena? Is an emergent theory just a way of repackaging the microscopic theory, or is it something truly novel? For that matter, is the behavior of the emergent theory derivable, even in principle, from the microscopic description, or does the underlying stuff literally act differently in the macroscopic context? A more provocative way of putting the same questions would be: are emergent phenomena *real*, or merely illusory?

As you might imagine, these questions lie front and center when we start talking about knotty issues such as the emergence of consciousness or free will. Sure, you think you're making a choice about whether to have that last slice of pizza or virtuously resist the temptation, but are you sure you really are? If the underlying laws of nature are deterministic, then isn't your volition simply an illusion?

But the independent reality of emergent phenomena is an important issue even when we stick to physics. Philip Anderson won the Nobel Prize in Physics in 1977 for his work on the electronic properties of materials. He is a "condensed matter" physicist—someone who thinks about materials, fluids, or other macroscopically tangible forms of matter here on Earth, as opposed to an astrophysicist, atomic physicist, or particle physicist. In the 1990s, when the US Congress was contemplating the fate of the Superconducting Super Collider particle accelerator, Anderson was called to testify as an expert in physics who was not directly involved in particle physics. He told the committee that the new machine would doubtless do good work, but any discoveries it would make would be utterly irrelevant to his own research. That was honest, and accurate, if a bit frustrating to the particle physicists who hoped the whole field would present a unified front. (Congress canceled the SSC in 1993; a competing machine, the Large Hadron Collider, was built in Europe, and went on to discover the Higgs boson in 2012.)

Anderson's comments were based on the fact that an emergent theory can be completely independent of more fine-grained comprehensive descriptions of the same system. The emergent theory is autonomous (it works by itself, without reference to other theories) and multiply realizable (many microscopic theories can lead to the same emergent behavior).

Anderson would be interested in questions about, for example, how current flows through a particular kind of ceramic. We know that the material is made of atoms, and we know the rules by which electricity and magnetism interact with those atoms. For the questions Anderson cares about, that's all we need to know. We can think of the theory of atoms, electrons, and their interactions as the emergent theory, and anything more fine-grained than that as a microscopic theory. The emergent theory has its own rules, independent of any purported lower levels. And it may very well be multiply realizable. Anderson doesn't need to worry about the quarks zipping about inside an atomic nucleus, or about the Higgs boson itself, and certainly not about superstring theory or anything that tries to give a more comprehensive microscopic description of matter. (For much of his work, he doesn't even need to know about atoms, as he is working at an even higher level of coarse-graining.)

Given this situation, condensed-matter physicists have long argued that we should think of emergent phenomena as truly new, not "merely" smeared-out versions of some deeper description. In 1972 Anderson published an influential article entitled "More Is Different," arguing that every one of the multiple overlapping stories we can tell about nature deserves to be studied and appreciated for its own sake, rather than focusing primarily on the most fundamental level. He has a point. A famous problem in condensed-matter physics is to find a successful theory of high-temperature superconductors, materials through which electrical current can flow without resistance. Everyone working on the problem believes that such materials are made out of ordinary atoms, obeying the ordinary microscopic rules; knowing that has been of essentially zero help in guiding us toward an understanding of why high-temperature superconductivity happens at all.

There are several different questions here, which are related to one another but logically distinct.

1. Are the most fine-grained (microscopic, comprehensive) stories the most interesting or important ones?
2. As a research program, is the best way to understand macroscopic phenomena to first understand microscopic phenomena, and then derive the emergent description?
3. Is there something we learn by studying the emergent level that we could not understand by studying the microscopic level, even if we were as smart as Laplace's Demon?
4. Is behavior at the macroscopic level incompatible—literally inconsistent with—how we would expect the system to behave if we knew only the microscopic rules?

Regarding question 1, it's obviously a subjective matter. If you're interested in particle physics, and your friend is interested in biology, neither is right or wrong; you're just different. Question 2 is a bit more practical, and the answer is fairly obvious: no. In almost all cases of interest, we might learn a little bit about higher levels by studying lower ones, but we'll learn more (and more quickly) by studying those higher levels themselves.

It's at question 3 where things become contentious. One point of view would say: if we completely understand the microscopic level, which has a domain of applicability that strictly contains that of the emergent theory, we know everything there is to know. Whatever question you have could, in principle, be translated into the microscopic language and answered there.

But "in principle" covers a multitude of sins here, or at least one very big sin. This perspective amounts to saying "You want to know if it will rain tomorrow? Just tell me the position and velocity of all the molecules in the Earth's atmosphere, and I'll get to calculating." Not only is that wildly unrealistic; it's also ignoring the fact that the emergent theory describes true features of the system that might be completely hidden from the microscopic point of view. You might have a self-contained and comprehensive theory of how things behave, but that doesn't mean you know everything; in particular, you don't know all of the useful ways of talking about the system. (Even if you know how every atom in a box of gas behaves, you might be blind to the important fact that the system can also be described as a fluid.) From that perspective—the correct one—we really do learn something new by studying emergent theories for their own sakes, even if all the theories are utterly compatible.

Then we have question 4, where all hell breaks loose.

We're now entering into the realm known as *strong emergence*. So far we've been discussing "weak emergence": even if the emergent theory gives you new understanding and an enormous increase in practicality in terms of calculations, in principle you could put the microscopic theory on a computer and simulate it, thereby finding out exactly how the system would behave. In strong emergence—if such a thing actually exists—that wouldn't be possible. When many parts come together to make a whole, in this view, not only should we be on the lookout for *new knowledge* in the form of better ways to describe the system, but we should contemplate *new behavior*. In strong emergence, the behavior of a system with many parts is not reducible to the aggregate behavior of all those parts, even in principle.

The notion of strong emergence is a bit puzzling, on the face of it. It starts by admitting that there is a sense in which a big macroscopic object, such as a person, is made up of smaller constituents, such as atoms. (In quantum mechanics, remember, this division into constituents isn't always possible, but that's not the subtlety that strong emergentists usually have in mind.) It further admits that there is a microscopic theory, one that will tell you how an atom will behave in any particular circumstance. But then it claims that there is an effect on that atom by the larger system of which it is a part—an effect that cannot be thought of as arising from all of the other atoms individually. The only way to think of it is as an effect of the whole on the individual parts.

I can imagine focusing on one particular atom that currently resides as part of the skin on the tip of my finger. Ordinarily, using the rules of atomic physics, I would think that I could predict the behavior of that atom using the laws of nature and some specification of the conditions in its surroundings—the other atoms, the electric and magnetic fields, the force due to gravity, and so on. A strong emergentist will say: No, you can't do that. That atom is part of you, a person, and you can't predict the behavior of that atom without understanding something about the bigger person-system. Knowing about the atom and its surroundings is not enough.

That is certainly a way the world could work. If it's how the world actually does work, then our purported microscopic theory of the atom is simply *wrong*. The nice thing about theories in physics is that they are very clear about what information is needed to predict the behavior of an object, and also clear about what the predicted behavior actually is. There's no ambiguity in what that atom is supposed to do, according to our best theory of physics. If there are situations in which the atom behaves otherwise, such as when it's part of the tip of my finger, then our theory is wrong and we have to do better.

Which is completely possible, of course. (Many things are possible.) In chapters 22 to 24 we'll dive more deeply into how our best theories of physics work, including the remarkably successful and unforgiving framework of quantum field theory. Within quantum field theory, there is no way for new forces or influences to play an important role in what atoms do in my body—or, more precisely, all of the possible ways this could happen have been ruled out by experiments. But it's always conceivable that quantum field theory itself is just wrong. There's no evidence that it's wrong, however, and very powerful experimental and theoretical reasons to think it's right, within a very wide domain of applicability. So we're allowed to contemplate alterations in this basic paradigm of physics—but we should be aware of how dramatically we are changing our best theories of the world, just in order to account for a phenomenon (human behavior) that is manifestly extremely complex and hard to understand.

We may or may not need to bite the bullet of strong emergence in order to understand the relationship between the atoms of which we are made and the consciousness we all experience. But it's our duty to figure out how they are related, given that both atoms and consciousness exist in the real world.

Or do they?

There is a continuum of possible stances toward the way that the different stories of reality fit together, with "strong emergence" (all stories are autonomous, even incompatible) on one end and "strong reductionism" (all stories reduce to one fundamental one) on the other. A strong reductionist would be someone who not only wants to relate macroscopic features of the world to some underlying fundamental description but also wants to go further by denying that elements of the emergent ontology even *exist*, under some appropriate definition of "exist." The real problem with consciousness, according to this school of thought, would be that there's no such thing. Consciousness is merely an illusion; it doesn't really exist. In the context of philosophy of mind, this hard-core flavor of reductionism is known as *eliminativism*, since its proponents want to eliminate talk of mental states entirely. (Naturally, there is a rich zoo of different types of eliminativism, each of which disagrees with the others about what should be eliminated and what should be kept.)

What is real, and what is not, doesn't seem like an intractable problem at first glance. The table in front of you is real; unicorns are not. But what if that table is made of atoms? Would it be fair to say that the atoms are real, but not the table?

That would be a certain construal of the word "real," limiting its applicability to only the most fundamental level of existence. It's not the most convenient definition we can imagine. One problem is that we don't, as yet, actually have a full theory of reality at its deepest level. If that were our standard for true existence, the only responsible attitude would be to say that nothing that human beings have ever contemplated is actually real. It's a philosophy with a certain Zen purity, but it's not very helpful if we would like to use the concept of "real" to distinguish certain phenomena from others. Wittgenstein would say that it doesn't make sense to talk that way.

A poetic naturalist has another way out: something is "real" if it plays an essential role in some particular story of reality that, as far as we can tell, provides an accurate description of the world within its domain of applicability. Atoms are real; tables are real; consciousness is undoubtedly real. (A similar view was put forward by Stephen Hawking and Leonard Mlodinow, under the label "model-dependent realism.")

Not everything is real, even by this permissive standard. Physicists used to believe in the "luminiferous aether," an invisible substance that filled all of space, and which served as a medium through which electromagnetic waves of light traveled. Albert Einstein was the first to have the courage to stand up and remark that the aether served no empirical purpose; we could simply admit that it doesn't exist, and all of the predictions of the theory of electromagnetism go through unscathed. There is no domain in which our best description of the world invokes the concept of luminiferous aether; it's not real.

Illusions are just mistakes, concepts that play no useful role in descriptions at any level of coarse-graining. When you are crawling across the desert sands, out of water and not completely in your right mind, and think you see a lush oasis with palm trees and a pond in the distance—that's an illusion (probably), in the sense that it's actually not there. But if you get lucky and it really *is* there, and you scoop up liquid water into your hand, that liquid is real, even if we have a more comprehensive way of talking that describes it in terms of molecules made of oxygen and hydrogen.

Consciousness is not an illusion, even if we think it is "just" an emergent way of talking about our atoms each individually obeying the laws of physics. If hurricanes are real—and it makes sense to think that they are—even though they are just atoms in motion, there is no reason why we should treat consciousness any differently. To say that consciousness is real isn't to say that it's something over and above the physical world; it's emergent, and it's also real, just like almost every other thing we've encountered in our lives.

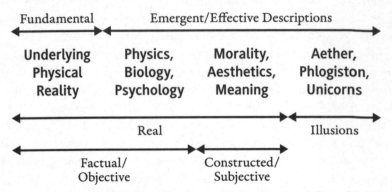

How poetic naturalism divides up "fundamental" versus "emergent/effective," "real" versus "illusion," and "objective" versus "subjective."

Describing our naturalism as "poetic" is helpful because there are other kinds of naturalism out there. There are austere forms of naturalism that try to eliminate everything in sight, and insist that the only "true" way of talking about the world is the deepest, most fundamental one. On the other side of the spectrum are augmented forms of naturalism, which hold that there is more to the world at a fundamental level than mere physical reality. This is a grab-bag category that would include those who believe mental properties are real and distinct from physical ones, or those who believe that moral principles are as objective and fundamental as the physical world.

Poetic naturalism sits in between: there is only one, unified, physical world, but many useful ways of talking about it, each of which captures an element of reality. Poetic naturalism is at least consistent with its own standards: it tries to provide the most useful way of talking about the world we have.

The most seductive mistake we can be drawn into when dealing with multiple stories of reality is to mix up vocabularies appropriate to different ways of talking. Someone might say, "You can't truly *want* anything, you're just a collection of atoms, and atoms don't have wants." It's true that atoms don't have wants; the idea of a "want" is not part of our best theory of atoms. There would be nothing wrong with saying "None of these atoms making up you want anything."

But it doesn't follow that *you* can't have wants. "You" are not part of our best theory of atoms either; you are an emergent phenomenon, meaning that you are an element in a higher-level ontology that describes the world at a macroscopic level. At the level of description where it is appropriate to talk about "you," it's also perfectly appropriate to talk about wants and feelings and desires. Those are all real phenomena in our best understanding of human beings. You can think of yourself as an individual human being, *or* you can think of yourself as a collection of atoms. Just not both at the same time, at least when it comes to asking how one kind of thing interacts with another one.

That's the ideal case, anyway. Following Galileo's lead of ignoring complications and searching for simplicity, physicists have developed formalisms in which the separation between different ways of talking—"effective field theories"—is precise and well-defined. Once we get beyond physics to the more nuanced and complex realms of biology and psychology, demarcating one theory from another becomes more difficult. We can talk about human beings coming down with an illness and becoming contagious, possibly passing on their disease to other people. "Illness" is a useful category in our vocabulary for describing human beings, with a reality all its own, independent of its microscopic underpinnings. But we know that there is a deeper level according to which that illness is a manifestation of, for example, a viral infection. We can't help but be sloppy and mix up our talk of people and illnesses and viruses into one big messy vocabulary.

Just as investigating dualities between different physical theories provides full employment for physicists, investigating how different vocabularies relate to one another and sometimes intermingle provides full employment for philosophers. For our purposes, we can leave that as homework for the ontologically fastidious, and leap into a different question: How do we go about constructing a set of ways to talk about our actual world?

14

Planets of Belief

Most people don't lose sleep worrying whether the world they see is basically real, or whether they're being tricked by an evil demon. We accept that what we see and hear reflects reality with at least some degree of reliability, and move on from there. This leaves us with a more subtle problem: how do we construct a comprehensive picture of how things work that is both reliable and consistent with our experience?

Descartes was looking for a "foundation" for justified belief. A foundation keeps a structure firmly rooted in solid ground. *Foundationalism* is the search for such solid ground, on which to erect the edifice of knowledge.

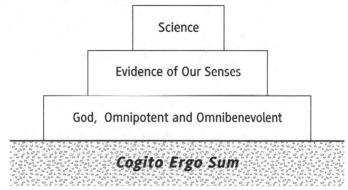

Knowledge as a series of beliefs resting on a secure foundation.

Let's take that metaphor more seriously than it perhaps deserves. On the scale of a human being, the ground beneath our feet is unquestionably solid and reliable. If we zoom out a bit, however, we see that the ground is simply part of the planet on which we live. And that planet, the Earth, isn't grounded on anything at all; it is moving freely through space, orbiting around the sun. The individual bits of matter that constitute the Earth aren't embedded in an unmoving structure; they are held together by their mutual gravitational force. All of the planets in the solar system formed gradually, as bits of rock and dust accreted together, each collection growing in influence and pulling together what remaining scraps of matter it could.

Without meaning to, we've discovered a much more accurate metaphor for how systems of belief actually work. Planets don't sit on foundations; they hold themselves together in a self-reinforcing pattern. The same is true for beliefs: they aren't (try as we may) founded on unimpeachable principles that can't be questioned. Rather, whole systems of belief fit together with one another, in more or less comfortable ways, pulled in by a mutual epistemological force.

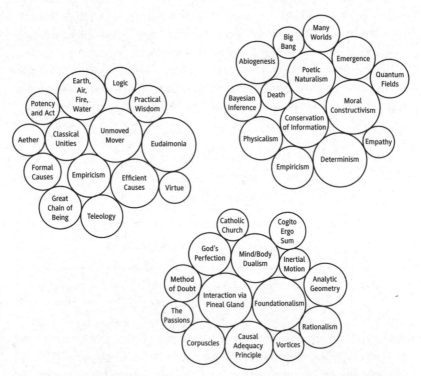

Knowledge as a set of beliefs held together by the "gravitational pull" of their mutual consistency. Parts of the planets of belief for Aristotle, Descartes, and a modern poetic naturalist.

In this picture, a *planet of belief* is much richer and more complex than simply an ontology. An ontology is a view about what really exists; a planet of belief contains all sorts of other convictions, including methods for understanding the world, a priori truths, derived categories, preferences, aesthetic and ethical judgments, and more. If you believe that two plus two equals four and chocolate ice cream is objectively better than vanilla, those are not parts of your ontology, but they are parts of your planet of belief.

No analogy is perfect, but the planets-of-belief metaphor is a nice way to understand the view known in philosophical circles as *coherentism*. According to this picture, a justified belief is one that belongs to a coherent set of propositions. This coherence plays the role of the gravitational pull that brings together dust and rocks to form real planets. A stable planet of belief will be one where all the individual beliefs are mutually coherent and reinforcing.

Some planets are not stable. People go through life with a very large number of beliefs, some of which may not be compatible with others, even if they don't recognize it. We should think of planets of belief as undergoing gradual but constant churning, bringing different beliefs into contact with one another, just as real planets experience convection in the mantle and plate tectonics near the surface. When two dramatically incompatible beliefs come into direct contact, it can be like highly reactive chemicals being mixed together, leading to an impressive explosion—possibly even blowing the entire planet apart, until a new one can be reassembled from different parts.

Ideally, we should be constantly testing and probing our planets of belief for inconsistencies and structural deficiencies. Precisely because they are floating freely through space, rather than remaining anchored on solid and immovable ground, we should always be willing to improve on our planets' composition and architecture, even to the point of completely jettisoning old beliefs and replacing them with better ones. The new information we receive through our observations is like the rain of meteors and comets that is perpetually falling on real planets, to be incorporated into our view of the world. Occasionally, there may even be an asteroid impact of such magnitude that the entire planet is destroyed. These instabilities, either from internal inconsistency or from an external shock, are more likely to happen to relatively young planets, ones that have not completely settled down, but we're all vulnerable.

The real problem is that we can imagine more than one stable planet— there can be multiple sets of beliefs that are consistent within the sets, but not among them. One person's planet might include the scientific method, as well as the belief that the universe is billions of years old; another's might include a belief in biblical literalism, as well as the belief that the world was created a few thousand years ago. If each planet consists of beliefs that are consistent with each other, how do we ever know which is *right*?

This is a real worry. People do hold beliefs that clash violently with the beliefs of others, even though they may seem consistent with other beliefs of their own. But there is reason to hope that the problem isn't insurmountable.

As a matter of empirical fact, there are a number of important, common beliefs that almost everyone shares. Most people believe that reason and logic play an important role in finding truth. They might disagree over whether those are uniquely powerful techniques, but very few people reject them outright. We also tend to share the goal of coming up with models of the world that provide accurate representations of what we actually observe. If you confront a young-Earth creationist who thinks that the world came into being 6,000 years ago with scientific evidence for a very old Earth and universe, their typical response is not "Oh, I don't believe in evidence and logic." Rather, they will attempt to account for that evidence within their belief system, for example, by explaining why God would have created the universe that way.

That's the way it's supposed to work, anyway. But mere "coherence" might seem like precious little on which to base a theory of truth. Abandoning the quest for a secure foundation in favor of a planet of belief is like moving from firm ground to a boat on choppy seas or a spinning teacup ride. It can make you dizzy, if not seasick. We are spinning through space, nothing to hold on to.

What rescues our beliefs from being completely arbitrary is that one of the beliefs in a typical planet is something like "true statements correspond to actual elements of the real world." If we believe that, and have some reliable data, and are sufficiently honest with ourselves, we can hope to construct belief systems that not only are coherent but also agree with those of other people and with external reality. At the very least, we can hold that up as a goal.

There is a crucial difference, in other words, between *stable* planets of belief, ones where all the different pieces attract one another in a consistent and coherent way, and *habitable* planets, ones where we could actually live. A habitable planet of belief necessarily includes some shared convictions about evidence and rationality, as well as the actual information we have gathered about the world. We can hope that people working in good faith will, after trying hard to understand reality the best they can, end up constructing planets of belief that are somewhat compatible with one another.

We shouldn't overestimate people's rationality or willingness to look at new evidence as objectively as possible. For better or for worse, planets eventually develop highly sophisticated defense mechanisms. When you realize that you are holding two beliefs that are in conflict with each other, psychologists refer to the resulting discomfort as *cognitive dissonance*. It's a sign that there is something not completely structurally sound about your planet of belief. Unfortunately, human beings are extremely good at maintaining the basic makeup of their planets, even under very extreme circumstances.

Leon Festinger, an American social psychologist who was the founder of cognitive-dissonance theory, and his collaborators once studied an apocalyptic cult led by a woman named Dorothy Martin (known to generations of psychology students by the pseudonym Marian Keech). Following Martin's lead, members of her group became convinced that the Earth was going to be destroyed on December 21, 1954, but that the true believers would be rescued by aliens the night before. The cult members were extremely serious; they quit their jobs, left their families, and huddled together to await the big day. Festinger was curious as to how they would react when—as his own planet of belief led him to surmise—nothing special happened on the appointed day. Would they, confronted with the unassailable fact that their leader's prophecy had been incorrect, change their minds about her mystical powers?

The day came and went—and afterward, the believers were more convinced of Martin's prophetic ability than ever. On the morning of the twenty-first, as it happened, Martin had conveyed a new vision: it was precisely the unflagging belief of their small group that had been able to prevent the Earth's destruction. Overjoyed, and very ready to believe, her followers doubled down on their commitment, and proceeded to try to spread the word of their insights as widely as possible.

Human beings are not nearly as coolly rational as we like to think we are. Having set up comfortable planets of belief, we become resistant to altering them, and develop *cognitive biases* that prevent us from seeing the world with perfect clarity. We aspire to be perfect Bayesian abductors, impartially reasoning to the best explanation—but most often we take new data and squeeze it to fit with our preconceptions.

It's worth highlighting two important cognitive biases that we can look to avoid as we put together our own planets. One is our tendency to give higher credences to propositions that we *want* to be true. This can show up at a very personal level, as what's known as *self-serving bias*: when something good happens, we think it's because we are talented and deserving, while bad things are attributed to unfortunate luck or uncontrollable external circumstances. At a broader level, we naturally gravitate toward theories of the world that somehow flatter ourselves, make us feel important, or provide us with comfort.

The other bias is our preference for preserving our planet of belief, rather than changing it around. This can also show up in many ways. *Confirmation bias* is our tendency to latch on to and highlight any information that confirms beliefs we already have, while disregarding evidence that may throw our beliefs into question. This tendency is so strong that it leads to the *backfire effect*—show someone evidence that contradicts what they believe, and studies show that they will usually come away holding their initial belief even more strongly. We cherish our beliefs, and work hard to protect them against outside threats.

Our need to justify our own beliefs can end up having a dramatic influence on what those beliefs actually are. Social psychologists Carol Tavris and Elliot Aronson talk about the "Pyramid of Choice." Imagine two people with nearly identical beliefs, each confronted with a decision to make. One chooses one way, and the other goes in the other direction, though initially it was a close call either way. Afterward, inevitably, they work to convince themselves that the choice they made was the right one. They each justify what they did, and begin to think there wasn't much of a choice at all. By the end of the process, these two people who started out almost the same have ended up on opposite ends of a particular spectrum of belief—and often defending their position with exceptionally fervent devotion. "It's the people who *almost* decide to live in glass houses who throw the first stones," as Tavris and Aronson put it.

We're faced with the problem that the beliefs we choose to adopt are shaped as much, if not more, by the beliefs we already have than by correspondence with external reality.

How can we guard ourselves against self-reinforcing irrationality? There is no perfect remedy, but there is a strategy. Knowing that cognitive biases exist, we can take that fact into account when doing our Bayesian inference. Do you want something to be true? That should count *against* it in your assignment of credences, not for it. Does new, credible evidence seem incompatible with your worldview? We should give it *extra* consideration, not toss it aside.

A Utopia of rationality might not be achievable by flawed human beings, but it's something to which we can aspire. Robert Aumann, an Israeli American mathematician who shared the Nobel Prize in Economics in 2005, was able to prove a wonderful mathematical theorem: two people, both acting rationally, who start with the same Bayesian prior credences for their beliefs, and who have access to the same information, including knowing what the other knows, *cannot disagree* about the updated credences for those beliefs. You might think that people can start with common priors but disagree about the likelihoods for observations being obtained, but Aumann's theorem shows that this can't happen if both share "common knowledge"—that is, when everyone knows what everyone else knows (and they all know that they all know it).

Aumann's "agreement theorem" sounds too good to be true, in part because it doesn't comport very well with actual human behavior. In the real world, people are not completely rational, they don't have common knowledge, they misinterpret one another, and they certainly don't start with the same priors. But it gives us hope that we could come to common agreement, even on very contentious issues, if we worked hard enough at it. Even wildly different priors will eventually be swamped by the process of updating if we collect enough evidence. If we try to be as honest as possible with others and with ourselves, we can hope to bring our planets of belief into closer alignment.

15

Accepting Uncertainty

L et's say you want to take a scientist down a peg, make her a bit flustered. Here's an easy way to do it: whenever she says that something is true, in her considered opinion as a scientist, just ask, "Can you really *prove* that?" If your adversary is a good scientist, but not trained in public relations, chances are very high that she will hem and haw, finding it difficult to give a straight answer. Science never proves anything.

A lot depends on our definition of "proof." Scientists will often have in their minds the kind of proof we have access to in mathematics or logic: a rigorous demonstration of the truth of a proposition, starting with some explicitly stated axioms. This differs in important ways from how we might hear "proof" used in casual conversation, where it's closer to "sufficient evidence that we believe something is true."

In a court of law, where precision is a goal but metaphysical certitude can never be attained, the flexible nature of proof is explicitly recognized by invoking different standards depending on the case. In a civil court, proving your case requires that a "preponderance of evidence" be on your side. In some administrative courts, "clear and convincing evidence" is required. And a criminal defendant is not considered to be proven guilty unless the case has been demonstrated "beyond a reasonable doubt."

None of these would impress a mathematician in the slightest; their first instinct would be to start thinking about the unreasonable doubts. Scientists, who have often taken a few math courses in their day, tend to have a similar idea about what constitutes proving something—and they know that it's not what they do for a living. So if a scientist says "Human activity is heating up the planet," or "The universe is billions of years old," or "The Large Hadron Collider is not going to make a black hole that will gobble up the Earth," all you have to do is innocently ask whether they can really prove it. Once they hesitate, you will have won a rhetorical victory. (You will not have made the world a better place, but that's your decision.)

Let's see the distinction more explicitly. Here is a mathematical theorem: There is no largest prime number. (Primes are whole numbers greater than zero that can be evenly divided by only one and themselves.) And here is a proof:

> Consider the set of all the prime numbers: {2, 3, 5, 7, 11, 13 . . . }. Suppose that there is a largest prime, p. Then there are only a finite number of primes. Now consider the number X that we obtain by multiplying together all of the primes from our list, exactly once each, and adding 1 to the result. Then X is clearly larger than any of the primes in our list. But it is not divisible by any of them, since dividing by any of them yields a remainder 1. Therefore either X itself must be prime, or it must be divisible by a prime number larger than any in our list. In either case there must be a prime larger than p, which is a contradiction. Therefore there is no largest prime.

Here is a scientific belief: Einstein's theory of general relativity (GR) accurately describes how gravity works, at least within the solar system, and at least to an extremely high accuracy. And here is the argument for it:

> GR incorporates both the principle of relativity (positions and velocities can be measured only relative to other objects) and the principle of equivalence (in small regions of space, gravity is indistinguishable from acceleration), both of which have been tested to very high precision. Einstein's equation of GR is the simplest possible non-trivial dynamic equation for the

curvature of spacetime. GR explained a preexisting anomaly—
the precession of Mercury—and made several new predictions,
such as deflection of light by the sun and the gravitational red-
shift, which have successfully been measured. Higher-precision
tests from satellites continue to constrain any possible devia-
tions from GR. Without taking GR effects into account, the
Global Positioning System would rapidly go out of whack, and
by including GR it works like a charm. All of the known alternatives are more complicated than GR, or introduce new free
parameters that must be finely tuned with experiment to avoid
contradiction. Furthermore, we can start from the idea of mass-
less graviton particles that interact with all sources of energy,
and show that the only complete version of such a theory leads
to GR and Einstein's equation. Although the theory is not successfully incorporated into a quantum-mechanical framework,
quantum effects are expected to be negligible in present-day experiments. In particular, quantum corrections to Einstein's
equation are expected to be unobservably small.

None of the details here is important; what matters is the difference in
underlying method. The mathematical proof is airtight; it's just a matter of
following the rules of logic. Given the assumptions, the conclusion necessarily follows.

The argument in favor of believing general relativity—a scientific one,
not a mathematical one—is of an utterly different character. It's abduction:
hypothesis testing, and accumulating better and better pieces of evidence,
seeking the best explanation of the phenomena. We throw a hypothesis out
there—gravity is the curvature of spacetime, governed by Einstein's
equation—and then we try to test it or shoot it down, while simultaneously
searching for alternative hypotheses. If the tests get better and better, and
the search for alternatives doesn't turn up any reasonable competitors, we
gradually start saying that the hypothesis is "right." There is no sharp, bright
line that we cross, at which the idea goes from being "just a theory" to being
"proven correct." When scientists observed the deflection of starlight
during a total eclipse of the sun, just as Einstein had predicted, that

didn't prove that he was right; it simply added to a growing pile of evidence in his favor.

It is an intrinsic part of this process that the conclusion didn't have to turn out that way. We could certainly imagine a world in which some more complicated theory than Einstein's was the empirically correct theory of gravity, or perhaps even one in which Newtonian gravity was correct. Deciding between the alternatives is not a matter of proving or disproving; its a matter of accumulating evidence past the point where doubt is reasonable, updating our credences along the way like good Bayesians. This is a fundamental difference between the kind of knowledge given to us by mathematics/logic/pure reason and the kind we get from science. The truths of math and logic would be true in any possible world; the things science teaches us are true about our world, but could have been false in some other one. Most of the interesting things it is possible to know are not things we could ever hope to "prove," in the strong sense.

Even when we do believe a theory beyond reasonable doubt, we still understand that it's an approximation, likely (or certain) to break down somewhere. There could very well be some new hidden field that we haven't yet detected that acts to slightly alter the true behavior of gravity from what Einstein predicted. And there is certainly something going on when we get down to quantum scales; nobody believes that general relativity is really the final word on gravity. But none of that changes the essential truth that GR is "right" in a certain well-defined regime. When we do hit upon an even better understanding, the current one will be understood as a limiting case of the more comprehensive picture.

These features of science—a form of knowledge gathering that we understand relatively well—apply more broadly. The basic recognition is that knowledge, like most things in life, is never perfect. Inspired by logically rigorous proofs of geometry, Descartes wanted to establish an absolutely secure, bedrock foundation for our understanding of the world. That's just not how knowledge of the world works.

Think about Bayes's Theorem: the credence we place in an idea after receiving some new information is the prior credence we started with for that idea, times the likelihood of obtaining that new information if our idea was correct. At first glance, it seems easy to achieve perfect certainty: if the likelihood for a particular outcome is exactly zero according to some idea, and we observe that outcome to occur, our credence in that idea gets set to zero.

But if we're being scrupulous, we shouldn't ever think that the likelihood of observing a particular outcome is precisely zero. You might think something like "In special relativity, particles never travel faster than light, so I have zero credence that I would ever observe a faster-than-light particle if special relativity were correct." The problem is that your observations could always be mistaken. Maybe you think you've seen a particle traveling faster than light, but instead your apparatus was faulty. This is always possible, no matter how careful you are. We should always imagine that there is some nonzero likelihood for absolutely any observation in absolutely any theory.

As a result, our credences never go all the way to zero—nor precisely to 100 percent, since there are always competing possibilities. And it's a good thing that credences never reach these points of absolute certainty; if they did, no amount of new evidence could ever change our minds. That's no way to go through life.

Not everyone agrees, of course. You may have heard that there is a long-running dispute about the relationship between "faith" and "reason." Some argue that there is perfect harmony between them, and indeed there have historically been many successful scientists and thinkers who have been extremely devout. Others argue that the very notion of faith is inimical to the practice of reason.

The discussion is complicated by the presence of multiple incompatible notions of what is meant by "faith." A dictionary might define it as "trust" or "confidence" in a belief, but it will go on to offer meanings along the lines of "belief without justification." The New Testament (Hebrews 11:1) says "Now faith is the substance of things hoped for, the evidence of things not seen." For many, faith is simply a firm conviction in their religious beliefs.

The word "faith" is highly charged, and this isn't the place to argue over how it should be defined. Let us merely note that sometimes faith is taken as something that is absolutely certain. Consider these statements from the Catechism of the Catholic Church:

- The faithful receive with docility the teachings and directives that their pastors give them in different forms.
- To obey (from the Latin *ob-audire*, to "hear or listen to") in faith is to submit freely to the word that has been heard, because its truth is guaranteed by God, who is Truth itself. Abraham is the model of such obedience offered us by Sacred Scripture. The Virgin Mary is its most perfect embodiment.
- Faith is certain. It is more certain than all human knowledge because it is founded on the very word of God who cannot lie.

It is this kind of stance—that there is a kind of knowledge that is certain, which we should receive with docility, to which we should submit—that I'm arguing against. There are no such kinds of knowledge. We can always be mistaken, and one of the most important features of a successful strategy for understanding the world is that it will constantly be testing its presuppositions, admitting the possibility of error, and trying to do better. We all want to live on a stable planet of belief, where the different parts of our worldview fit together harmoniously; but we want to avoid being sucked into a black hole of belief, where our convictions are so strong that we can never escape, no matter what kind of new insight or information we obtain.

You will sometimes hear the claim that even science is based on a kind of "faith," for example, in the reliability of our experimental data or in the existence of unbreakable physical laws. That is wrong. As part of the practice of science, we certainly make *assumptions*—our sense data is giving us roughly reliable information about the world, simple explanations are preferable to complex ones, we are not brains in vats, and so forth. But we don't have "faith" in those assumptions; they are components of our planets of belief, but they are always subject to revision and improvement and even, if necessary, outright rejection. By its nature, science needs to be completely open to the actual operation of the world, and that means that we stand ready to discard any idea that is no longer useful, no matter how cherished and central it may once have seemed.

Because we should have nonzero credences for ideas that might seem completely unlikely or even crazy, it becomes useful to distinguish between "knowing" and "knowing with absolute logical certainty." If our credence for some proposition is 0.0000000001, we're not absolutely certain it's wrong—but it's okay to proceed as if we know it is.

When the Large Hadron Collider particle accelerator in Geneva began operation in 2008, a fuss was raised by people who had heard that the LHC might create black holes that would ultimately destroy the Earth, ending all life as we know it. Sure, the physicists gave assurances that such an occurrence was extremely unlikely. But they couldn't *prove* that it wasn't going to happen. And with consequences as drastic as these, can it ever be worth taking the risk, no matter how unlikely the outcome is supposed to be?

One possible response to such people would be: Consider going home tonight and cooking some pasta for dinner. But before you open the lid on that jar of marinara sauce, ask yourself: What if a freak mutation inside the jar has created a deadly pathogen that will be released if and only if you open the lid, spreading through the world and killing all forms of life? Clearly that would be bad; just as clearly, it seems very unlikely. But you can't *prove* that it won't happen. There's a chance, even if it's very small.

The resolution is to admit that some credences are so small that they're not worth taking seriously. It makes sense to act as if we know those possibilities to be false.

So we take "I believe x" not to mean "I can prove x is the case," but rather "I feel it would be counterproductive to spend any substantial amount of time and effort doubting x." We can accumulate so much evidence in favor of a theory that maintaining skepticism about it goes from being "prudent caution" to being "crackpottery." We should always be open to changing our beliefs in the face of new evidence, but the evidence required might need to be so overwhelmingly strong that it's not worth the effort to seek it out.

We are left with, not absolute proof of anything, but a high degree of confidence in some things, and greater uncertainty in others. That's both the best we can hope for and what the world does as a matter of fact grant us. Life is short, and certainty never happens.

16

What Can We Know about the World
without Actually Looking at It?

O ur most direct, tangible, verifiable connection to the world around
us is through our senses. We see things, touch them, and come to
understand something about them. But there are times when we
seem to experience reality at a deeper level, without the intermediation of
our senses. How are we to account for such experiences as we try to understand the big picture?

The first time I visited London, wandering around one evening with no
real plans, I noticed a poster advertising a concert at St Martin-in-the-
Fields, a church near Trafalgar Square. It's a famous place, especially in
classical-music circles, but at that moment its primary virtue was that it was
nearby, and a concert seemed to qualify as the kind of cultural enrichment
that young people were supposed to seek out when traveling abroad.

It was more than that. The concert was by candlelight: electricity extin-
guished, the expansive nave was illuminated by the soft flicker of hundreds
of small flames. The musicians played selections from Bach and Haydn,
sonorous notes reverberating through the shadowy space. Locals and tour-
ists alike huddled in overcoats, partaking both of the immediate moment
and of the larger sweep of history—musical, architectural, sacred. The
vaulted ceilings evoked the night sky, and the cadence of the music played
off the human rhythms of breaths and heartbeats. Perhaps for the regular
attendees of the concert series it was just another pleasant night out; for me
it was a transcendent experience.

"Transcendent," from the Latin *transcendere,* "climb over, surpass," is a word we attach to experiences that seem to reach beyond our mundane physical situation. A wide variety of circumstances can earn the label. For some, transcendence occurs when your spirit comes into direct contact with the divine. For Christians it might involve the witness of the Holy Spirit, while for Hindus or Buddhists it can refer to escaping the material world in favor of a higher spiritual reality. Individuals can experience transcendence through prayer, meditation, solitude, or even psychoactive drugs such as ayahuasca or LSD. It could simply be a matter of letting one's self be lost in a particularly moving piece of music, or in the love of one's family.

Many of us have had such experiences, though disputes arise over whose have been "truly" transcendent. They can play an important role in who we are, helping us achieve peace or joy, even guide us in making important decisions. For our present purposes, we want to know what transcendent experiences imply about the structure of the world. Do they arise from the behavior of the atoms and neurons in our physical brains, or should we think of such moments as indications of contact with a numinous realm, something truly beyond the physical? What, in other words, does transcendence teach us about ontology?

Behind these questions lurks an even bigger issue. Science advances by observation and experiment: we pose hypotheses about how the world works and then test them by collecting new information and performing the appropriate Bayesian updating. But is that the *only* way to learn about the world? Isn't it at least conceivable that we could come to knowledge of reality in ways other than the scientific, using methods other than hypothesis testing and collecting data? Certainly, throughout history, people have thought that they've gained understanding through revelation, spiritual practice, or other nonempirical methods. The possibility needs to be taken seriously.

Science, even broadly construed, is certainly not the only way that we can come to acquire new knowledge. The obvious exceptions are mathematics and logic.

While math is lumped together with science in many school curricula—and while they certainly enjoy a close and mutually beneficial relationship—at heart they are completely different endeavors. Math is all about proving things, but the things that math proves are not true facts about the actual world. They are the implications of various assumptions. A mathematical demonstration shows that *given* a particular set of assumptions (such as the axioms of Euclidean geometry or of number theory), certain statements inevitably follow (such as the angles inside a triangle adding up to 180 degrees, or there being no largest prime number). In this sense, logic and mathematics can be thought of as different aspects of the same underlying strategy. In logic, as in math, we start with axioms and derive results that inevitably follow from them. Though we casually speak of "logic" as a single set of results, it is actually a procedure for inferring conclusions from axioms. There are different possible sets of axioms from which one can draw logical conclusions, just as there are different sets of axioms one could use in geometry or number theory.

The statements we can prove based on explicitly stated axioms are known as *theorems*. But "theorem" doesn't imply "something that is true"; it only means "something that definitely follows from the stated axioms." For the conclusion of the theorem to be "true," we would also require that the axioms themselves be true. That's not always the case; Euclidean geometry is a marvelous edifice of mathematical results, and certainly useful in many real-world situations, but Einstein helped us see that the actual geometry of the world obeys a more general set of axioms, invented by Bernhard Riemann in the nineteenth century.

We can think of the difference between math and science in terms of possible worlds. Math is concerned with truths that would hold in any possible world: given these axioms, these theorems will follow. Science is all about discovering the *actual* world in which we live. Working scientists might find it useful to occasionally consider non-real worlds (like ones with no friction, or a different number of dimensions of space) for purposes of improving their intuition, but among all the possible worlds, it's the one real world that they ultimately care about. There are possible worlds in which space is flat and Euclid's axioms are true, and other possible worlds in which space is curved and those axioms are false; but in every possible world, Euclid's axioms imply that the interior angles of a triangle add up to 180 degrees.

The way that science goes about narrowing down our world from an infinite number of possible ones is pretty clear: by looking at it. Performing observations and experiments, gathering data, and using that to increase our credence in the useful, explanatory theories.

Science is sometimes described as adhering to *methodological naturalism*: choosing only to consider explanations that are grounded in the natural world, and to discount from the start possible interventions by non-natural phenomena. This characterization is even used by its supporters, in part for political and strategic reasons. The United States has long been plagued by arguments over the teaching of creationism (biological species were created by God) versus that of Darwin's theory of natural selection. An approach called Intelligent Design has been put forward as a "scientific" version of creationism, under the theory that it could therefore be taught as science rather than as religion. Opponents of creationism sometimes countered this argument by appealing to the principle of methodological naturalism; by their lights, the reference in Intelligent Design to a supernatural creator immediately rendered it nonscientific. No less an authority than the National Academy of Sciences wrote,

> Because science is limited to explaining the natural world by means of natural processes, it cannot use supernatural causation in its explanations. Similarly, science is precluded from making statements about supernatural forces because these are outside its provenance.

Not really. Science should be interested in determining the *truth*, whatever that truth may be—natural, supernatural, or otherwise. The stance known as methodological naturalism, while deployed with the best of intentions by supporters of science, amounts to assuming part of the answer ahead of time. If finding the truth is our goal, that is just about the biggest mistake we can make.

Fortunately, it's also an inaccurate characterization of what science actually is. Science isn't characterized by methodological naturalism but by methodological *empiricism*—the idea that knowledge is derived from our experience of the world, rather than by thought alone. Science is a technique, not a set of conclusions. The technique consists of imagining as many different ways the world could be (theories, models, ways of talking) as we possibly can, and then observing the world as carefully as possible.

This broad characterization includes not only the obviously recognized sciences like geology and chemistry but social sciences like psychology and economics, and even subjects such as history. It's not a bad description of how many people typically figure things out about the world, albeit in a somewhat less systematic way. Nevertheless, science shouldn't be simply identified with "reason" or "rationality." It doesn't include math or logic, nor does it address issues of judgment, such as aesthetics or morality. Science has a simple goal: to figure out what the world actually is. Not all the possible ways it could be, nor the particular way it should be. Just what it is.

There's nothing in the practice of science that excludes the supernatural from the start. Science tries to find the best explanations for what we observe, and if the best explanation is a non-natural one, that's the one science would lead us to. We can easily imagine situations in which the best explanation scientists could find would reach beyond the natural world. The Second Coming could occur; Jesus could return to Earth, the dead could be resurrected, and judgment could be passed. It would be a pretty dense set of scientists indeed who, faced with the evidence of their senses in such a situation, would stubbornly insist on considering only natural explanations.

The relationship between science and naturalism is not that science *presumes* naturalism; it's that science has provisionally *concluded* that naturalism is the best picture of the world we have available. We lay out all of the ontologies we can think of, assign some prior credences to them, collect as much information we can, and update those credences accordingly. At the end of the process, we find that naturalism gives the best account of the evidence we have, and assign it the highest credence. New evidence could lead to future adjustments in our credences, but right now naturalism is well out ahead of the alternatives.

Science uses the strategy of empiricism, learning about the world by looking at it. There is a countervailing tradition: *rationalism*, the idea that we can come to true knowledge of the world by methods other than through our sensory experience.

"Rationalism" sounds like a good idea; who doesn't want to be rational? But this particular use of the word refers to learning about the world by reason alone, without any help from observation. There are a number of different ways it could happen: we could be equipped with innate knowledge, we could reason about how things are on the basis of incontrovertible metaphysical principles, or we could be gifted with insight through spiritual or other nonphysical means. A close look reveals that none of these is a very reliable way to learn about our world.

None of us comes to life as a blank slate. We have intuitions, instincts, built-in heuristics for dealing with our environment, developed over the long course of evolution—or perhaps, one might believe, planted there by God. The mistake is to think of any of those ideas as "knowledge." Some might be correct, but how would we know? Just as assuredly, some of our natural instincts about the world often turn out to be wrong. The only good reason we have for trusting any supposedly innate ideas is that we test them against experience.

A related route to rationalism is based on the belief that the world has an underlying sensible or logical order, and from this order we can discern a priori principles that simply have to be true, without any need to check up on them by collecting data. Examples might include "for every effect there is a cause," or "nothing comes from nothing." One motivation for this view is our ability to abstract from individual things we see in the world to universal regularities that are obeyed more widely. If we were thinking deductively, like a mathematician or logician, we would say that no collection of particular facts suffices to derive a general principle, since the very next fact might contradict the principle. And yet we seem to do that all the time. This has prompted people like Gottfried Wilhelm Leibniz to suggest that we must secretly be relying on a kind of built-in intuition about how things work.

Perhaps we are. The best way of knowing whether we are is to test that belief against the data, and adjust our credences appropriately.

※

John Calvin, an influential theologian of the Protestant Reformation, suggested that human beings possess an ability known as the *sensus divinatis*, a capacity to directly sense the divine. The notion has been taken up in contemporary discussion by theologian Alvin Plantinga, who goes on to suggest that the sense is shared by all human beings, but that it is faulty or silent in atheists.

Is it possible that God exists, and communicates with human beings in ways that circumvent our ordinary senses? Absolutely. As Plantinga correctly points out, *if* theism is true, then it makes perfect sense to think that God would implant knowledge of his existence directly into human beings. If we are already convinced that God is real and cares about us, there would be good reason to believe that we could learn about God through nonsensory means, such as prayer and contemplation. Theism and this flavor of rationalism could, under these assumptions, be parts of a fully coherent planet of belief.

What that doesn't do is help us decide whether theism actually is true. We have two competing propositions: one is that God exists, and that transcendental experiences represent (at least in part) moments when we are closer to divinity; the other is naturalism, which would explain such experiences the same way it would explain dreams or hallucinations or other impressions that arise from a combination of sensory input and the inner workings of the physical brain. To decide between them, we need to see which one coheres better with other things we believe about the world.

One way that inner, personal spiritual experiences would count as genuine evidence against naturalism would be if it were possible to demonstrate that such mental states—feelings of being in touch with something greater, of being outside one's own body, dissolving the boundaries of self, communicating with nonphysical spirits, participating in a kind of cosmic joy—did not, or could not, arise from ordinary material causes. Like many questions about consciousness and perception, this one is somewhat open, though there is an increasing amount of research that draws direct connections between apparently spiritual experiences and biochemistry in the brain.

The author Aldous Huxley, in his nonfiction book *The Doors of Perception*, describes his experiences with the psychoactive drug mescaline, including "sacramental vision." Similar drugs, such as peyote and ayahuasca, have long been used to induce spiritual states, especially by Native

Americans, and related effects have been noted in association with LSD and psilocybin (magic mushrooms). Huxley felt that mescaline acted to enhance his consciousness, removing filters that shielded his mind from a greater awareness. He would return to psychedelics repeatedly in his life, including at the very end, when he asked his wife, Laura, to inject him with LSD to help alleviate the extreme pain caused by laryngeal cancer. Afterward, Laura reported that his doctors had never seen a patient with that kind of cancer, usually marked by violent convulsions, spend their final moments with so little pain and struggle.

Recent neuroscience indicates that Huxley may have been on the right track about the filtering effects of mescaline. We tend to think of psychedelics as stimulating visions and sensations, but work by Robin Carhart-Harris and David Nutt used functional magnetic resonance imaging (fMRI) to argue that these drugs actually work to suppress neuronal activity in parts of the brain that act as filters. Some parts of our brain, it turns out, are constantly buzzing with images and sensations, which other parts then work to suppress in order to maintain the coherence of our conscious self. The detailed mechanism is unclear, but there are indications that some hallucinogens help activate a certain receptor for serotonin, a neurotransmitter that helps regulate our moods. Psychedelics, in this picture, don't conjure up new hallucinations but simply allow us to consciously perceive what is already bouncing around inside our brains.

It proves nothing about whether we *also* have feelings and visions as a result of a direct connection to a spiritual reality. Perhaps certain drugs have effects that mimic those of genuine transcendent experiences, without actually explaining them away. Perhaps, indeed, drugs or direct physical influences on the brain can open us up to such experiences and bring us into contact with a broader reality. On the other hand, there might be simple and elegant explanations for transcendent experiences that don't lean on a non-natural world in any way.

Given the profound and deeply personal nature of prayer, meditation, and contemplation, it can seem frivolous or diminishing to relate them to psychedelics or the activity of neurons, or even to dispassionate scientific investigation of any sort. But if we want to undertake our journey to the best possible understanding of the world with the intellectual honesty it deserves, we always have to question our beliefs, consider alternatives, and compare them with the best evidence we can gather. It may be the case that transcendent experiences arise from a direct connection with a higher level of reality, but the only way to know is to weigh that idea against what we learn from the world by looking at it.

17

Who Am I?

All of this discussion about emergence and overlapping vocabularies and domains of applicability isn't merely arid philosophizing. It cuts to the very essence of who we are.

Consider an issue that is central to our self-conception: gender and sexuality. As I am typing these words, societies across the world are going through dizzying changes in how they think about this topic. One indication of the change is the shifting status of same-sex marriages. In the United States, the Defense of Marriage Act, which defined "marriage" as far as the federal government was concerned as the union of one man and one woman, was passed overwhelmingly in 1996. The House Judiciary Committee affirmed that the act was intended "to express moral disapproval of homosexuality." By 2013, the Supreme Court had declared that definition unconstitutional, so that the federal government would recognize same-sex marriages that had been sanctioned by any of the states; two years later, the Supreme Court found that it was unconstitutional for individual states to ban the practice, effectively legalizing it nationwide. Thus the United States caught up with Canada, Brazil, much of Europe, and other countries that had already legalized same-sex marriages. Meanwhile, there are still a large number of countries where same-sex relationships are subject to imprisonment, even the death penalty.

If marriage is a contentious issue, gender identity is even more challenging. As social mores are changing, an increasing number of people who identify as a gender different from their biological sex are deciding to accept that aspect of who they are, rather than hiding it or fighting to suppress it. Some transgender people choose to undergo medical procedures to alter their anatomical makeup, while others do not; either way, their psychological affiliation with the gender they identify with can be just as strong as that of "cisgender" people (those whose gender identity agrees with their biological sex). You will always remember the first time that a friend who you've known for years as a woman, and referred to using pronouns "she" and "her," requests to be thought of from now on as a man, using pronouns "he" and "him."

After Ben Barres, a professor of neurobiology at Stanford, gave a well-received seminar at a conference, one of the scientists in the audience remarked, "Ben Barres's work is much better than his sister's." Except that Barres didn't have a sister; the scientist was thinking of Barres himself, who had previously been a woman known as Barbara Barres. It was the same work that was being judged—it just seemed more impressive coming from a man. Our opinion of a person is greatly affected by what sex we perceive them to be.

Whether you are forward-thinking about such things or staunchly traditionalist, it can be a difficult transition to get used to. How can a person you know, or thought you knew, as a man, suddenly just declare that she's a woman? That's like deciding one day that you are eight feet tall. There are some things you just don't get to decide; they simply are what they are. Right?

Part of how we respond to people who are different from us depends on basic features of our own social orientation and frame of mind. Some people have a fundamental live-and-let-live attitude, or are committed social liberals, and make a point of accepting an individual's right to declare who they are. Others tend to be more naturally wary or judgmental, and frown upon behavior that seems unconventional to them.

But there is something deeper here than mere personal attitudes: there is a question of ontology. What categories do you take to "really exist," to play a central role in how the world is organized?

For many people, the concepts of "male" and "female" are deeply rooted in the fabric of the world. There is a natural order of things, and these concepts are an indelible part of it. If eliminativism is the urge to declare as many things illusory as possible, its opposite is *essentialism*: the tendency to take certain categories as immovable features of the bedrock of reality. At the current moment in history, most people are essentialists about gender, but things are changing.

Religious doctrine is a wellspring of essentialism. Consider how the National Catholic Bioethics Center talks about "Gender Identity Disorder" (italics in original):

> We are either male or female persons, and nothing can change that . . . Persons seeking such operations are clearly uncomfortable with who they *really* are . . .
>
> A person can change what genitalia they have, but not one's sex. Receiving hormones of the opposite sex and removing genitalia are not sufficient to change one's sex. Sexual identity is not reducible to hormonal levels or genitalia but is an objective fact rooted in the specific nature of the person . . .
>
> A person's sex identity is not determined by one's subjective beliefs, desires or feelings. It is a function of his or her *nature*. Just as there are geometrical givens in a geometrical proof, sexual identity is an ontological given.

It would be hard to find a more straightforward declaration of gender essentialism, asserting that a person's gender is a function of their "nature," part of "who they really are."

Religion isn't the only source of such a stance. The notion of "Gender Identity Disorder," as a diagnosed condition of people whose gender identity disagrees with their biological sex, first appeared in the *Diagnostic and Statistical Manual* of the American Psychiatric Association in 1980. Long before that, surgical procedures and hormone therapies were used on children who didn't look or feel the way their doctors judged that they should. Only in 2013 was the official APA diagnosis changed to "gender dysphoria," used to refer to psychological discontent with one's own condition, rather than a mismatch with a purportedly objective judgment of what one's sex "really" is.

Poetic naturalism sees things differently. Categories such as "male" and "female" are human inventions—stories we tell because it helps us make sense of our world. The basic stuff of reality is a quantum wave function, or a collection of particles and forces—whatever the fundamental stuff turns out to be. Everything else is an overlay, a vocabulary created by us for particular purposes. Therefore, if a person has two X chromosomes and identifies as male, what of it?

That doesn't mean we should simply eliminate gender, either. A person who is biologically male but identifies as a woman isn't thinking to themselves, "Male and female are just arbitrary categories, I can be whatever I want." They're thinking, "I'm a woman." Just because a concept is invented by human beings, it doesn't imply that it's an illusion. Saying, "I am a woman," or just knowing it, is absolutely useful and meaningful.

This can sound reminiscent of the old postmodern slogan that "reality is socially constructed." There's a sense in which that's true. What's socially constructed are the ways we talk about the world, and if a particular way of talking involves concepts that are useful and fit the world quite accurately, it's fair to refer to those concepts as "real." But we can't forget that there is a single world underlying it all, and there's no sense in which the underlying world is socially constructed. It simply *is*, and we take on the task of discovering it and inventing vocabularies with which to describe it.

People who think that transgenderism is a violation of the natural order sometimes like to use a slippery-slope argument: If gender and sexuality are up for grabs, what about our basic identity as human beings? Is our *species* socially constructed?

There is, indeed, a condition known as "species dysphoria." It is analogous to gender dysphoria but is characterized by a conviction that the subject belongs to a different species. Someone might think that, despite their nominal human form, they are actually a cat, or a horse. Others go further, identifying with species that don't actually exist, like dragons or elves.

Even for the relatively open-minded, a certain grumpiness tends to kick in when confronted with species dysphoria: "If poetic naturalism means that I have to pretend to go along with my crazy teenage nephew who thinks he's a unicorn, I'm going back to my comfortable species essentialism, thank you very much."

The question, however, is whether a particular way of talking about the world is *useful*. And usefulness is always relative to some purpose. If we're being scientists, our goal is to describe and understand what happens in the world, and "useful" means "providing an accurate model of some aspect of reality." If we're interested in a person's health, "useful" might mean "helping us see how to make a person more healthy." If we're discussing ethics and morality, "useful" is closer to "offering a consistent systematization of our impulses about right and wrong."

So poetic naturalism doesn't automatically endorse or condemn someone who thinks they are a dragon, or for that matter someone who thinks they are male or female. Rather, it helps us understand what questions we should ask: What vocabulary gives us the most insight into how this person is thinking and feeling? What helps us understand how they can be happy and healthy? What is the most useful way of conceptualizing this situation? We can certainly imagine thinking through these questions in good faith, and at the end concluding with "Sorry, Kevin. You're not a unicorn."

The real lives of people whose self-conceptions do not match those that society would like them to have can be extremely challenging, and their obstacles are highly personal. No amount of academic theorizing is going to solve those problems with a simple gesture. But if we insist on talking about such situations on the basis of outdated ontologies, chances are high that we'll end up doing more harm than good.

18

Abducting God

Everyone knows Friedrich Nietzsche proclaimed that God is dead. It's one of the few sentences in the history of philosophy that you can buy on T-shirts and bumper stickers. Or if snappy comebacks are more your style, you can also find NIETZSCHE IS DEAD—GOD.

But many people assume that Nietzsche was celebrating God's supposed demise, which isn't really accurate. Although he wasn't denying it, he was certainly worried about the consequences. The famous quip appears in a short parable entitled "The Madman," where Nietzsche's title character runs crying through a marketplace filled with unbelievers.

> The madman jumped into their midst and pierced them with his eyes. "Whither is God?" he cried; "I will tell you. *We have killed him*—you and I . . .
>
> "Do we not feel the breath of empty space? Has it not become colder? Is not night continually closing in on us? Do we not need to light lanterns in the morning? Do we hear nothing as yet of the noise of the gravediggers who are burying God? Do we smell nothing as yet of the divine decomposition? Gods, too, decompose. God is dead. God remains dead. And we have killed him."

Neither Nietzsche nor his fictional madman are happy about the death of God; if anything, they're trying to wake people up to what it really means.

Starting in the nineteenth century, it began to sink in to a growing number of people that the comforting certainties of the old order were beginning to crumble away. As science developed a unified view of nature that exists and evolves without any outside support, many cheered the triumphs of human knowledge. Others saw a dark side to the new era.

Science can help us live longer, or journey to the moon. But can it tell us what kind of life to live, or account for the feeling of awe that overcomes us when we contemplate the heavens? What becomes of meaning and purpose when we can't rely on gods to provide them?

Thinking about God in a rigorous way is not an easy task. He seems to be reluctant to reveal himself very explicitly in the operation of the world. We can debate about the legitimacy of reported miracles, but most of us will grant that they are rare at best. People may feel that they have an inner, personal experience of the divine—but that's not the kind of evidence that is convincing to people other than the experiencer.

For another thing, people don't agree about God. He's a notoriously slippery notion. To some people, God is very much a *person*—an omniscient, omnipotent, omnibenevolent being who created the universe and cares deeply about the fate of human beings, individually and collectively. Others prefer to think of a more abstract notion of God, as something closer to an explanatory *idea* that plays a crucial role in accounting for our world.

What all theists—people who believe in God—tend to agree on is that God is absolutely important. One of the most significant features of someone's ontology is whether or not it includes God. It's the biggest part of the big picture. So, slippery notion or not, deciding how to think about God is something we simply have to do.

Remember that there are two parts to Bayesian reasoning: coming up with prior credences before any evidence is in, and then figuring out the likelihood of obtaining various kinds of information under the competing ideas. When it comes to God, both of these steps are enormously problematic. But we don't have any choice.

For the sake of keeping things simple, let's divide all of the possible ways of thinking about God into just two categories: theism (God exists) and atheism (no, he doesn't). These are catchall terms for a variety of possible beliefs, but we're illustrating general principles here. For the sake of being definite, let's imagine we're talking about God as a person, as some kind of enormously powerful being who is interested in the lives of humans.

What should our priors be for theism and atheism? We could argue that atheism is simpler: it has one fewer conceptual category than theism does. Simple theories are good, so that suggests our prior for atheism should be higher. (If atheism doesn't actually account for the universe we see, that prior will become irrelevant, as the corresponding likelihoods will be very small.) On the other hand, even though God is a separate category from the physical world, we might hope to explain features of the world using that hypothesis. Explanatory power is a good thing, so that might argue in favor of a greater prior for theism.

Let's call it a wash. You are entitled to your own priors, but for purposes of this discussion let's imagine that the prior credences for theism and atheism are about equal. Then all of the heavy lifting will be done by the likelihoods—how well the two ideas do in accounting for the world we actually see.

Here is where things get interesting. What we're supposed to do is to imagine, as fairly as possible, what the world would probably look like according to either of our two possibilities, and then compare it to what it actually is like. That's really hard. Neither "theism" nor "atheism," by itself, is an extremely predictive or specific framework. We can imagine many possible universes that would be compatible with either idea. And our considerations are contaminated by the fact that we actually do know quite a bit about the world. That's a considerable bias to try to overcome.

Take the problem of evil. Why would a powerful and benevolent God, who presumably could simply stop humans from being evil, nevertheless allow it in the world? There are many possible responses to this question. A common one relies on free will: perhaps to God, it is more important that humans be free to choose according to their own volition—even if they end up choosing evil—than to coerce them into being uniformly good.

Our job, however, isn't simply to reconcile the data (the existence of evil) with the theory (theism). It's to ask how the data changes our credences for each of the two competing theories (theism and atheism).

So imagine a world that is very much like ours, except that evil does not exist. People in this world are much like us, and seem able to make their own choices, but they always end up choosing to do good rather than evil. In that world, the relevant data is the absence of evil. How would that be construed, as far as theism is concerned?

It's hard to doubt that the absence of evil would be taken as very strong evidence in favor of the existence of God. If humanity simply evolved according to natural selection, without any divine guidance or interference, we would expect to inherit a wide variety of natural impulses—some for good, some for not so good. The absence of evil in the world would be hard to explain under atheism, but relatively easy under theism, so it would count as evidence for the existence of God.

But if that's true, the fact that we do experience evil is unambiguously evidence *against* the existence of God. If the likelihood of no evil is larger under theism, then the likelihood of evil is larger under atheism, so evil's existence increases our credence that atheism is correct.

Put in those terms, it's easy to come up with features of our universe that provide evidence for atheism over theism. Imagine a world in which miracles happened frequently, rather than rarely or not at all. Imagine a world in which all of the religious traditions from around the globe independently came up with precisely the same doctrines and stories about God. Imagine a universe that was relatively small, with just the sun and moon and Earth, no other stars or galaxies. Imagine a world in which religious texts consistently provided specific, true, nonintuitive pieces of scientific information. Imagine a world in which human beings were completely separate from the rest of biological history. Imagine a world in which souls survived after death, and frequently visited and interacted with the world of the living, telling compelling stories of life in heaven. Imagine a world that was free of random suffering. Imagine a world that was perfectly just, in which the relative state of happiness of each person was precisely proportional to their virtue.

In any of those worlds, diligent seekers of true ontology would quite rightly take those aspects of reality as evidence for God's existence. It follows, as the night the day, that the absence of these features is evidence in favor of atheism.

How strong that evidence is, is another question entirely. We could try to quantify the overall effect, but we're faced with a very difficult obstacle: theism isn't very well defined. There have been many attempts, along the lines of "God is the most perfect being conceivable," or "God is the grounding of all existence, the universal condition of possibility." Those sound crisp and unambiguous, but they don't lead to precise likelihoods along the lines of "the probability that God, if he exists, would give clear instructions on how to find grace to people of all times and cultures." Even if one claims that the notion of God itself is well defined, the connection between that concept and the actuality of our world remains obscure.

One could try to avoid the problem by denying that theism makes any predictions at all for what the world should be like—God's essence is mysterious and impenetrable to our minds. That doesn't solve the problem—as long as atheism does make predictions, evidence can still accumulate one way or the other—but it does ameliorate it somewhat. Only at a significant cost, however: if an ontology predicts almost nothing, it ends up explaining almost nothing, and there's no reason to believe it.

There are some features of our world that count as evidence in favor of theism, just as some features are evidence for atheism. Imagine a world in which nobody had thought of the concept of God—the idea had simply never occurred. Given our definition of theism, that's a very unlikely world if God exists. It would seem a shame for God to go to all the trouble to create the universe and humankind, and then never let us know about his existence. So it's perfectly reasonable to say that the simple fact that people think about God counts as some evidence that he is real.

That's a somewhat whimsical example, but there are more serious ones. Imagine a world with physical matter, but in which life never arose. Or a universe with life, but no consciousness. Or a universe with conscious beings, but ones who found no joy or meaning in their existence. At first glance, the likelihoods of such versions of reality would seem to be higher under atheism than under theism. Much of the task of the rest of this book is to describe how these features are quite likely in a naturalistic worldview.

There's not much to be gained by rehearsing all of the arguments for and against theism here. What matters more is understanding the basis for making progress on this and similar questions. We lay out our prior credences, determine the likelihoods for different things to happen under each competing conception of the world, and then update our credences on the basis of what we observe. That's just as true for the existence of God as it is for the theory of continental drift or the existence of dark matter.

It all sounds very tidy, but we are fallible, finite, biased humans. Someone will argue that a universe with a hundred billion galaxies is exactly what God would naturally create, while someone else will roll their eyes and ask whether that expectation was actually put forward before we went out and discovered the galaxies in our telescopes.

All we can hope to do is to survey our own planets of belief, recognize our biases, and try to correct for them the best we can. Atheists sometimes accuse religious believers of being victims of wishful thinking—believing in a force beyond the physical world, a higher purpose to existence, and especially a reward after death, simply because that's what they want to be true. This is a perfectly understandable bias, one we would be wise to recognize and try to take into consideration.

But there are biases on both sides. Many people may be comforted by the idea of a powerful being who cares about their lives, and who determines ultimate standards of right and wrong behavior. Personally, I am not comforted by that at all—I find the idea extremely off-putting. I would rather live in a universe where I am responsible for creating my own values and living up to them the best I can, than in a universe in which God hands them down, and does so in an infuriatingly vague way. This preference might unconsciously bias me against theism. On the other hand, I'm not at all happy that my life will come to an end relatively soon (cosmically speaking), with no hope for continuing on; so that might bias me toward it. Whatever biases I may have, I need to keep them in mind while trying to objectively weigh the evidence. It's all any of us can hope to do from our tiny perch in the cosmos.

PART THREE

ESSENCE

19

How Much We Know

When I was twelve years old, I was fascinated by psychic powers. Who wouldn't be? It's a provocative notion, to be able to reach out and push things around, hear what other people are thinking, or tell the future, all just by using your mind.

I read everything I could find about ESP, telekinesis, clairvoyance, precognition—the whole gamut of mental abilities that stretched beyond the ordinary. I was a big fan of comic books, where all the heroes were endowed with superpowers, but also of science-fiction and fantasy stories, not to mention straightforwardly "scientific" accounts of what purported to be evidence for human capabilities beyond the normal. I wanted to penetrate the mystery, figure out how this kind of thing could really work. I loved mind-bending ideas, and what's more mind-bending than the possibility that the mind itself can actually bend things?

I was also a young scientist at heart. So eventually I decided on the obvious course of action—I would perform my own experiments.

We had a spare room in the ground floor of our house. There I was with the door closed, the rest of my family occupied elsewhere. (I didn't say I was an especially courageous young scientist.) I started with small things like dice and coins, placed carefully on a smooth tabletop. Then I just . . . *thought* at them. I concentrated as hard as I could, trying to push the little trinkets across the table with the sheer force of my mind. Sadly, nothing. I switched to easier targets: tiny scraps of paper that shouldn't require as much force to get moving. In the end I had to admit it: maybe some people were able to push things around just by thinking, but I wasn't one of them.

As experiments go, this wasn't the most careful one ever performed. But it was convincing to me at the time. I gave up on the idea that I could move things around with my mind, and became pretty skeptical of anyone else who claimed to have such powers. I didn't lose my fascination for mind-bending ideas, or penetrating deep mysteries. I still wish it were true that I could move objects by thinking at them. It would be really useful, not to mention scientifically fascinating.

A great deal of investigation, more professional than mine, has gone into evaluating the possibility of psychic or paranormal phenomena. J. B. Rhine, a professor at Duke University, famously carried out a long series of tests that concluded that psychic powers were real. His studies were extremely controversial; many attempts to replicate them failed, and Rhine was criticized for having lax protocols that would allow subjects to cheat on his tests. Today, parapsychology is not taken seriously by most academics. The magician and skeptic James Randi has offered a million dollars to anyone who can demonstrate such abilities under controlled conditions; many have tried to claim the prize, but to date no one has succeeded.

And nobody ever will succeed. Psychic powers—defined as mental abilities that allow a person to observe or manipulate the world in ways other than through ordinary physical means—don't exist. We can say that with confidence, even without digging into any controversies about this or that academic study.

The reason is simple: what we know about the laws of physics is sufficient to rule out the possibility of true psychic powers.

That's a very strong claim. And more than a little bit dangerous: the trash heap of history is populated by scientists claiming to know more than they really do, or predicting that they will know almost everything any day now:

> "[We are] probably nearing the limit of all we can know about astronomy."
>
> —Simon Newcomb, 1888

"The more important fundamental laws and facts of physical science have all been discovered."

—Albert Michelson, 1894

"Physics, as we know it, will be over in six months."

—Max Born, 1927

There is a 50 percent chance that "we would find a complete unified theory of everything by the end of the century."

—Stephen Hawking, 1980

My claim is different. (That's what everyone says, of course—but this time it really is.) I'm not claiming that we know everything, or anywhere close to it. I'm claiming that we know *some things*, and that those things are enough to rule out some other things—including bending spoons with the power of your mind. The reason we can say that with confidence relies heavily on the specific form that the laws of physics take. Modern physics not only tells us that certain things are true; it comes with a built-in way of delineating the limits of that knowledge—where our theories cease to be reliable. To see how that works, in this section we'll dig into the rules by which contemporary physics says the universe operates.

My twelve-year-old self wasn't really being overly optimistic, given his knowledge at the time. The idea that our minds can reach out and influence or observe the outside world seems completely plausible. We see things in one place affecting things far away every day. I pick up a remote control, push some buttons, and my TV comes to life and changes the channel. I pick up a phone and suddenly I'm talking to someone thousands of miles away. It's obvious that invisible forces can fly across great distances through the power of technology—why not through the power of the mind?

The human mind is a mysterious thing. It's not that we know nothing about it; wise people have been contemplating the mind's workings for thousands of years, and modern psychology and neuroscience have added considerably to our understanding. Still, it's fair to say that there are more looming questions than settled facts. What is consciousness? What happens when we dream? How do we make decisions? How do we record memories? How do emotions and feelings interact with our rational thoughts? Where do experiences of awe and transcendence come from?

So why not psychic powers? We should be properly skeptical, and try to determine through careful testing whether any particular claim actually holds up to scrutiny. Wishful thinking is a powerful force, and it makes sense to guard against it. But it's important to be honest about what we know and what we don't. On the face of it, reading minds or bending spoons doesn't seem any crazier than talking over a telephone, and maybe less crazy than many of the triumphs of modern technology.

There is a wide gap between admitting that we don't know everything about how the mind works and remembering that whatever it does, it needs to be compatible with the laws of nature. There are things we don't understand about, for example, treating the common cold. But there is no reason to think that cold viruses are anything other than particular arrangements of atoms obeying the rules of particle physics. And that knowledge puts limits on what those viruses can possibly do. They cannot teleport from one person's body to another one, nor can they spontaneously turn into antimatter and cause explosions. The laws of physics don't tell us everything we might want to know about how viruses work, but they undoubtedly tell us some things.

Those same laws tell us that you can't see around corners, or levitate through sheer force of will. All of the things you've ever seen or experienced in your life—objects, plants, animals, people—are made of a small number of particles, interacting with one another through a small number of forces. By themselves, those particles and forces don't have the capability of supporting the psychic phenomena that so fascinated my twelve-year-old self. More important, we know that there aren't new particles or forces out there yet to be discovered that would support them. Not simply because we haven't found them yet, but because we definitely would have found them if they had the right characteristics to give us the requisite powers. We know enough to draw very powerful conclusions about the limits of what we can do.

We never know anything about the empirical world with absolute certainty. We must always be open to changing our theories in the face of new information.

But we can, in the spirit of the later Wittgenstein, be sufficiently confident in some claims that we treat the matter as effectively settled. It's possible that at noon tomorrow, the force of gravity will reverse itself, and we'll all be flung away from the Earth and into space. It's *possible*—we can't actually prove it won't happen. And if surprising new data or an unexpected theoretical insight forces us to take the possibility seriously, that's exactly what we should do. But until then, we don't worry about it.

Psychic powers are like that. There's no harm in doing careful laboratory tests to search for people with the ability to read minds or push things around through telekinesis. But there's no real point, since we know such abilities aren't real, just as we know that gravity won't reverse tomorrow.

David Hume, writing in *An Enquiry Concerning Human Understanding*, considered the question of how we should treat claims of miraculous events, defined as "a violation of the laws of nature." His answer was Bayesian in spirit: we should accept such a claim only if it would be harder to disbelieve it than to believe it. That is, the evidence should be so overwhelming that it would strain our credulity more to deny it than to accept that the laws we thought governed the world have in fact been violated. The same holds for psychic phenomena: as long as the evidence in favor of them is weaker than our evidence in favor of the laws of physics (as it surely is), our credence in their existence should be extremely low.

None of which is to say that science is finished, or that there aren't things we have yet to understand. Every scientific theory we have is one way of talking about the world, one particular story we tell with a certain domain of applicability. Newtonian mechanics works pretty well for baseballs and rocket ships; for atoms, it breaks down and we need to invoke quantum mechanics. Yet we still use Newtonian mechanics where it works. We teach it to our students, and we use it to send spaceships to the moon. It's "correct," as long as we understand the domain in which it's applicable. And no future discovery will suddenly make us think that it is incorrect in that domain.

Right now we have a certain theory of particles and forces, the Core Theory, that seems indisputably accurate within a very wide domain of applicability. It includes everything going on within you, and me, and everything you see around you right this minute. And it will continue to be accurate. A thousand or a million years from now, whatever amazing discoveries science will have made, our descendants are not going to be say-ing "Ha-ha, those silly twenty-first-century scientists, believing in 'neu-trons' and 'electromagnetism.'" Hopefully by then we will have better, deeper concepts, but the concepts we're using now will still be legitimate in the appropriate domain.

And those concepts—the tenets of the Core Theory, and the framework of quantum field theory on which it is based—are enough to tell us that there are no psychic powers.

Many people still believe in psychic phenomena, but they are for the most part dismissed in respectable circles of thought. The same basic story holds for other tendencies we sometimes have to appeal to extraphysical aspects of what it means to be human. The position of Venus in the sky on the day you were born does not affect your future romantic prospects. Con-sciousness emerges from the collective behavior of particles and forces, rather than being an intrinsic feature of the world. And there is no immate-rial soul that could possibly survive the body. When we die, that's the end of us.

We are part of the world. Comprehending how the world works, and what constraints that puts on who we are, is an important part of under-standing how we fit into the big picture.

20

The Quantum Realm

T he history of science is sometimes told—for dramatic effect, if not always in the interests of accuracy—as a story of revolutions. We had the Copernican revolution in astronomy, and the Darwinian revolution in biology. Physics has witnessed two revolutions that transformed the very foundations of the discipline: Newtonian mechanics, which describes the classical world, and quantum mechanics.

There's a story that Chinese premier Zhou Enlai was asked in 1972 about his opinion of the impact of the French Revolution, and he replied, "It's too early to say." Sounds too good to be true, and it is. An interpreter later admitted that, given how the question was phrased, it is clear that Zhou was thinking of the student riots of 1968, not the revolution of 1789.

If they had been talking about the quantum revolution of the 1920s, on the other hand, the quip would have been entirely appropriate. In 1965, physicist Richard Feynman opined, "I think I can safely say that nobody understands quantum mechanics," and the sentiment is equally applicable today. For a theory that has seen unparalleled empirical success at predicting and accounting for the outcomes of high-precision experiments, the embarrassing truth is that physicists cannot claim to have a very good understanding of what the theory actually *is*. Or at least, if some people know what it is, their views are not widely shared by their colleagues.

But we shouldn't exaggerate the mysteriousness of quantum mechanics just for effect. We understand an enormous amount about the theory—otherwise we wouldn't be able to make those predictions that have been checked to amazing precision. Give a well-trained physicist a well-posed question about what quantum mechanics predicts in some specific situation, and they will come up with the uniquely correct answer. But the essence of the theory, its final correct formulation and its ultimate ontology, are still very much in dispute.

This is unfortunate, because where misunderstanding dwells, misuse will not be far behind. No theory in the history of science has been more misused and abused by cranks and charlatans—and misunderstood by people struggling in good faith with difficult ideas—than quantum mechanics. We need to get as clear a view as possible of what the theory says and doesn't say, since it is the deepest and most fundamental picture of the world we now have. Quantum mechanics has direct implications for many issues that confront us as we try to make sense of our human experience of the world: determinism, causality, free will, the origin of the universe itself.

Let's start with the part of quantum mechanics that everyone agrees on: what you will see when you observe a system.

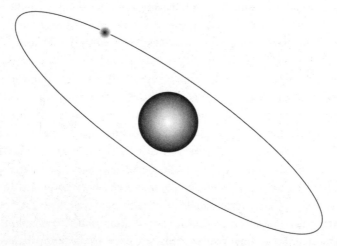

Consider a hydrogen atom. That's the simplest kind of atom there is; its nucleus is a single proton, and there is a single electron bound to it. When we visualize it in our head, we tend to imagine the electron orbiting around the proton much like a planet in the solar system orbits around the sun. This is the "Rutherford model" of the atom.

It's also wrong, and here's why. Electrons are electrically charged, which means they interact with electric and magnetic fields. When you shake an electron, it emits electromagnetic waves—that's the origin of much of the light you actually see in your daily life, whether it's from the sun or from an incandescent bulb. Some electrons were heated up, started shaking, and lost energy by radiating light. In our hydrogen atom, that orbiting electron carries a certain amount of energy, depending on how close it is to the proton— the closer it gets, the less energy it has. So an electron that is far away from the proton, but still bound to it, has a relatively large energy. And it's being "shaken," simply by the fact that it's orbiting around. We therefore expect the electron to give off light and in the process lose energy and spiral closer and closer to the proton. (We expect the same thing for planets moving around the sun, which lose energy by gravitational radiation—but gravity is such a weak force that the net effect is negligible.)

When should this process stop? In a Newtonian world, the answer is simple: when the electron is sitting right on top of the proton. Every electron orbiting around every nucleus of every atom should very rapidly spiral to the center, so that every atom in the universe should collapse to the size of a nucleus in less than a billionth of a second. There should be no molecules, no chemistry, no tables, no people, no planets.

That would be bad. Also, it's not what happens in the actual world.

We can get an idea about what does happen by considering cases when the electron in the hydrogen atom actually does lose energy by giving off an electromagnetic wave. When you collect the emitted light, you notice something funny right off the bat: you only ever see certain discrete wavelengths. Newtonian mechanics predicts that we should see all sorts of waves with any wavelength you can imagine. What we observe, instead, is only certain allowed wavelengths emitted at each transition.

That means the electron in the atom can't just be in any old orbit. There must only be some special orbits it can be in, with fixed amounts of energy. The reason we observe only certain wavelengths in the emitted light is that the electrons are not gently spiraling inward but spontaneously leaping from one allowed orbit to another, emitting a packet of light to make up the difference in energy between them. The electron is doing "quantum jumps."

Okay. Electrons don't orbit atomic nuclei with any energy they like, as classical mechanics would have it. For some reason, they stick to certain allowed orbits, with fixed energies. That seems to be a fact of enormous significance, apparently incompatible with the Newtonian worldview that had been utterly entrenched in the structure of physics. But the data should always overrule our expectations; if certain fixed electron orbits are what we have to imagine in order to explain the stability of tables and other objects made of atoms, let's go with it.

The next question is: What makes an electron skip from one allowed orbit to another? When does it happen? How does it know that it's time? Does the state of the electron contain information other than simply what orbit it's in?

It took quite a bit of genius and hard work to figure out the answers to these questions. Physicists were forced to throw out what we mean by the "state" of a physical system—the complete description of its current situation—and replace it with something utterly different. What is worse, we had to reinvent an idea we thought was pretty straightforward: the concept of a *measurement* or *observation*.

We all think we know what those terms mean, but in classical mechanics there's nothing all that special about them. We can measure anything we want about the system, as accurately as we would like, at least in principle. Not so in quantum mechanics. First off, there are only certain things we can measure at any one experiment. We can measure the location of a particle, for example, or we can measure its velocity; but we can't measure both at the same time. And when we do make those measurements, only certain results are allowed, depending on the physical circumstances. If we measure the location of an electron, for example, it could be anywhere; but if we measure its energy when it is orbiting inside an atom, only certain discrete values will ever be obtained. (That's where the word "quantum" comes from, since in the early days of the field, physicists were extremely interested in how electrons behaved in atoms; but not all observables have discrete possible outcomes, so the name is something of a misnomer.)

In classical mechanics, if you know the state of the system, you can predict with certainty what any measurement outcome will be. In quantum mechanics, the state of a system is a *superposition* of all the possible measurement outcomes, known as the "wave function" of the system. The wave function is a combination of every result you could get by doing an observation, with different weights for each possibility. The state of an electron in an atom, for example, will be a superposition of all the allowed orbits with fixed energies. The superposition representing a given quantum state might be heavily concentrated on one specific outcome—the electron might be almost perfectly localized in an orbit with some particular energy—but in principle every possible measurement outcome can be part of the quantum state.

Quantum mechanics is a profound change from classical mechanics, whereby the outcomes of experiments are not perfectly predictable, even if we know the state exactly. Quantum mechanics tells us the *probability* that, upon observing a quantum system with a specified wave function, we will see any particular outcome. We don't lack perfect predictability because we have incomplete information about the system; it's just the best quantum mechanics allows us to do.

This quantum probability is very different from ordinary classical uncertainty. Think once again of playing poker. At the end of a certain hand, your opponent makes a big bet, and you need to decide whether your hand can beat theirs. You don't know what their hand is, but you know what the possibilities are: nothing, a pair, three of a kind, and so forth. Given their behavior so far in the hand, and the odds that they received certain cards to start, you can be a good Bayesian and assign different probabilities to the various hands they could have. Quantum states sound kind of like that, but they are crucially different. In the (classical) game of poker, you don't know what your opponent has, but they have *something* definite. When we say that a quantum state is a superposition, we don't mean "it could be any one of various possibilities, we're not sure which." We mean "it is a weighted combination of all those possibilities at the same time." If you could somehow play "quantum poker," your opponent would really have some combination of each of the possible hands all at once, and their hand would become one specific alternative only once they turned over the cards for you to look at them.

If it all makes your brain hurt, you're not alone. Quantum mechanics took a long time to be put together, and we're still arguing about what it all means.

Consider a billiard ball on a table. Ordinarily, you might think there is something called "the location of the ball." In quantum mechanics, there's no such thing. Were you to observe the ball in order to determine its location, you would indeed *see* it located in one place or another. But when you are not looking, the ball has no location; it has a wave function, which is a superposition of every possible location it could be. It's a bit like a literal wave, sitting on top of the table; where the wave is highest, there's the largest chance of seeing the ball were you to look. If you knew what that wave function was ahead of time, you could predict the probability it would be in one location or any other. For big, real-world objects like billiard balls, the wave function is typically very strongly peaked around one particular position on the table. As that "most likely" position evolves over time, it obeys the rules of classical mechanics, just as Newton and Laplace thought. But there is a chance that when you look at it, you'll see it somewhere else.

This situation is unsatisfying, to put it mildly. Quantum mechanics, at least the way we teach it to physics majors taking their first college courses in the subject, says that there are two completely different ways that the state of a system evolves over time.

One kind of evolution happens when we're not observing the system. Then there's an equation that the wave function obeys—the Schrödinger equation, after Austrian physicist Erwin Schrödinger, who later became famous for torturing cats in thought experiments. (Not real cats, it should be emphasized.) Here it is, in its most general form:

$$i\hbar \partial_t |\Psi\rangle = \hat{H} |\Psi\rangle.$$

It's quite beautiful in its way. The symbol $|\Psi\rangle$ represents the quantum state. The left-hand side of the equation asks "How is the state changing over time?" The right-hand side provides an answer, by doing a certain operation on the state itself. It's parallel to Newton's famous "force equals mass times acceleration," in which forces determine how the system changes through time.

Evolution according to the Schrödinger equation is very much like the evolution of a state in classical mechanics. It is smooth, reversible, and completely deterministic; Laplace's Demon would have no problem predicting what the state would be in the past and future. If that were all we had to the story, quantum mechanics wouldn't be problematic.

But there is also an entirely different way the quantum state can evolve, according to the textbook treatment: namely, when it is observed. In that case, we teach our undergraduates, the wave function "collapses," and we obtain some particular measurement outcome. The collapse is sudden, and the evolution is nondeterministic—knowing what the state was before, you can't perfectly predict what the state will be afterward. All you have are probabilities.

Despite the appearance of probabilities, the predictions of quantum mechanics can be extraordinarily precise. For example, we can measure the strength of the electromagnetic interaction by one kind of experiment, such as how an atom recoils when it emits a photon. Then we can use that measurement to predict the outcome of a different experiment, such as how fast electrons precess in a magnetic field. Finally, we can compare that prediction to an actual observation. The resulting agreement is breathtakingly good:

Observation/Prediction = 1.000000002.

The observed and predicted values aren't exactly the same, both because of experimental error and because of theoretical approximations. But the lesson is clear: quantum mechanics isn't some loosey-goosey, anything-goes kind of operation. It is relentlessly specific and unforgiving.

21

Interpreting Quantum Mechanics

What really bothers us about quantum mechanics is that the word "observer" appears in the theory at all.

What counts as an "observer" or an "observation" anyway? Does a microscope count, or does a conscious human being have to be using it? What about a squirrel, or a video camera? What if I just glance at the thing rather than observing it closely? When exactly does the "wave function collapse" take place? (So you're not kept in suspense, almost no modern physicist thinks that "consciousness" has anything whatsoever to do with quantum mechanics. There are an iconoclastic few who do, but it's a tiny minority, unrepresentative of the mainstream.)

Together these issues are known as the *measurement problem* of quantum mechanics. After fretting about it for decades, physicists still don't agree on how to address it.

They have ideas. One approach is to suggest that while the wave function plays an important role in predicting experimental outcomes, it doesn't actually represent physical reality. It might be that there is a deeper way of describing the world, in addition to the wave function, in terms of which the evolution would be in principle completely predictable. This possibility is sometimes called the "hidden variables" approach, since it suggests that we just haven't yet pinpointed the real way to best describe the state of a quantum system. If such a theory is true, it would have to be nonlocal—parts of the system would have to directly interact with parts at other locations in space.

An even more radical approach is to simply deny the existence of an underlying reality altogether. This would be an *antirealist* approach to quantum mechanics, since it treats the theory as merely a bookkeeping device for predicting the outcomes of future experiments. If you ask an antirealist what aspect of the current universe that knowledge is *about*, they will tell you that it's not a sensible question to ask. There is, in this view, no underlying "stuff" that is being described by quantum mechanics; all we are ever allowed to talk about is the outcomes of experimental measurements.

Antirealism is a pretty dramatic step to take. It seems to have been advocated, however, by no less of an authority than Niels Bohr, the grandfather of quantum mechanics. His views were described as "There is no quantum world. There is only an abstract physical description. It is wrong to think that the task of physics is to find out how nature *is*. Physics concerns what we can *say* about nature."

Perhaps the biggest problem with antirealism is that it's hard to see how it could be a position that one holds with perfect consistency. It's one thing to say that our understanding of nature is incomplete; it's another thing entirely to say that there is no such thing as nature. For one thing, who is it that's doing the saying? Even Bohr, in the quote above, speaks of what we can say "about nature." That would seem to imply that there's something called "nature" that we can say things about.

Fortunately, we have not yet exhausted our possibilities. The simplest possibility is that the quantum wave function isn't a bookkeeping device at all, nor is it one of many kinds of quantum variables; the wave function simply represents reality directly. Just as Newton or Laplace would have thought of the world as a set of positions and velocities of particles, the modern quantum theorist can think of the world as a wave function, full stop.

The difficulty with this robust brand of straightforward quantum realism is the measurement problem. If everything is just wave function, what makes states "collapse," and why is the act of observation so important?

A resolution was suggested in the 1950s by a young physicist named Hugh Everett III. He proposed that there is only one piece of quantum ontology—the wave function—and only one way it ever evolves—via the Schrödinger equation. There are no collapses, no fundamental division between system and observer, no special role for observation at all. Everett proclaimed that quantum mechanics fits perfectly comfortably into a deterministic Laplacian view of the world.

But if that's right, why does it seem to us that wave functions collapse when we observe them? The trick, in modern language, can be traced to a feature of quantum mechanics called *entanglement*.

In classical mechanics, we can think of every different piece of the world as having its own state. The Earth is moving around the sun with a particular position and velocity, and Mars has a position and velocity of its own. Quantum mechanics tells a different story. There is not a wave function for the Earth, another one for Mars, and so on through all of space. There is only one wave function for the entire universe at once—what we call, with no hint of modesty, the "wave function of the universe."

A wave function is simply a number we assign to every possible measurement outcome, like the position of a particle, such that the number tells us the probability of obtaining that outcome. The probability is given by the wave function squared; that's the famous Born rule, after German physicist Max Born. So the wave function of the universe assigns a number to every possible way that objects in the universe could be distributed through space. There's one number for "the Earth is here, and Mars is over there," and another number for "the Earth is at this other place, and Mars is yet somewhere else," and so on.

The state of Earth can therefore be entangled with the state of Mars. For big macroscopic things like planets this possibility isn't realized in a demonstrable way, but for tiny things like elementary particles it happens all the time. Say we have two particles, Alice and Bob, each of which could be spinning either clockwise or counterclockwise. The wave function of the universe could assign a 50 percent probability to Alice spinning clockwise and Bob counterclockwise, and another 50 percent to Alice spinning counterclockwise and Bob clockwise. We have no idea what answer we would get were we to measure the spin of either particle; but we know that once we measure one of them, the other is definitely spinning the other way. They are entangled with each other.

Everett says that we should take the formalism of quantum mechanics at face value. Not only is the system you're going to observe described by a wave function, but *you* are described by a wave function yourself. That means that you can be in a superposition. When you make a measurement of a particle to see whether it's spinning clockwise or counterclockwise, Everett suggests, the wave function doesn't collapse into one possibility or the other. It evolves smoothly into an entangled superposition, part of which has "the particle is spinning clockwise" and "you saw the particle spinning clockwise," while the other of which has "the particle is spinning counterclockwise" and "you saw the particle spinning counterclockwise." Both parts of the superposition actually exist, and they continue to exist and evolve as the Schrödinger equation demands.

At last, then, we have a candidate for a final answer to the critical ontological question "What is the world, really?" It is a quantum wave function. At least until a better theory comes along.

Everett's bare-bones approach to quantum mechanics—just wave functions and smooth evolution, no new variables or unpredictable collapses or denials of objective reality—has been dubbed the *Many-Worlds Interpretation*. The two parts of the wave function of the universe, one in which you saw the particle spinning clockwise and the other in which you saw it spinning counterclockwise, subsequently evolve completely independently of each other. There is no future communication or interference between them. That's because you and the particle become entangled with the rest of the universe, in a process known as *decoherence*. The different parts of the wave function are different "branches," so it's convenient to say that they describe different worlds. (There's still one "world" in the sense of "the natural world," described by the wave function of the universe, but there are many different branches of that wave function, and they evolve independently, so we call them "worlds." Our language hasn't yet caught up to our physics.)

There's a lot to love about the Everett/Many-Worlds approach to quantum mechanics. It is lean and mean, ontologically speaking; there's just the quantum state and its single evolution equation. It's perfectly deterministic, even though individual observers can't tell which world they are in before they actually look at it, so there is necessarily some probabilistic component when it comes to people making predictions. And there's no difficulty in explaining things like the measurement process, or any need to invoke conscious observers to carry out such measurements. Everything is just a wave function, and all wave functions evolve in the same way.

There are, of course, an awful lot of universes.

Many people object to Many-Worlds because they simply don't like the idea of all of those universes out there. Especially unobservable universes— the theory predicts them, but there's no practical way of ever seeing them. This is not a very thoughtful objection. If our best theory predicts that something is true, we should place a relatively high Bayesian credence that it actually is true, until a better theory comes along. If you have some visceral or a priori bad feeling about multiple universes, then by all means work on better formulations of quantum mechanics. But a bad feeling is not a principled stance.

The secret to making your peace with Many-Worlds is to appreciate that the approach doesn't start with the formalism of quantum mechanics and add in a preposterously big multiverse. All those other universes are already there, at least potentially, in the formalism. Quantum mechanics describes individual objects as being in superpositions of different measurement outcomes. The wave function of the universe automatically includes the possibility that the whole universe is in such a superposition, which we then choose to talk about as "multiple worlds." It's all the other versions of quantum mechanics that have to work to *get rid of the extra worlds*—by changing the dynamics, or adding in new physical variables, or denying the existence of reality itself. But you gain nothing in explanatory or predictive power, and have unnecessarily made a simple framework more elaborate— at least as Everettians see things.

Which isn't to say that there aren't very good reasons to be concerned about Everettian quantum mechanics. According to Everett, the branching of the wave function into different parallel worlds isn't an objective feature; it's simply a convenient way of talking about the underlying reality. But what exactly determines the best way of drawing the line between universes? Why do we see the emergence of a reality that is well approximated by the rules of classical mechanics? These are perfectly respectable questions—though ones that seem quite answerable to partisans of Many-Worlds.

There are two important things to take away from this discussion, as far as the big picture is concerned. One is that, while we don't have a finished understanding of quantum mechanics at a fundamental level, there is nothing we know about it that necessarily invalidates determinism (the future follows uniquely from the present), realism (there is an objective real world), or physicalism (the world is purely physical). All of these features of the Newtonian/Laplacian clockwork universe can easily still hold true in quantum mechanics—but we don't know for sure.

The other important takeaway is a feature common to all interpretations of quantum mechanics: what we see when we look at the world is quite different from how we describe the world when we're not looking at it. As human knowledge has progressed over the centuries, we have occasionally been forced to dramatically rearrange our planets of belief to accommodate a new picture of the physical universe, and quantum mechanics certainly qualifies as that. In a sense it is the ultimate unification: not only does the deepest layer of reality not consist of things like "oceans" and "mountains"; it doesn't even consist of things like "electrons" and "photons." It's just the quantum wave function. Everything else is a convenient way of talking.

22

The Core Theory

Quantum mechanics is, as far as we currently know, the way the universe works. But quantum mechanics isn't a specific theory of the world; it's a framework within which particular theories can be constructed. Just as classical mechanics includes the theory of planets moving around the sun, or the theory of electricity and magnetism, or even Einstein's theory of general relativity, there are an enormous number of particular physical models that qualify as "quantum-mechanical." If we want to know how the world really works, we need to ask, "The quantum-mechanical theory *of what*?"

Your first guess might be "particles and forces." When we talk about atoms, for example, the central nucleus is a collection of particles called *protons* and *neutrons*, while orbiting around the nucleus are particles called *electrons*. The protons and neutrons are bound to each other by a force (the nuclear force), and the electrons are bound to the nucleus by a different force (electromagnetism), and everything pulls toward everything else because of yet another force (gravitation). Particles and forces are reasonable guesses for what the world is made of, the fundamental stuff that the quantum theory of reality describes.

And that's almost true, but not quite. Our best theory of the world—at least in the domain of applicability that includes our everyday experience—takes unification one step further, to say that both particles and forces arise out of *fields*. A field is kind of the opposite of a particle; while a particle has a specific location in space, a field is something that stretches all throughout space, taking on some particular value at every point. Modern physics says that the particles and the forces that make up atoms all arise out of fields. That viewpoint is called *quantum field theory*. It's quantum field theory that gives us confidence that we can't bend spoons with the power of our minds, and that we know all of the pieces of which you and I are made.

And what are the fields made of? There isn't any such thing. The fields are the stuff that everything else is made of. There could always be a deeper level, but we haven't found it yet.

It's easy enough to accept that the forces of nature arise from fields filling space. It was our old friend Pierre-Simon Laplace who first showed that Newton's theory of gravity could be thought of as describing a "gravitational potential field" that was pushed around by, and in turn pulled back on, objects moving through the universe. Electromagnetism, the theory put together in the nineteenth century by Scottish physicist James Clerk Maxwell and his contemporaries, provides a unified description of electric and magnetic fields.

But what about the particles? Particles and fields seem like they're diametrically opposed to each other—particles live at one spot, while fields live everywhere. Surely we're not going to be told that a particle like an electron comes out of some "electron field" filling space?

That is exactly what you are going to be told. And the connection is provided by quantum mechanics.

The fundamental feature of quantum mechanics is that what we see when we look at something is different from how we describe the thing when we're not looking at it. When we measure the energy of an electron orbiting a nucleus, we get a definite answer, and that answer is one of a specific number of allowed outcomes; but when we're not looking at it, the state of the electron is generally a superposition of all those possible outcomes.

Fields are exactly the same way. According to quantum field theory, there are certain basic fields that make up the world, and the wave function of the universe is a superposition of all the possible values those fields can take on. If we observe quantum fields—very carefully, with suffi-ciently precise instruments—what we see are individual particles. For elec-tromagnetism, we call those particles "photons"; for the gravitational field, they're "gravitons." We've never observed an individual graviton, be-cause gravity interacts so very weakly with other fields, but the basic struc-ture of quantum field theory assures us that they exist. If a field takes on a constant value through space and time, we don't see anything at all; but when the field starts vibrating, we can observe those vibrations in the form of particles.

There are two basic kinds of fields and associated particles: bosons and fermions. Bosons, such as the photon and graviton, can pile on top of each other to create force fields, like electromagnetism and gravity. Fermions take up space: there can only be one of each kind of fermion in one place at one time. Fermions, like electrons, protons, and neutrons, make up the ob-jects of matter like you and me and chairs and planets, and give them all the property of solidity. As fermions, two electrons can't be in the same place at the same time; otherwise objects made of atoms would just collapse to a microscopic size.

The ordinary stuff out of which you and I are made, as well as the Earth and everything you see around you, only really involves three matter particles and three forces. Electrons in atoms are bound to the nucleus by electromagnetism, and the nucleus itself is made of protons and neu-trons held together by the nuclear force, and of course everything feels the force of gravity. Protons and neutrons, in turn, are made out of two kinds of smaller particles: up quarks and down quarks. They are held together by the strong nuclear force, carried by particles called gluons. The "nuclear force" between protons and neutrons is a kind of spillover of the strong nuclear force. There's also a weak nuclear force, carried by W and Z bosons, which lets other particles interact with a final kind of fermion, the neu-trino. And the four fermions (electron, neutrino, up and down quarks) are just one generation out of a total of three. Finally, in the background lurks the Higgs field, responsible for giving masses to all the particles that have them.

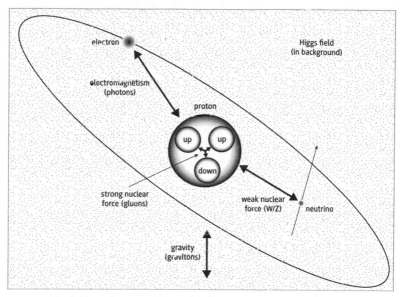

The fields, and associated particles, that make up our everyday world.

The basic collection of fields and their associated particles is illustrated in the figure, a more sophisticated version of the illustration of a hydrogen atom from chapter 20. The two heavier generations of fermions aren't included, as they tend to decay away extremely quickly. The particles we've shown here are the only ones that stick around long enough to make up everyday objects; the full set is discussed in the Appendix.

Physicists divide our theoretical understanding of these particles and forces into two grand theories: the *standard model of particle physics*, which includes everything we've been talking about except for gravity, and *general relativity*, Einstein's theory of gravity as the curvature of spacetime. We lack a full "quantum theory of gravity"—a model that is based on the principles of quantum mechanics, and matches onto general relativity when things become classical-looking. Superstring theory is one promising candidate for such a model, but right now we just don't know how to talk about situations where gravity is very strong, like near the Big Bang or inside a black hole, in quantum-mechanical terms. Figuring out how to do so is one of the greatest challenges currently occupying the minds of theoretical physicists around the world.

But we don't live inside a black hole, and the Big Bang was quite a few years ago. We live in a world where gravity is relatively weak. And as long as the force is weak, quantum field theory has no trouble whatsoever describing how gravity works. That's why we're confident in the existence of gravitons; they are an inescapable consequence of the basic features of general relativity and quantum field theory, even if we lack a complete theory of quantum gravity.

The domain of applicability of our present understanding of quantum gravity includes everything we experience in our everyday lives. There is, therefore, no reason to keep the standard model and general relativity separate from each other. As far as the physics of the stuff you see in front of you right now is concerned, it is all very well described by one big quantum field theory. Nobel Laureate Frank Wilczek has dubbed it the *Core Theory*. It's the quantum field theory of the quarks, electrons, neutrinos, all the families of fermions, electromagnetism, gravity, the nuclear forces, and the Higgs. In the Appendix we lay it out in a bit more detail. The Core Theory is not the most elegant concoction that has ever been dreamed up in the mind of a physicist, but it's been spectacularly successful at accounting for every experiment ever performed in a laboratory here on Earth. (At least as of mid-2015—we should always be ready for the next surprise.)

In the previous chapter we concluded that "what the world is" is a quantum wave function. A wave function is a superposition of configurations of stuff. The next question is "What is the stuff that the wave function is a function *of*?" The answer, as far as the regime of our everyday life is concerned, is "the fermion and boson fields of the Core Theory."

※

We don't need nearly all of the Core Theory to describe almost all of our everyday lives. The heavier fermions decay away very quickly. The Higgs field lurks in the background, but to make an actual Higgs *boson*—the particle that you see when the Higgs field starts vibrating—requires a $10-billion particle accelerator like the Large Hadron Collider in Geneva, and even then the particle decays in about a zeptosecond. Neutrinos are all around us, but the weak nuclear force is so weak that they are very hard to detect. The sun is emitting neutrinos like mad, so that about a hundred trillion of them pass through your body every second, but I suspect you've never noticed.

Almost all of human experience is accounted for by a very small number of ingredients. The various atomic nuclei that we find in the elements of the periodic table; the electrons that swirl around them; and two long-range forces through which they all interact, gravity and electromagnetism. If you want to describe what goes on in rocks and puddles, pineapples and armadillos—that's all you need. And gravity, let's face it, is pretty simple. Everything pulls on everything else. All of the real structure and complexity we see in the world come from electrons (and the fact that they can't lie on top of each other) interacting with nuclei and with other electrons.

There are exceptions, of course. The weak nuclear force plays an important role in nuclear fusion, which powers the sun, so we wouldn't want to do without that. Muons, which are the heavier cousins of electrons, can be produced when cosmic rays hit the Earth's atmosphere, and may be involved in the rate at which DNA mutates, and therefore in the evolution of life. These and other phenomena are important to keep track of—and the Core Theory does a fantastic job accounting for them. But the vast majority of life is gravity and electromagnetism pushing around electrons and nuclei.

We can be confident that the Core Theory, accounting for the substances and processes we experience in our everyday life, is *correct*. A thousand years from now we will have learned a lot more about the fundamental nature of physics, but we will still use the Core Theory to talk about this particular layer of reality. From the perspective of poetic naturalism, there is one story of reality we can tell with confidence, in a well-defined domain of applicability. We can't be metaphysically certain of this; it's not something we can prove mathematically, since science never proves things. But in any good Bayesian accounting, it seems overwhelmingly likely to be true. The laws of physics underlying everyday life are completely known.

23

The Stuff of Which We Are Made

Quantum field theory is an immensely powerful framework. If Godzilla and the Hulk had a baby, and that baby was a framework describing a certain kind of physical theory, that baby would be quantum field theory.

"Powerful" doesn't mean "capable of smashing cities to rubble." (Although quantum field theory is that, since it's the only way we have of describing one kind of particle transforming into another one, which is a crucial part of nuclear reactions and therefore nuclear weapons.) When we're talking about scientific theories, powerful actually means *restrictive*—a powerful theory is one in which there are many things that simply cannot happen. The power we're talking about here is the ability to start with very few assumptions and draw conclusions that are reliable and wide-ranging in their scope. Quantum field theory doesn't knock down buildings lying in its path; it knocks down our speculations about what kinds of things can happen in physical reality.

The claim we're making is pretty audacious:

> **Claim:** The laws of physics underlying everyday life are completely known.

An assertion like that invites a great deal of skepticism. It's bombastic, self-congratulatory, and it doesn't seem that hard to think of plausible ways in which our understanding could be dramatically incomplete. It sounds an awful lot like all the many times throughout history when some great thinker or another boasted that the quest for perfect knowledge was nearly complete. Every one of which turned out to be hilariously premature.

But we're not claiming that all the laws of physics are known, only a restricted set that suffices to describe what happens at the level underlying everyday life. Even that sounds pretty presumptuous. Surely there must be all sorts of ways to add new particles or forces to the Core Theory that could be important to everyday-level physics, or for that matter new kinds of phenomena that fall outside the scope of quantum field theory entirely. Right?

Not so. The situation now really is different from the way it has ever been at previous moments in the history of science. Not only do we have a successful theory, but we also know how far that theory can be extended before it ceases to be reliable. That's just how powerful quantum field theory is.

The logic behind our audacious claim is simple:

1. Everything we know says that quantum field theory is the correct framework for describing the physics underlying everyday life.
2. The rules of quantum field theory imply that there can't be any new particles, forces, or interactions that could be relevant to our everyday lives. We've found them all.

Could quantum field theory not apply in the appropriate regime? Of course. As good Bayesians, we know better than to set our credences all the way to zero even for the most extreme options. In particular, quantum field theory could fail to completely describe human behavior, since *physics* could fail to describe human behavior. There could be a miraculous intervention, or some inherently nonphysical phenomenon that affects the behavior of physical matter. No amount of scientific progress will ever rule that out entirely. What we can do is show that physics by itself is fully up to the task of accounting for what we see.

Einstein's *special relativity* (as opposed to general relativity) is the theory that melds space and time together and posits the speed of light as an absolute limit on the universe. Let's say you want to invent a theory that simultaneously embraces these three ideas:

1. Quantum mechanics
2. Special relativity
3. Sufficiently separated regions of space behave independently from one another

Nobel laureate Steven Weinberg has argued that every theory that fits these requirements will *look* like a quantum field theory at (relatively) long distances and low energies—say, anything bigger than a proton. No matter what happens at the ultimate, most fundamental and comprehensive level of nature, in the regime that humans can probe, the world will be well described by quantum field theory.

If we are interested in describing the everyday low-energy world around us, therefore, and we want to stick purely to physics, we should work in the framework of quantum field theory.

Let's accept the idea that quantum field theory works in the everyday regime, and ask why there couldn't be undiscovered particles that are relevant to the everyday world.

First, we need to establish that there can't be real, tangible particles buzzing around and bumping into us, somehow affecting the behavior of the particles we know about. Then we'll have to assure ourselves that there aren't any *virtual* particles or new interactions that could likewise affect the particles we see. In quantum field theory, virtual particles are ones that quickly flick in and out of existence as quantum fluctuations, affecting what regular particles do without ever being observed themselves. We'll look at this second issue in the next chapter, and for the moment focus on the possibility of real particles.

The reason why we know there are no new fields or particles that play an important role in the physics underlying our everyday lives is a crucial property of quantum field theory known as *crossing symmetry*. This amazing feature helps us be sure that certain kinds of particles do not exist; otherwise we would have found them already. Crossing symmetry basically says that if one field can interact with another one (for example, by scattering off of it), then the second field can *create* particles of the first one under the right conditions. It can be thought of as the quantum-field-theory analogue of the principle that every action implies a reaction.

Consider a new particle X that you might suspect leads to subtle but important physical effects in the everyday world, whether it's the ability to bend spoons with your mind or consciousness itself. That means that the X particle must interact with ordinary particles like quarks and electrons, either directly or indirectly. If it didn't, there would be no way for it to have any effect on the world we directly see.

Interactions between particles in quantum field theory can be visualized by the lovely mechanism of *Feynman diagrams*. Think of an X particle bouncing off of an electron by the exchange of some other new particle, Y. From left to right in the diagram, an X and an electron came in, exchanged a Y particle, then went off on their own ways.

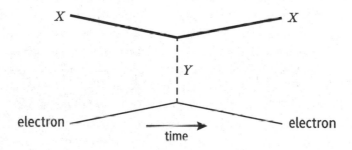

The diagram isn't just a picture of what can happen; it's associated with a number, which tells us how strong the interaction is—in this case, how likely an X is to scatter off an electron. Crossing symmetry says that for every such process, there is another process of the same strength, obtained by rotating the diagram by ninety degrees, and switching any lines that changed directions from particle to antiparticles. One result of crossing symmetry is shown in the next figure.

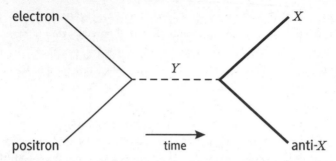

A diagram representing the annihilation of an electron and a positron (antiparticle of an electron) into a Y particle, which then decays into an X and an
anti-X. This diagram is related to the previous one by crossing symmetry.

In field theory, every particle has an antiparticle with the opposite electric charge. The antiparticle of an electron is a particle called the positron,
which is positively charged. Crossing symmetry says that the first process,
scattering of an X off an electron, implies the existence of a related process
in which an electron and positron annihilate to create one of our X particles as well as its antiparticle.

Here is the payoff. We have smashed electrons and positrons together,
often and with great care. From 1989 to 2000, a particle accelerator called
the Large Electron-Positron Collider (predecessor of today's Large Hadron
Collider) operated underground outside Geneva. Within its experiments,
electrons and positrons collided at enormous energies, and physicists kept
extremely careful track of everything that came out. They were hoping with
all their hearts to find new particles; discovering new particles, especially
unexpected ones, is what keeps particle physics exciting. But they didn't see
any. Just the known particles of the Core Theory, produced in great numbers.

The same has been done for protons smashing into antiprotons, and various
other combinations. The verdict is unambiguous: we've found all of the
particles that our best current technology enables us to find. Crossing symmetry assures us that, if there were any particles lurking around us that
interact with ordinary matter strongly enough to make a difference to the
behavior of everyday stuff, those particles should have easily been produced
in experiments. But there's nothing there.

There are probably more particles yet to be found. They just won't be relevant to our everyday world. The fact that we haven't yet found such particles tells us a great deal about what properties they must have; that's the power of quantum field theory. Any particle that we haven't yet detected must have one of the following features:

1. It could be so very weakly interacting with ordinary matter that it is almost never produced; or—
2. It could be extremely massive, so that it takes collisions at energies even higher than what our best accelerators can achieve in order to make it; or—
3. It could be extremely short-lived, so that it gets made but then almost immediately decays away into other particles.

If any particle we haven't yet found lasted long enough and interacted with ordinary matter with sufficient strength that it could possibly affect the physics of everyday goings-on, we would have produced it in experiments by now.

One as-yet-undiscovered particle we believe exists is dark matter. Astronomers, studying the motions of stars and galaxies as well as the large-scale structure of the universe, have become convinced that most matter is "dark"—some kind of new particle that is not part of the Core Theory. The dark-matter particle must be quite long-lived, or it would have decayed away long ago. But it cannot interact strongly with ordinary matter, or it would have already been found in one of the many dark-matter detection experiments that physicists are currently running. Whatever the dark matter is, it certainly plays no role in determining the weather here on Earth, or anything having to do with biology, consciousness, or human life.

There is an apparent loophole in this analysis. There is a particle that we think exists but have never directly detected: the graviton. It is light and stable enough to be produced, but gravity is such a weak force that any gravitons we might make in a particle accelerator will be swamped by the huge number of other particles produced. And yet, gravity does affect our everyday lives.

The basic reason why gravity matters to us is that it is a long-range force that accumulates—the more stuff you have causing the gravity, the stronger its influence is. (That's not necessarily true for electromagnetism, for example, since positive and negative charges can cancel out; gravity always just adds up.) So while we have no hope of making or detecting an individual graviton by smashing two particles together, the combined gravitational effect of the whole Earth creates a noticeable amount of gravitational force.

Is it possible that some other force takes advantage of this loophole—it would be weak if we look at just a few particles, but could accumulate if we had a lot of matter working together? Absolutely—and physicists have been looking for such a "fifth force" for many years now. They haven't found one.

The search for new forces is greatly abetted by the fact that ordinary objects are made only of three kinds of particles: protons, neutrons, and electrons. Another feature of quantum field theory is that you can't turn the forces from individual particles on and off; the associated fields are always there. You can create macroscopic forces by arranging positive and negative charges in the right way, as in an electromagnet, but particle by particle the fields are always present. So we just have to look for forces between those three kinds of particles. Physicists have done precisely that: constructing impeccably precise experiments that bring objects of different compositions close together and then apart again, searching for any hint of an influence outside the known forces of nature.

The results, as of 2015, are shown schematically in the figure. Any possible force between two given kinds of particles is parameterized by two numbers: how strong it is, and the distance over which it reaches. (Gravity and electromagnetism are "long-range" forces, stretching essentially infinitely far; the strong and weak nuclear forces have very short ranges, smaller than individual atoms.) It's easiest to measure forces that are strong, and that reach over long distances. Those are the possible forces that we've already ruled out.

The result is that, if a new force stretches for more than a tenth of a centimeter—which it would have to, if you wanted to use it to bend spoons or reach from Saturn to the time and place of your birth—it would have to be substantially weaker than the force of gravity. That doesn't sound so weak, but keep in mind that gravity is extraordinarily feeble; every time you jump in the air, the puny electromagnetic forces in your body are overcoming the combined gravitational force of the entire Earth. To say that a force

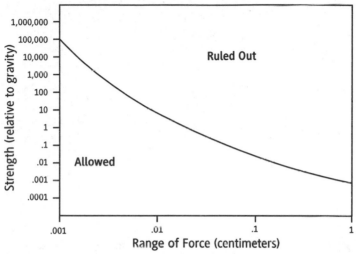

A rough guide to experimental constraints on new forces that could affect ordinary matter. To have escaped detection thus far, a new force must either be sufficiently weak or operate only over a very short range.

is as weak as gravity is to say that it is about one billionth of a billionth of a billionth of a billionth the strength of electromagnetism. An even weaker force would be completely negligible in everyday circumstances.

Here in our daily environment, the world of people and cars and houses, we have a complete inventory of the particles and forces and interactions that are strong enough to have any noticeable effect on anything. That's a tremendous intellectual achievement, one of which the human race can be justifiably proud.

24

The Effective Theory of the Everyday World

ll of this talk of particles and quantum fields can seem almost infinitely far away from the human side of the big picture—the cares and concerns of our personal and social lives. But we are made of particles and fields that obey the ironclad laws of physics. Everything we want to think about human beings has to be compatible with the nature and behavior of the pieces of which we are made, even if those pieces don't tell the whole story. Understanding what those particles and fields are and how they interact with one another is a crucial part of comprehending what it means to be human.

The constraints provided by quantum mechanics and relativity make quantum field theory an extremely restrictive and unforgiving framework. We can use that rigidity to map out how well we've tested the Core Theory, the specific set of fields and interactions that governs our local environment. The answer is: really well. Enough to be convinced that we know what the relevant particles and fields are in this regime, and any new discoveries will involve phenomena that only manifest themselves elsewhere—at higher energies, shorter distances, more extreme conditions.

But how do we know, even if we can't directly see new particles or fields, that they can't exert some subtle but important influence on the particles that we do see? The answer can be traced to another feature of quantum fields: an idea called *effective field theory*. In quantum field theory, the modifier "effective" doesn't mean something like "does a good job fitting the data." Rather, an effective theory is an emergent approximation to a deeper theory. A kind of approximation that is specific, reliable, and well controlled—all due to the power of quantum field theory.

Given some physical system, there are some things you care about, and some you don't. An effective theory is one that models only those features of the system that you care about. The features you don't care about are too small to be noticed, or moving back and forth in ways that everything just averages out. An effective theory describes the macroscopic features that emerge out of a more comprehensive microscopic description.

Effective theories are extremely useful in a wide variety of situations. When we talked about describing the air as a gas rather than as a collection of molecules, we were really using an effective theory, since the motions of the individual molecules didn't concern us. Think about the Earth moving around the sun. The Earth contains approximately 10^{50} different atoms. It should be nearly impossible to describe how something so enormously complex moves through space—how could we conceivably keep track of all of those atoms? The answer is that we don't have to: we have to keep track of only the single quantity we are interested in, the location of the Earth's center of mass. Whenever we talk about the motion of big macroscopic objects, we're almost always implicitly using an effective theory of their center-of-mass motion.

The idea of an effective theory is ubiquitous, but really comes into its glory when we're dealing with quantum fields. That's because of an insight due to Nobel laureate Kenneth Wilson, who thought deeply about the "field" nature of quantum field theory.

Wilson focused on a fact well-known to physicists: if you have a vibrating field, you can always break those vibrations up into a certain contribution at each different wavelength. That's what we're doing when we pass a beam of light through a prism and decompose it into different colors; red light is a long-wavelength vibration in the electromagnetic field, blue light is a short-wavelength vibration, and so on for all the colors in between. In quantum mechanics, short-wavelength vibrations are oscillating faster, and therefore have more energy, than long-wavelength ones. The things we care about are the low-energy, long wavelength vibrations; those are the ones that are easy to make and observe in our everyday lives (unless your everyday life exposes you to particle accelerators or high-energy cosmic rays).

So, Wilson says, quantum field theory comes automatically equipped with a very natural way to create effective theories: keep track of only the long-wavelength/low-energy vibrations in the fields. The short-wavelength/high-energy vibrations are still there, but as far as the effective theory is concerned, all they do is affect how the long-wavelength vibrations behave. Effective field theories capture the low-energy behavior of the world, and by particle-physics standards, everything we see in our daily lives is happening at low energies.

For example, we know that protons and neutrons are made out of up quarks and down quarks, held together by gluons. The quarks and gluons, zipping around at high energies inside the protons and neutrons, are short-wavelength field vibrations. We don't need to know anything about them to talk about protons and neutrons and how they interact with each other. There is an effective field theory of protons and neutrons that works perfectly well, as long as we don't zoom in so closely that we can see the individual quarks and gluons.

This simple example highlights important aspects of how effective theories work. For one thing, notice that the actual entities we're talking about—the ontology of the theory—can be completely different in the effective theory from that of a more comprehensive microscopic theory. The microscopic theory has quarks; the effective theory has protons and neutrons. It's an example of emergence: the vocabulary we use to talk about fluids is completely different from that of molecules, even though they can both refer to the same physical system.

Two features characterize how wonderfully simple and powerful effective field theories are. First, for any one effective theory, there could be many different microscopic theories that give rise to it. That's multiple realizability in the context of quantum physics. Consequently, we don't need to know all the microscopic details to make confident statements about macroscopic behavior. Second, given any effective theory, the kinds of dynamics it can have are generally extremely limited. There simply aren't that many different ways that quantum fields can behave at low energies. Once you've told me what particles are in your theory, all I need to do is measure a few parameters like their masses and interaction strengths, and the theory is completely specified. It's like the planets orbiting the sun; it doesn't make a single whit of difference that Jupiter is a hot gas giant and Mars is a cold rocky planet; they both move on orbits such that their centers of mass are obeying Newton's laws.

This is why we're so confident the Core Theory is basically correct in its domain of applicability. Even if there were something utterly different at the microscopic level—not a field theory at all, perhaps not even space or time as we understand them—the emergent effective theory would still be an ordinary field theory. The fundamental stuff of reality might be something wholly distinct from anything any living physicist has ever imagined; in our everyday world, physics will still work according to the rules of quantum field theory.

All of which is enormously frustrating if you're a physicist who wants to construct a Theory of Everything, but the flip side is that we have a really good handle on the Theory of Some Low-Energy Things—in particular, the kinds of things we encounter in our everyday lives.

We know that the Core Theory isn't the final answer. It doesn't account for the dark matter that dominates the matter density of the universe, and neither does it describe black holes or what happened at the Big Bang.

We can, therefore, imagine improving it by adding some as-yet-unknown "new physics," which would be enough to account for astrophysical and cosmological phenomena. Then we can describe the domains of applicability of various theories in the kind of Venn diagrams we looked at in chapter 12. Astrophysics needs more than the Core Theory, but our everyday experience is well within its domain of applicability.

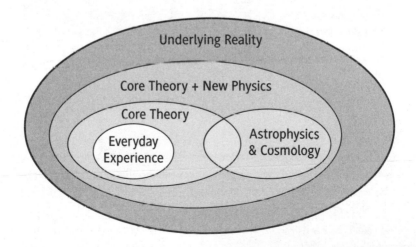

Another way of conveying the same idea is to think about which phenomena depend on which other phenomena—what *supervenes* on what, as the philosophers would say. This is shown in the next figure. Astrophysical phenomena depend on the Core Theory, but also on new physics. And everything, of course, depends on the same underlying reality. But crucially, the emergent phenomena we see in our everyday lives do *not* depend on dark matter or other new physics. Moreover, they only depend on underlying reality through their dependence on the Core Theory particles and interactions. That's the power of effective field theory. All sorts of microscopic quantum-gravitational craziness could be breaking out deep within the underlying reality, but none of that matters for the behavior of chairs and cars and central nervous systems; it's all subsumed in the effective field theory of the Core Theory.

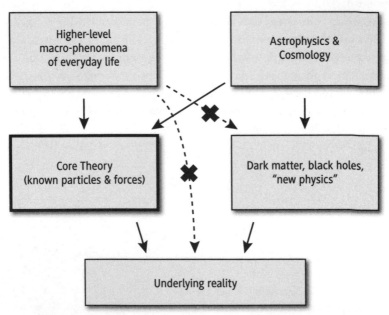

Different ways of talking about the world, and how they relate to each other. Solid arrows indicate how one theory depends on another; for example, astrophysics depends on the Core Theory and also on dark matter and dark energy. Dashed arrows show dependencies that could have existed but don't; everyday life does not depend on dark matter, and depends on underlying reality only through the Core Theory.

The strength of effective field theory is what allows us to assert "This time is different" when we make our audacious claim that the laws of physics underlying everyday life are completely known. When Newton and Laplace contemplated the glory of classical mechanics, they may very well have considered the possibility that it would someday have to be superseded by more comprehensive theories.

And eventually it was—by special relativity, general relativity, and quantum mechanics. Newtonian theory is a good approximation in a certain domain of applicability, but ultimately it breaks down and we need a better description of reality.

What's new is that Newton and Laplace, even if they had thought of their ideas as only accurate in a certain regime, had no way of knowing how far that regime extended. Newtonian gravity works very well for the Earth or Venus; it eventually starts breaking down when we consider the orbit of Mercury, whose tiny precession became some of the strongest evidence in favor of Einstein's general relativity. But Newton would have had no idea how far his theory might be accurate.

With effective field theory, however, that's exactly what we have. An effective field theory describes everything that happens to a certain set of fields, as long as the energies are lower than a certain cutoff, and distances are larger than a certain lower limit (as set by experiment). Once we have the parameters of the effective theory pinned down, we know what will happen to our fields in any experiment we can imagine within its domain of applicability, even if we haven't done that experiment yet.

It's this special feature of quantum field theory that gives us the confidence to make such audacious claims about the scope of our knowledge.

There are a million ways to misinterpret "The laws of physics underlying everyday life are completely known." While it's an undeniably bold claim, it would be easy to mistake it for something even more grandiose than it actually is, and then dismiss that exaggerated claim. It certainly does not imply that we know all of physics.

Nor does it, by any wild stretch of the imagination, imply that we know *how everything works* at the level of the everyday. Nobody in their right mind thinks that we have, or are close to having, complete theories of biology or neuroscience or the weather, or for that matter of the flow of electricity through ordinary materials. Those phenomena need to be *compatible* with the Core Theory, but the phenomena themselves are emergent. As we discussed in chapter 12, understanding emergent phenomena is a matter of discovering new knowledge—finding those patterns (where they exist) that allow us to describe simple behaviors out of many underlying moving parts. Sometimes the simple demand of compatibility with an underlying theory tells us a great deal, as in the case of planets moving around the sun. Conservation of momentum immediately tells us that the Earth won't go careening off in a random direction; the absence of long-range forces other than gravity and electromagnetism tells us that you can't bend spoons with your mind. But for the most part, there is a wide gap between knowing a theory at one level and knowing the emergent theories that are related to it by coarse-graining.

The success of the Core Theory, and our understanding of its domain of applicability, thanks to the principles of effective field theory, implies that there is an enormous presumption (a high Bayesian credence) in favor of understanding macroscopic phenomena in terms that are compatible with the underlying laws of physics. There can always be exceptions. But as David Hume would have said, if you believe that any one particular case is a true example of the Core Theory being violated, your evidence in favor of it needs to be strong enough to overcome the enormous amounts of evidence to the contrary.

Even accepting that science never proves anything and that surprises are always possible, there are still some small loopholes in our arguments that the laws of physics underlying everyday life are completely known. It would be intellectually dishonest not to acknowledge them, so here we go.

The most straightforward loophole would be if quantum field theory were just flat-out wrong in the domain that includes everyday life. For example, if there were physical effects that stretched from one particle to another, but not via anything like a quantum field. This seems very unlikely, on general grounds; once you accept the basic principles of relativity and quantum mechanics, you are more or less forced into accepting quantum field theory. In regions where gravity is strong, like the Big Bang and black holes, field theory may very well break down. There aren't any black holes in your living room, happily. But for the sake of completeness, we should admit that it's always a possibility.

The second possible loophole, arguably more plausible than the first, is the looming problem that we don't fully understand quantum mechanics. It's *possible* that we have in hand all of the basic pieces of quantum ontology (wave functions, the Schrödinger evolution equation), and the foundational work that remains is to interpret how that formalism describes the real world. In that case, this loophole closes with a slam. Indeed, in all of the most popular approaches to quantum mechanics, there really isn't any loophole here at all; there's no place in quantum dynamics for the general principles of effective field theory to be violated.

But because we don't all agree on the correct formulation of quantum mechanics, it's conceivable that none of the most popular alternatives is correct. We can imagine that the correct theory of quantum mechanics will ultimately tell us that wave functions don't really collapse randomly, for example; perhaps there are subtle features of quantum measurement that have thus far eluded experimental detection, but will end up playing an important role in how we come to understand biology or consciousness. It's possible.

Another loophole is the possibility that "new physics" lurks not in new dynamic laws but in something we don't yet appreciate about the initial conditions of the universe. A kind of prearrangement, rather than predestination. The early universe seems to have been a very simple, low-entropy place, which means (following Boltzmann's definition of entropy) there aren't many states it could have been in. But it's at least conceivable that it was in a very special state featuring extremely subtle correlations that work to influence our world today. We have no direct reason to believe that's true, but it deserves a place on our list of loopholes.

Finally, there is the manifest loophole that describing the world in terms of physics alone might not be good enough. There might be more to reality than the physical world. We'll leave serious discussion of that possibility for chapter 41.

The most likely scenario for future progress is that the Core Theory continues to serve as an extremely good model in its domain of applicability while we push forward to understand the world better at the levels above, below, and to the side. We used to think that atoms consisted of a nucleus and some electrons orbiting around it; now we know that the nucleus is made of protons and neutrons, which are in turn made of quarks and gluons. But we didn't stop believing in nuclei when we learned about protons and neutrons, and we didn't stop believing in protons and neutrons when we learned about quarks and gluons. Likewise, even after another hundred or thousand years of scientific progress, we will still believe in the Core Theory, with its fields and their interactions. Hopefully by then we'll be in possession of an even deeper level of understanding, but the Core Theory will never go away. That's the power of effective theories.

25

Why Does the Universe Exist?

I fell in love with the universe at an early age. Lying in bed at night, ready to fall asleep, I'd often be thinking about the expansion of space, and what things were like back near the Big Bang, and what other kinds of universes could exist—until I would come to the thought: What if our universe hadn't existed at all? What if there were simply *nothing*? That would be it. No sleep for me that night.

These are classic questions, and behind them lurks a conviction that the existence of the universe demands some kind of explanation. In a 1697 essay entitled "On the Ultimate Origin of Things," Gottfried Leibniz—whom we remember as the proponent of the Principle of Sufficient Reason and the Principle of the Best, as well as the coinventor of calculus—argued that we should be somewhat surprised that anything exists at all. Nothingness, after all, is simpler than any one particular existing thing ever could be; there is only one nothing, and many kinds of something. More recently, British philosopher Derek Parfit has sympathized, saying that "it can seem astonishing that anything exists."

Just because these questions are common, it doesn't mean they're the right ones to ask. Sidney Morgenbesser, a much-beloved professor of philosophy at Columbia University, renowned for his aphoristic wisdom, was once asked, "Why is there something, rather than nothing?"

"If there were nothing," Morgenbesser immediately replied, "you'd still be complaining."

Beyond the worries and the witticisms, there are two interesting questions facing us, similar-sounding but different in important ways.

1. *Could* the universe, possibly, simply exist? Can we at least imagine reasonable scenarios in which the universe simply *is*, all by itself, or is it necessary to imagine something outside the universe in order to account for its existence?
2. What is the *best explanation* for the existence of the universe? If we need to invoke something outside the universe to account for its existence, what is that thing? And is it better or simpler to not invoke anything additional at all?

Following Aristotle, the fact that the universe exists is often cited as evidence in favor of the existence of God. The universe is specific and contingent, the argument goes; it could easily have been otherwise. So there must be something that explains the universe, and then something that explains that thing, and so on through the chain of reasons. To avoid diving down a rabbit hole of infinite regress, we need to invoke a necessary being— one that must exist and could not have been otherwise, and therefore requires no explanation. And that being is God.

Poetic naturalists don't like to talk about necessities when it comes to the universe. They prefer to lay all the options on the table, then try to figure out what our credences should be in each of them. Maybe there is an ultimate explanation; maybe there is an infinite chain of explanations; maybe there is no final explanation at all. The progress of modern physics and cosmology has sent a fairly unequivocal message: there's nothing wrong with the universe existing without any external help. Why it exists the particular way it does, rather than some other way, is worth exploring.

Let's start with the relatively straightforward, science-oriented question: could the universe exist all by itself, or does it need something to bring it into existence?

As Galileo taught us, one of the foundational features of modern physics is that objects can move, and tend to do so, without any need for an external cause or mover. Roughly speaking, the same goes for the universe.

The scientific question to ask isn't "What caused the universe?" or "What keeps the universe going?" All we want to know is "Is the existence of the universe compatible with unbroken laws of nature, or do we need to look beyond those laws in order to account for it?"

This question is complicated by the fact that we don't know what the ultimate laws of nature actually are. Consider an issue that is inextricably tied to why the universe exists: has it existed forever, or did it come into existence at some particular moment, presumably the Big Bang?

Nobody knows. If we were Pierre-Simon Laplace, who believed in the classical physics of Newton and scoffed at the idea that God would ever interfere in the workings of nature, the answer would be easy: the universe exists forever. Space and time are fixed and absolute, and it doesn't really matter what happens to the stuff that is moving around inside space. Time stretches from the infinite past to the infinite future. Of course you are always welcome to consider other theories, but in unmodified Newtonian physics the universe has no beginning.

Then in 1915 along comes Einstein and his theory of general relativity. Space and time are subsumed into a four-dimensional spacetime, and spacetime is not absolute—it is dynamic, stretching and twisting in response to matter and energy. Not long thereafter, we learned that the universe is expanding, which led to the prediction of a Big Bang singularity in the past. In classical general relativity, the Big Bang is the very first moment in the history of the universe. It is the beginning of time.

Then in the 1920s we stumbled across quantum mechanics. The "state of the universe" in quantum mechanics isn't simply a particular configuration of spacetime and matter. The quantum state is a superposition of many different classical possibilities. This completely changes the rules of the game. In classical general relativity, the Big Bang is the beginning of spacetime; in quantum general relativity—whatever that may be, since nobody has a complete formulation of such a theory as yet—we don't know whether the universe has a beginning or not.

There are two possibilities: one where the universe is eternal, one where it had a beginning. That's because the Schrödinger equation of quantum mechanics turns out to have two very different kinds of solutions, corresponding to two different kinds of universes.

One possibility is that time is fundamental, and the universe changes as time passes. In that case, the Schrödinger equation is unequivocal: time is infinite. If the universe truly evolves, it always has been evolving and always will evolve. There is no starting and stopping. There may have been a moment that looks like our Big Bang, but it would have only been a temporary phase, and there would be more universe that was there even before the event.

The other possibility is that time is not truly fundamental, but rather emergent. Then, the universe can have a beginning. The Schrödinger equation has solutions describing universes that don't evolve at all: they just sit there, unchanging.

You might think that's simply a mathematical curiosity, irrelevant to our actual world. After all, it seems pretty obvious that time does exist, and that it's passing all around us. In a classical world, you'd be right. Time either passes or it doesn't; since time seems to pass in our world, the possibility of a timeless universe isn't very physically relevant.

Quantum mechanics is different. It describes the universe as a superposition of various classical possibilities. It's like we take different ways a classical world could be and stack them on top of each other to create a quantum world. Imagine that we take a very specific set of ways the world could be: configurations of an ordinary classical universe, but at different moments in time. The whole universe at 12:00, the whole universe at 12:01, the whole universe at 12:02, and so on—but at moments that are much closer together than a minute apart. Take those configurations and superimpose them to create a quantum universe.

That's a universe that is not evolving in time—the quantum state itself simply *is*, unchanging and forever. But in any one part of the state, it *looks like* one moment of time in a universe that is evolving. Every element in the quantum superposition looks like a classical universe that came from somewhere, and is going somewhere else. If there were people in that universe, at every part of the superposition they would all think that time was passing, exactly as we actually do think. That's the sense in which time can be emergent in quantum mechanics. Quantum mechanics allows us to consider universes that are fundamentally timeless, but in which time emerges at a coarse-grained level of description.

And if that's true, then there's no problem at all with there being a first moment in time. The whole idea of "time" is just an approximation anyway.

I'm not making this up—this kind of scenario is exactly what was contemplated by physicists Stephen Hawking and James Hartle back in the early 1980s, when they helped pioneer the subject of "quantum cosmology." They showed how to construct a quantum state of the universe in which time isn't truly fundamental, and in which the Big Bang represents the beginning of time as we know it. Hawking went on to write *A Brief History of Time*, and become the most famous scientist of the modern age.

The idea of the universe having a beginning—whether time is fundamental or emergent—suggests to some people that there must be something that brought it into being, and typically that something is identified with God. This intuition is codified in the *cosmological argument* for God's existence, an idea that traces its lineage back at least as far as Plato and Aristotle. In recent years it has been championed by theologian William Lane Craig, who puts it in the form of a syllogism:

1. Whatever begins to exist, has a cause.
2. The Universe begins to exist.
3. Therefore, the Universe had a cause.

As we've seen, the second premise of the argument may or may not be correct; we simply don't know, as our current scientific understanding isn't up to the task. The first premise is false. Talking about "causes" is not the right vocabulary to use when thinking about how the universe works at a deep level. We need to be asking ourselves not whether the universe had a cause but whether having a first moment in time is compatible with the laws of nature.

As we go through our lives, we don't see random objects popping into existence. It might be forgivable to think that, at least with a high degree of credence, the universe itself shouldn't simply pop into existence. But there are two very substantial mistakes lurking beneath that innocent-sounding idea.

The first mistake is that saying that the universe *had a beginning* is not the same as saying it *popped into existence*. The latter formulation, which is natural from an everyday point of view, leans heavily on a certain way of thinking about time. For something to pop into existence implies that at an earlier moment it was not there, and at a later moment it was. But when we're talking about the universe, that "earlier" moment simply does not exist. There is not a moment in time where there is no universe, and another moment in time where there is; all moments in time are necessarily associated with an existing universe. The question is whether there can be a first such moment, an instant of time prior to which there were no other instants. That's a question our intuitions just aren't up to addressing.

Said another way: even if the universe has a first moment of time, it's wrong to say that it "comes from nothing." That formulation places into our mind the idea that there was a state of being, called "nothing," which then transformed into the universe. That's not right; there is no state of being called "nothing," and before time began, there is no such thing as "transforming." What there is, simply, is a moment of time before which there were no other moments.

The second mistake is to assert that things don't simply pop into existence, rather than asking *why* that doesn't happen in the world we experience. What makes me think that, despite my best wishes, a bowl of ice cream is not going to pop into existence right in front of me? The answer is that it would violate the laws of physics. Those include conservation laws, which say certain things remain constant over time, such as momentum and energy and electric charge. I can be fairly confident that a bowl of ice cream isn't going to materialize in front of me because that would violate the conservation of energy.

Along those lines, it seems reasonable to believe that the universe can't simply begin to exist, because it's full of stuff, and that stuff has to come from somewhere. Translating that into physics-speak, the universe has energy, and energy is conserved—it's neither created nor destroyed.

Which brings us to the important realization that makes it completely plausible that the universe could have had a beginning: as far as we can tell, every conserved quantity characterizing the universe (energy, momentum, charge) is exactly zero.

It's not surprising that the electric charge of the universe is zero. Protons have a positive charge, electrons have an equal but opposite negative charge, and there seem to be equal numbers of them in the universe, adding up to a total charge of zero. But claiming that the *energy* of the universe is zero is something else entirely. There are clearly many things in the universe that have positive energy. So to have zero energy overall, there would have to be something with negative energy—what is that?

The answer is "gravity." In general relativity, there is a formula for the energy of the whole universe at once. And it turns out that a uniform universe—one in which matter is spread evenly through space on very large scales—has precisely zero energy. The energy of "stuff" like matter and radiation is positive, but the energy associated with the gravitational field (the curvature of spacetime) is negative, and exactly enough to cancel the positive energy in the stuff.

If the universe had a nonzero amount of some conserved quantity like energy or charge, it couldn't have an earliest moment in time—not without violating the laws of physics. The first moment of such a universe would be one in which energy or charge existed without any previous existence, which is against the rules. But as far as we know, our universe isn't like that. There seems to be no obstacle in principle to a universe like ours simply beginning to exist.

To the question of whether the universe could possibly exist all by itself, without any external help, science offers an unequivocal answer: sure it could. We don't yet know the final laws of physics, but there's nothing we know about how such laws work that suggests the universe needs any help to exist.

For questions like this, however, the scientific answer doesn't always satisfy everyone. "Okay," they might say, "we understand that there can be a physical theory that describes a self-contained universe, without any external agent bringing it about or sustaining it. But that doesn't explain *why it actually does exist*. For that, we have to look outside science."

Sometimes this angle of attack appeals to fundamental metaphysical principles, which are purportedly more foundational even than the laws of physics, and cannot be sensibly denied. In particular, the pre-Socratic Greek philosopher Parmenides put forward the famous maxim *ex nihilo, nihil fit*—"out of nothing, nothing comes." Even Lucretius, the Roman poet who was closer to modern naturalism than almost anyone else in the ancient world, subscribed to a similar belief. According to this line of thought, it doesn't matter if physicists can cook up self-contained theories in which the cosmos has a first moment of time; those theories must necessarily be incomplete, since they violate this cherished principle.

This is perhaps the most egregious example of begging the question in the history of the universe. We are asking whether the universe could come into existence without anything causing it. The response is "No, because nothing comes into existence without being caused." How do we know that? It can't be because we have never seen it happen; the universe is different from the various things inside the universe that we have actually experienced in our lives. And it can't be because we can't imagine it happening, or because it's impossible to construct sensible models in which it happens, since both the imagining and the construction of models have manifestly happened.

In the *Stanford Encyclopedia of Philosophy*, an online resource written and edited by professional philosophers, the entry on "Nothingness" starts by asking, "Why is there something rather than nothing?" and immediately answering, "Well, why not?" That's a good answer. There is no reason why the universe couldn't have had a first moment in time, nor is there any reason it couldn't have lasted forever, even without the benefit of any external causal or sustaining influences. Our job, as always, is to ask how well competing theories account for the information we accumulate as we observe the actual universe.

Our job, in other words, is to move from the first question, "Can the universe simply exist?" (yes, it can) to the second, harder one: "What is the best explanation for the existence of the universe?"

The answer is certainly "We don't know." Understanding that time may be emergent, and that the laws of physics are perfectly compatible with the universe having a first moment of time, might help explain *how* the universe came to be, but it says essentially nothing about *why*. It says nothing about why we have these particular laws of physics at all. Why quantum mechanics rather than classical mechanics? Why do we seem to have three dimensions of space and one of time, and the particular zoo of particles and forces we have discovered?

It's possible that some of these have partial answers within a larger physical context. Modern theories of gravity, for example, envision scenarios in which the number of dimensions of spacetime can be different in different parts of the universe. Perhaps there is some dynamic mechanism that picks out 4 as a special number.

But that can't be the entire answer. *Why* would there be such a dynamic mechanism in the first place? Physicists sometimes fantasize about discovering that the laws of physics are somehow unique—that these are the only ones there possibly could have been. That's probably an unrealistic pipe dream. It's not hard to imagine all sorts of different possible ways the laws of physics could have been. Perhaps the universe could have been classical, rather than quantum. Perhaps the universe could be a lattice, like a checkerboard, with bits flipping from on to off as time passes in discrete units. Perhaps the sum total of reality could have been a single point, lacking either space or time. Perhaps there could be a universe that had no regularities at all, one where there would be nothing we would recognize as a "law of physics."

There may be no ultimate answer to the "Why?" question. The universe simply is, in this particular way, and that's a brute fact. Once we figure out how the universe behaves at its most comprehensive level, there will not be any deeper layers left to discover.

Theists think they have a better answer: God exists, and the reason why the universe exists in this particular way is because that's how God wanted it to be. Naturalists tend to find this unpersuasive: Why does God exist? But there is an answer to that, or at least an attempted one, which we alluded to at the beginning of this chapter. The universe, according to this line of reasoning, is contingent; it didn't have to exist, and it could have been otherwise, so its existence demands an explanation. But God is a *necessary* being; there is no optionality about his existence, so no further explanation is required.

Except that God isn't a necessary being, because there are no such things as necessary beings. All sorts of versions of reality are possible, some of which have entities one would reasonably identify with God, and some of which don't. We can't short-circuit the difficult task of figuring out what kind of universe we live in by relying on a priori principles.

It's important to be fair to both sides. Given a conventional understanding of what is meant by "God," the fact that the universe exhibits regularities at all, and in particular that it exhibits regularities that allow for the existence of human beings, seems to have a higher likelihood under theism than under naturalism. A caring deity is more likely to produce hospitable conditions than a brute-fact cosmos. If the existence of a universe governed by physical laws were the only piece of information we had, that piece of evidence would tilt us in the direction of theism.

It's not the only piece of evidence we have, of course. As we saw in chapter 18, naturalists find many aspects of the universe that do not fit well at all with theism, and count heavily against it. The theistic side of the argument would be much stronger if it extended beyond "God would have wanted a hospitable universe to exist, and here we are" to specific aspects of the physical world, especially ones we haven't yet discovered. If you want to claim that the properties of our kind of universe provide evidence for God's existence, you need to believe that you understand God's motivations well enough to say that it's more likely God would have created this kind of universe rather than some other kind. And if *that's* true, it's natural to ask for even more. How many galaxies would God have wanted to create? What would God have made the dark matter consist of?

There may be answers to these questions, either in naturalism or in theism. Or we may have to live with simply accepting the universe the way it is. What we can't do is demand explanations that the universe may not be able give us.

26

Body and Soul

I n another world, just slightly different from ours, the woman we know as Princess Elisabeth of Bohemia might have been an influential and celebrated philosopher or scientist. Instead, her ideas come to us primarily from her correspondence with the great thinkers of her age, especially René Descartes. Known as virtuous and pious, in her later years she served as an active leader of a major convent in Saxony. But she was most distinguished by her freedom of thought and questioning intellect, which led her to challenge one of Descartes's most famous positions: mind-body dualism, the idea that the mind or soul is an immaterial substance distinct from the body. If that were true, she insisted on knowing, how did the two substances communicate with each other?

These days we would say it this way: our bodies are made of atoms, which are in turn made of particles, and those particles obey the equations of the Core Theory. If you want to say that the mind is a separate substance, not just a way of talking about the collective effect of all those particles, how does that substance interact with the particles? How are the equations of the Core Theory incorrect, and how should we improve them?

In the early seventeenth century, the Holy Roman Empire was a loose confederation of city-states centered in modern-day Germany. One of the most influential of them was the Electoral Palatinate, a group of municipalities scattered along the Rhine. Elisabeth Simmern van Pallandt was born there

Elisabeth of the Palatinate, Abbess of Herford Abbey and
Princess of Bohemia, 1618–1680.

in 1618, daughter of Frederick V, Elector Palatine, and Elizabeth Stuart,
who herself was the daughter of James I of England. Elisabeth's upbringing
seems tumultuous from our perspective, although perhaps it was a typical
Central European royal childhood back in those days.

Elisabeth didn't grow up in Bohemia. After a short and unsuccessful
stint as the ruling couple of Bohemia, her parents sought refuge in the
Netherlands. Elisabeth was raised for a while by her grandmother in Hei-
delberg, before moving to The Hague at the age of nine with other members
of her exiled family. Through the upheaval she managed to obtain a wide-
ranging education, including philosophy, astronomy, mathematics, juris-
prudence, history, and classical languages, for which her fluency earned her
the nickname "the Greek" among her brothers and sisters. Her father died
when she was twelve, leaving her in the hands of an uninterested mother

who would tease Elisabeth for her earnest, studious demeanor. Her life at home was probably not made any smoother by her penchant for valuing honesty over courtly manners.

Despite not living an easy or luxurious life by princess standards, Elisabeth managed to be active and engaged both intellectually and politically. She was committed to social justice, befriending and supporting William Penn and other influential Quakers, notwithstanding the theological differences they may have had with her own Calvinism. She received one recorded offer of marriage, to the elderly King Wladyslaw IV of Poland, whom she had never met in person. The Polish Diet wouldn't let the match go forward unless Elisabeth converted to Catholicism, which she refused to do, so the wedding was called off.

In 1667 she entered the convent of Herford Abbey, where she eventually rose to the station of abbess. Elisabeth wasn't the retiring sort of nun, but rather was an active philanthropist and humanitarian, offering the abbey as a place of refuge for anyone persecuted for reasons of conscience, as well as essentially governing the surrounding town. She died in 1680, having become gravely ill, but not before putting her affairs in order and writing a letter of farewell to her sister Louise.

In our actual world, René Descartes certainly succeeded in becoming an influential and celebrated philosopher and scientist. As we have seen, he delved deep into skepticism of the physical world, ultimately relying on his belief in his own existence (and in God's) to pull himself up by his bootstraps. But at the moment our concern is with Descartes's mind-body dualism.

It was in the *Meditations on First Philosophy*, the same work in which he established his own existence, that Descartes argued for the idea that the mind is independent of the body. It's not a completely crazy thing to think. Both living organisms and nonliving objects clearly have "matter" in them, but conscious creatures are manifestly different in some important way from non-conscious lumps of stuff. The mind or the soul seems, at very first glance, to be something quite different from the body itself.

Descartes's argument was pretty simple. He'd already established that we can doubt the existence of many things, even the chair we are sitting on.

So there's no real problem doubting the existence of your own body. But you can't doubt the existence of your mind—you think, therefore your mind must really exist. And if you can doubt the existence of your body but not your mind, they must be two different things.

The body, Descartes went on to explain, works like a machine, having material properties and obeying the laws of motion. The mind is an entirely separate kind of entity. Not only is it not made of material stuff; it doesn't even have a specific location on the material plane. Whatever the mind is, it's something very different from tables and chairs, something that occupies an utterly distinct realm of existence. We label this view *substance dualism*, since it claims that mind and body are two distinct kinds of substance, not merely two different aspects of one underlying kind of stuff.

But the mind and body interact with each other, of course. Certainly our minds communicate with our bodies, nudging them to perform this or that action. Descartes felt that the interaction also went the other way: our bodies can influence our minds. This was a minority position at the time, although it also seems fairly unobjectionable at first glance. When we stub a toe, it's the body that is first affected, but our minds certainly experience the pain. For a Cartesian dualist, minds and bodies coexist in an ongoing dance of influence and response.

Elisabeth read Descartes's *Meditations* in 1642, soon after they were first published. She was intrigued, but skeptical. Fortunately for her, (1) Descartes was himself living in the Netherlands at the time, and (2) she was a princess. Before too long she was able to bring up her worries with the philosopher himself.

Elisabeth's father had died in 1631, leaving her mother, Elizabeth Stuart, as the head of an indebted and unruly family. She would frequently host salons that entertained politicians, scientists, artists, and adventurers. Descartes attended one such event, at which Elisabeth was present, but the studious young woman didn't muster the courage to engage the famous thinker in direct conversation. She did afterward speak of her interest in Descartes's recent writings to a mutual friend, who passed word along to him.

Having royal allies is always a good thing, even if the family is out of power and relatively poor. Accordingly, on his next visit to The Hague, Descartes once again stopped by the house of the exiled queen of Bohemia. Elisabeth, as fate would have it, wasn't in at the time. A few days later, however, he received a letter from her, the beginning of a correspondence that would last until his death in 1650.

Elisabeth's letters combine a mastery of formal etiquette with an intellectual's impatient distaste for beating around the bush. After a few polite preliminaries, she dives into the problems she has with Descartes's mind/body dualism. Her writing is urgent and pointed:

> How can the soul of a man determine the spirits of his body so as to produce voluntary actions (given that the soul is only a thinking substance)? For it seems that all determination of movement is made by the pushing of a thing moved, either that it is pushed by the thing which moves it or it is affected by the quality or shape of the surface of that thing. For the first two conditions, touching is necessary, for the third extension. For touching, you exclude entirely the notion that you have of the soul; extension seems to me incompatible with an immaterial thing. This is why I ask you to give a definition of the soul more specific than the one you gave in your *Metaphysics*.

It's a question that cuts to the heart of the mind/body split. You say that mind and body act on each other, fine. But how, exactly? What precisely happens?

It's not simply a matter of "We don't know this part of the story, but we'll figure it out eventually." Elisabeth was presumably not a physicalist, someone who believes that the world is made purely of physical stuff. Not many people were in 1643. She was a pious Christian, and most likely had no trouble believing there was more to life than the immediately apparent world. But she was also scrupulously honest, and could not understand how an immaterial mind was supposed to push around the material body. When something pushes something else, the two things need to be located at the same place. But the mind isn't "located" anywhere—it's not part of the physical plane. Your mind has a thought, such as "I've got it—*Cogito, ergo sum.*" How is that thought supposed to lead to the body lifting a pen and committing those words to paper? How is it even conceivable that something with no extent or location could influence an ordinary physical object?

Descartes's initial response was at once both fulsomely flattering and somewhat patronizing. He wanted to remain in the princess's favor, but at first he didn't take her question all that seriously, offering a halfhearted suggestion that "mind" was somewhat like "heaviness," though not really. His argument was the following (roughly paraphrased):

- We want to know how an immaterial substance such as the soul can influence the motion of a physical object like the body.
- Well, "heaviness" is an immaterial quality, not a physical object itself. And yet we often speak as if it has an effect on what happens to physical objects—"I couldn't lift that package because it was too heavy." That is, we attribute causal powers to it.
- Of course, he quickly notes, mind is not exactly like that, because mind actually *is* a separate kind of substance. Nevertheless, perhaps the way the mind influences the body is somehow analogous to the way we say heaviness influences objects, even though one is a true substance and the other is not.

If you're confused, you should be, since Descartes's story makes no sense. Ironically, though, it's close to correct. To a poetic naturalist, "mind" is simply a way of talking about the behavior of certain collections of physical matter, just as "heaviness" is. The problem is that Descartes is nobody's naturalist. His burden was to explain how something nonphysical could influence something physical, and he proffered an explanation that utterly failed to do so.

Elisabeth was not impressed. In her subsequent letters she continued to press him on the issue, explaining that she knew perfectly well what heaviness was, but couldn't fathom how it was supposed to help her understand the interactions of physical bodies and immaterial minds. She asks why a mind that is completely independent of the body could be so affected by it—why, for example, "the vapors" are able to affect our capacity for reasoning.

Descartes never offered a satisfactory answer. He believed that the mind's relationship to the body was not like that of a captain to his ship, with the mind pushing around the material object; rather, the two were "tightly joined" and "mingled together." And that mingling occurred, he hypothesized, in a very particular anatomical location: the pineal gland, a tiny part of the vertebrate brain that (we now know) produces the hormone melatonin, responsible for our sleep rhythms. He focused on that specific organ because it seemed to be the only part of the human brain that was unified rather than split bicamerally, and he believed that the mind only experienced one thought at a time. Descartes suggested that the pineal gland was a physical object that could be moved both by the "animal spirits" of the body, and by the immaterial soul itself, serving to mediate influences between the two.

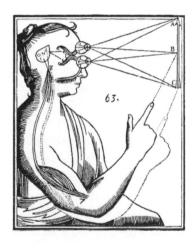

An illustration of the role of the pineal gland, from Descartes's *Treatise of Man*. (Illustration by René Descartes)

The suggestion that the pineal gland serves as "principal seat of the soul" never really caught on, even among thinkers who were otherwise sympathetic to Cartesian dualism. People continued to try to understand how the mind and body could interact. Nicolas Malebranche, a French philosopher who was born just a few years before Elisabeth and Descartes began their correspondence, suggested that God was the *only* causal agent in the world, and that every mind/brain interaction was mediated by God's intervention. As Isaac Newton later noted in a discussion of vision, "To determine by what modes or actions light produceth in our minds the phantasm of colour is not so easie."

How an immaterial soul might interact with the physical body remains a challenging question for dualists even today, and indeed it has grown enormously more difficult to see how it might be addressed. While Elisabeth pointed out some of the difficulties with the idea, she didn't offer an incontrovertible argument that souls and bodies cannot interact in any possible way. She simply noted a crucial difficulty with the dualistic worldview: it's hard to see how something immaterial could affect the motion of something material. Religious believers will sometimes point to an aspect of naturalism that hasn't yet been fully explicated, such as the origin of the universe or the nature of consciousness, and insist that naturalism is therefore defeated; such arguments are rightly derided as "God of the gaps" reasoning, finding evidence for the divine in the gaps in our physical understanding. Likewise, the inability of Descartes and his successors to explain how souls and bodies interact doesn't undermine dualism once and for all; to pretend otherwise would be indulging in "naturalism of the gaps."

It does highlight the difficulties that dualism must face. Today, those difficulties are larger than anything Descartes would have imagined. Modern science knows a lot more about the behavior of matter than seventeenth-century science did. The Core Theory of contemporary physics describes the atoms and forces that constitute our brains and bodies in exquisite detail, in terms of a rigid and unforgiving set of formal equations that leaves no wiggle room for intervention by nonmaterial influences. The way we talk about immaterial souls, meanwhile, has not risen to that level of sophistication. To imagine that the soul pushes around the electrons and protons and neutrons in our bodies in a way that we haven't yet detected is certainly conceivable, but it implies that modern physics is profoundly wrong in a way that has so far eluded every controlled experiment ever performed. How should we modify the Core Theory equation (shown in the Appendix) to allow for the soul to influence the particles in our body? It's a substantial hurdle to leap.

For the moment, Elisabeth's questions remain unanswered. Twentieth-century British philosopher Gilbert Ryle criticized what he called "the dogma of the Ghost in the Machine." As Ryle saw it, thinking of the mind as a separate kind of thing from the body was one big mistake, not just in how the mind works but in what it fundamentally is. We certainly don't have a comprehensive understanding of how matter in motion gives rise to thought and feeling. But from what we do understand, that seems like a much simpler task than making sense of how the mind could be a completely distinct category of existence.

Another strategy for the would-be dualist is to give up on straightforward Cartesian "substance dualism," in which mind and matter are two distinct substances, and go for something more subtle. *Property dualism* is the idea that there's only one kind of stuff—matter—but it has both physical properties and mental properties. We can imagine how Princess Elisabeth might have reacted to this idea: "So how do the mental properties affect the physical ones?" We'll tackle this question in greater depth, but it's not hard to see how the move to property dualism merely pushes the issue back a step rather than actually resolving it.

Besides her insistent questioning on the mind/body interaction question, Elisabeth had a profound influence on Descartes's later work. They corresponded about technical scientific issues, as this paragraph of hers demonstrates:

> I believe that you will justly retract the opinion you have of my understanding once you find out that I do not understand how quicksilver is formed, both so full of agitation and so heavy, contrary to the definition you have given of heaviness. And also when the body E, in the figure on page 255, presses it when it is above, why does it resist this contrary force when it is below, any more than air does in leaving a ship which it has been pressing?

Most importantly, she forcefully argued to Descartes that he was too aloof and disinterested in his moral and ethical philosophy, and needed to take greater account of everyday human reality and "the passions" (what we might today think of as "emotions"). His last published work, dedicated to Elisabeth, was entitled *The Passions of the Soul*, and can be thought of as a response to her prompting.

Elisabeth was a devoted Christian of the late Reformation, not a modern-day naturalist. It is her attitudes and methodology, not her beliefs, that make her a hero for this book. She was not content to posit an attractive picture of the world, such as mind/body dualism, and move on from there without further questioning. How would it work? How does this move that? How would we know? Good questions to be asking, no matter how you ultimately view the fundamental nature of reality.

27

Death Is the End

One of the most impressive properties of the Core Theory of the physics underlying everyday life is its *rigidity*. We specify a particular physical situation, such as a configuration of atoms and ions in a neuron in your brain, and the theory predicts with magnificent accuracy how that situation will evolve. At the microscopic scale, quantum mechanics implies that individual measurement outcomes are expressed in probabilities rather than certainties, but those probabilities are unambiguously fixed by the theory, and when we aggregate many particles the overall behavior becomes fantastically predictable (at least in principle, to a Laplace's Demon–level intellect). There are no vague or unspecified pieces waiting to be filled in; the equations predict how matter and energy behave in any given situation, whether it's the Earth revolving around the sun, or electrochemical impulses cascading through your central nervous system.

This rigidity makes the modern version of Princess Elisabeth's question enormously more pressing than it had been in the seventeenth century. Whether you are a physicalist who believes that there is nothing to us other than the particles of the Core Theory, or someone who thinks that there is some crucial nonphysical component to a human being, everyone admits that the particles are *part* of who we are. If you want to say there is something else, you have to explain how that something else interacts with the particles. How, in other words, the Core Theory is incomplete, and has to change.

To address this issue seriously, we wouldn't necessarily need to have a

"Soul Theory" that is as rigorous and well developed as the Core Theory of physics. We would, however, need to be specific and quantitative about how the Core Theory could possibly be changed. There needs to be a way that "soul stuff" interacts with the fields of which we are made—with electrons, or photons, or something. Do those interactions satisfy conservation of energy, momentum, and electric charge? Does matter interact back on the soul, or is the principle of action and reaction violated? Is there "virtual soul stuff" as well as "real soul stuff," and do quantum fluctuations of soul stuff affect the measurable properties of ordinary particles? Or does the soul stuff not interact directly with particles, and merely affect the quantum probabilities associated with measurement outcomes? Is the soul a kind of "hidden variable" playing an important role in quantum ontology?

If you want to be a dualist and believe in an immaterial soul that plays any role whatsoever in who we are as human beings, these questions are not optional. We're not rigging the game by demanding a full-blown mathematical theory of the soul itself; we're simply asking how the soul is supposed to affect the mathematical theory of the quantum fields that we already have.

Put aside for the moment the possibility of an immaterial soul, or other nonphysical effects that could influence our lives here on Earth. Let's consider the most straightforward construal of our present state of knowledge: the Core Theory underlies everything we witness in our everyday lives, including ourselves. What are the consequences of that picture for our human capacities, as well as for how we think about our place in the cosmos?

We've already alluded to the most obvious repercussion of the Core Theory: you can't bend spoons with your mind. Actually you can, but only by the traditional method: sending signals from your brain, down your arms, to your hands, which then pick up the spoon and bend it.

The argument is simple. Your body, including your brain, is made up of only a few particles (electrons, up quarks, and down quarks), interacting through a few forces (gravity, electromagnetism, and the strong and weak nuclear forces). If you're not going to reach out and touch the spoon with your hands, any influence you have on it is going to have to come through one of the four forces. It won't be through one of the nuclear forces, since those reach only over microscopically small distances. And it won't be through gravity, since gravity is far too weak. (If you didn't know about the Core Theory, you might think you could imagine simply increasing the strength of gravity, or otherwise manipulating it. In the real world, that won't work. A collection of particles, such as your brain, creates a very predictable gravitational field, determined by its total energy. We don't live in a science-fiction movie.)

We're left with electromagnetism. Unlike gravity, the potential electromagnetic force from your body actually is strong enough to bend spoons—indeed, that's what happens when you use your hands. All of chemistry is essentially due to electromagnetic forces acting on electrons and ions (atoms that are charged by having more or fewer electrons than protons). To greatly simplify a complex biological process, muscle contraction occurs when calcium ions provoke one kind of protein (myosin) into pulling on another kind of protein (actin), using energy stored in adenosine triphosphate (ATP) molecules. It's an interplay between a relatively modest collection of electrons, ions, and electromagnetic fields, but it's enough to provide the necessary oomph to bend a spoon as you will.

We might imagine that a brain could be able to somehow focus electromagnetic energy in such a way as to create forces on distant objects without actually touching them. While the brain is chock-full of charged particles, for the most part the electric field associated with them cancels out because there are an equal number of positively charged protons and negatively charged electrons. Conceivably, those particles could move about and arrange themselves in the right way to create an electric or magnetic field that could bend a spoon. (Charged particles at rest are surrounded by electric fields, while charged particles in motion generate magnetic fields in addition.) Something like that, after all, happens with radio transmitters and receivers: signals are sent when charged particles in motion create electromagnetic waves, which then start charges moving inside the receivers.

Having the brain function as a kind of electromagnetic tractor beam would not violate the laws of physics, but it doesn't work for more mundane reasons. The brain itself is subtle and complicated, so we could imagine generating a large electromagnetic field. But once generated, that field would be a blunt instrument. Spoons are not subtle and complicated; they are just inert pieces of metal. Not only would any brain-produced electromagnetic field have no special reason to home in on a spoon in the desired way; it would be incredibly easy to notice for other reasons. Every metallic object in the vicinity would go flying around in response to this force field, and it would be straightforward to measure it using conventional methods. Needless to say, no such field has ever been detected, while quite a few illusions that give the impression of magical spoon bending have been unmasked.

The same goes for phenomena such as astrology. The only fields that could possibly reach from another planet to Earth are gravity and electromagnetism. Gravity, again, is simply too weak to have any effect; the gravitational force caused by Mars on objects on Earth is comparable to that of a single person standing nearby. For electromagnetism the situation is even clearer; any electromagnetic signals from other planets are swamped by more mundane sources.

There's nothing wrong with doing elaborate double-blind studies to look for parapsychological or astrological effects, but the fact that such effects are incompatible with the known laws of physics means that you would be testing hypotheses that are so extremely unlikely as to render it hardly worth the effort.

There is a much more profound implication of accepting the Core Theory as underlying the world of our everyday experience. Namely: there is no life after death. We each have a finite time as living creatures, and when it's over, it's over.

The reasoning behind such a sweeping claim is even more straightforward than the argument against telekinesis or astrology. If the particles and forces of the Core Theory are what constitute each living being, without any immaterial soul, then the information that makes up "you" is contained in the arrangement of atoms that makes up your body, including your brain. There is no place for that information to go, or any way for it to be preserved, outside your body. There are no particles or fields that could store it and take it away.

This perspective can seem strange, because on the surface there appears to be some kind of "energy" or "force" associated with being alive. It certainly seems as if, when something dies, there is some *thing* that is no longer present. Where, it seems natural to ask, does the energy associated with life go when we die?

The trick is to think of life as a *process* rather than a substance. When a candle is burning, there is a flame that clearly carries energy. When we put the candle out, the energy doesn't "go" anywhere. The candle still contains energy in its atoms and molecules. What happens, instead, is that the process of combustion has ceased. Life is like that: it's not "stuff"; it's a set of things happening. When that process stops, life ends.

Life is a way of talking about a particular sequence of events taking place among atoms and molecules arranged in the right way. That wasn't always so obvious; the nineteenth century saw the flowering of a doctrine known as *vitalism*, according to which life is associated with a certain kind of spark or energy, labeled by French philosopher Henri Bergson as *élan vital* (life force). This idea has since gone the way of other similar nineteenth-century doctrines that posited new substances that we now recognize as simply ways of talking about the motions of ordinary matter. "Phlogiston," for example, was supposed to be a kind of element that was contained within flammable bodies, and released during the process of combustion. Today we know that combustion is simply a rapid chemical reaction in which molecules combine with oxygen. Similarly, "caloric" was a hypothetical fluid that represented the heat contained in a body, which would flow from hotter objects to colder ones. Now we understand heat as a measure of the energy contained in the random thermal motions of atoms and molecules.

Over and over, something that we once thought of as a distinct kind of substance has been revealed to be a particular property of ordinary matter in motion. Life is no different.

People have put forward direct evidence for life after death, in the form of near-death experiences or even cases of reincarnation. Often it is claimed that patients near death saw things that they couldn't possibly have seen, or that young children remember events from past lives that they couldn't have known about. Upon closer inspection, the large majority of such testimony proves to be less dramatic than originally suggested. One famous case is that of Alex Malarkey (his actual name, honest), who wrote the book *The Boy Who Came Back from Heaven* with his father, Kevin. After reaching bestseller status and being made into a TV movie, Alex admitted that his tale of visiting heaven and meeting Jesus during a near-death experience was a thorough fabrication.

No cases of claimed afterlife experiences have been subject to careful scientific protocols. People have tried; several studies have been conducted trying to find evidence for out-of-body experiences in patients who have near-death encounters. Researchers will visit hospital rooms and, without specific knowledge on the part of patients or medical staff, hide some kind of visual stimulus in a place where the patient would have to be floating freely of their own body to see it. To date, there has been no case where such a stimulus has been clearly seen.

When judging the veracity of such claims, we need to weigh them against the scientific knowledge we have acquired in much more controlled conditions. It's possible that the known laws of physics are dramatically wrong in such a way as to allow human consciousness to persist after the death of the physical body; however, it is also possible that people under the extreme conditions of nearly dying are likely to hallucinate, and that reports of prior lives are exaggerated or faked. Each of us must choose our priors and update our credences the best we can.

It might seem wrongheaded to draw such sweeping conclusions about human capacities and limitations from something as narrow and esoteric as quantum field theory. Quantum fields, however, are indisputably part of who we are. If they are *all* of who we are, we should have no problem drawing implications of that fact for our lives. If there is something in addition to the quantum fields, it is reasonable to seek an understanding of (and evidence for) that something that is just as precise and rigorous and reproducible as the one we have for field theory.

If we are collections of interacting quantum fields, the implications are enormous. It's not just that we can't bend spoons, and not even that our lives truly end when we die. The laws of physics governing those fields are resolutely impersonal and non-teleological. Our status as parts of the physical universe implies that there is no overarching purpose to human lives, at least not any inherent in the universe beyond ourselves. The very notion of a "person" is ultimately a way of talking about certain aspects of the underlying reality. It's a good way of talking, and we have good reason to take seriously all of the ramifications of that description, including the fact that human beings have individual purposes and can make decisions for themselves. It's when we start imagining powers or behaviors that contradict the laws of physics that we go astray.

If the world we see in our experiments is just a tiny part of a much bigger reality, the rest of reality must somehow act upon the world we do see; otherwise it doesn't matter very much. And if it does act upon us, that implies a necessary alteration in the laws of physics as we understand them. Not only do we have no strong evidence in favor of such alterations; we don't even have any good proposals for what form they could possibly take.

The burden for naturalists, meanwhile, is to show that a purely physical universe made of interacting quantum fields is actually able to account for the macroscopic world of our experience. Can we understand how order and complexity arise in a world without transcendent purpose, even in the face of increasing disorder as implied by the second law of thermodynamics? Can we make sense of consciousness and our inner experience without appealing to substances or properties beyond the purely physical? Can we bring meaning and morality to our lives, and speak sensibly about what is right and what is wrong?

Let's see if we can.

PART FOUR

COMPLEXITY

28

The Universe in a Cup of Coffee

W illiam Paley, a British clergyman writing at the turn of the nine-teenth century, invited you to imagine a walk through one of Britain's picturesque heaths. Suddenly your reverie is interrupted when you stub your toe against a stone. You would be annoyed, thought Paley, but what you wouldn't do is start wondering where such a stone could possibly have come from. Stones are the kinds of things one naturally expects to come across while walking through fields.

Now imagine instead that you notice a pocket watch lying on the ground during your walk. Here you have a puzzle—how did it get there? Not a difficult puzzle, admittedly; presumably someone dropped it while out on a walk similar to your own. Paley's point was that you would never imagine that the watch would just have been sitting there since time immemorial. A stone is a simple lump of material, but a watch is an intricate and purposeful mechanism. It is clear that someone must have made it; a watch implies a watchmaker.

And so it is, continues Paley, with so many things in nature. What we observe in the form of living creatures in the natural world, he argued, is "every manifestation of design"—not only complexity but structures that are obviously attuned to some specific purpose. Nature, he concluded, requires a watchmaker. A Designer, whom Paley identified with God.

It's an argument worth considering. If you found a watch lying on the ground, you would indeed surmise that someone had designed it. And there are specific mechanisms inside our bodies that, for example, help us tell time. (Among them is a protein, cleverly named CLOCK, whose production plays a crucial role in regulating our daily circadian rhythm.) The human body is much more complex than a mechanical watch. Concluding that biological organisms are designed doesn't seem like much of a leap.

We might be cautious about where exactly we should be leaping. David Hume, in his *Dialogues Concerning Natural Religion*, argued fairly compellingly—and even before Paley had popularized the "watchmaker analogy" version of the argument from design—that there is a substantial difference between "a designer" and our traditional notion of God. Paley's argument nevertheless has a good deal of persuasive power, and continues to be popular today.

Immanuel Kant, writing in 1784, mused, "There will never be a Newton for the blade of grass." Sure, you can invent unbending mechanistic rules governing the motions of planets and pendulums, but to account for the living world, you need to go beyond mindless patterns. There must be something that accounts for the purposive nature of living creatures.

These days we know better. We even know who the Newton for the blade of grass turned out to be: his name was Charles Darwin. In 1859, Darwin published *On the Origin of Species by Means of Natural Selection*, in which he laid out the basis for the modern theory of evolution. The great triumph of Darwin's theory was not only to account for the history of life as revealed in the fossil record, but to do so without invoking any kind of purpose or external guidance—"design without a designer," as biologist Francisco Ayala has labeled it.

Essentially every working professional biologist accepts the basic explanation provided by Darwin for the existence of complex structures in biological organisms. In the famous words of Theodosius Dobzhansky, "Nothing in biology makes sense except in the light of evolution." But evolution happens within a larger context. Darwin takes as his starting point creatures that can survive, reproduce, and randomly evolve, and then shows how natural selection can act on those random changes to produce the illusion of design. So where did those creatures come from in the first place?

Our goal over the next few chapters is to address the origin of complex structures—including, but not limited to, living creatures—in the context of the big picture. The universe is a set of quantum fields obeying equations that don't even distinguish between past and future, much less embody any long-term goals. How in the world did something as organized as a *human being* ever come to be?

The short answer comes in two parts: entropy and emergence. Entropy provides an arrow of time; emergence gives us a way of talking about collective structures that can live and evolve and have goals and desires. First we'll focus on entropy.

The role of entropy in the development of complexity seems counterintuitive at first. The second law of thermodynamics says that the entropy of isolated systems increases over time. Ludwig Boltzmann explained entropy to us: it's a way of counting how many possible microscopic arrangements of the stuff in a system would look indistinguishable from a macroscopic point of view. If there are many ways to rearrange the particles in a system without changing its basic appearance, it's high-entropy; if there are a relatively small number, it's low-entropy. The Past Hypothesis says that our observable universe started in a very low-entropy state. From there, the second law is easy to see: as time goes on, the universe goes from being low-entropy to high-entropy, simply because there are more ways that entropy can be high.

Increasing entropy isn't incompatible with increasing complexity, but it can seem that way because of how we sometimes translate the technical terms into informal speech. We say that entropy is "disorderliness" or "randomness," and that it always increases in isolated systems (such as the universe). If the general tendency of stuff is to grow more random and disorganized, it might seem strange that highly organized subsystems come into being without any guiding force working behind the scenes.

There's a common response to this worry, which is perfectly correct but doesn't quite get at the underlying concern. It goes like this: "The second law is a statement about the growth of entropy in *isolated* systems, ones that don't interact with an external environment. In *open* systems, exchanging energy and information with the outside world, of course entropy can go down. The entropy of a bottle of wine goes down when you put it in a refrigerator because its temperature goes down, and the entropy of your room goes down when you clean it up. None of that violates the laws of physics, since the total entropy is still going up—refrigerators expel heat from the back, and human beings sweat and grunt and radiate as they clean up a room."

While it addresses the letter of the concern, this response sidesteps its spirit. The emergence of complex structures on a place like the surface of the Earth is completely compatible with the second law, and it is silly to suggest otherwise. The Earth is an extremely open system, radiating into the universe and increasing its total entropy all the time. The problem is, while that explains why organized systems *can* come into being here on Earth, it doesn't explain why they actually *do*. A refrigerator lowers the entropy of its contents, but only by making them colder, not by making them more intricate or complex. And rooms can be cleaned, but in our experience it seems to require exactly what Paley was talking about: an external intelligence to do the work. Rooms don't spontaneously clean themselves, even if we allow them to interact with the environment.

We still need to understand how and why the laws of physics brought about complex, adaptive, intelligent, responsive, evolving, caring creatures like you and me.

What do we mean by "simple" or "complex," and how do they relate to entropy? Intuitively, we associate complexity with low entropy, and simplicity with high entropy. After all, if entropy is "randomness" or "disorganization," that sounds like the opposite of how we think about the intricate mechanisms we find in a wristwatch or an armadillo.

Our intuition here is a bit off. Think of mixing cream into coffee, inside a glass mug. Since we're doing a physics experiment rather than a morning ritual, let's do it first by gently putting the cream on top of the coffee, and only then mixing them together with a spoon. (The spoon is an external influence, but not a guided or intelligent one.)

At the beginning, the system is low-entropy. There are relatively few ways to rearrange the atoms in the cream and coffee without changing its macroscopic appearance; we could swap individual cream molecules amongst themselves, or individual molecules in the coffee, but once we started exchanging cream with coffee, our glass mug would look different. At the end, everything is mixed together and the entropy is relatively high. We could exchange any bit of the mixture with any other bit and the system would look essentially the same. Entropy has gone up throughout the process, just as the second law would lead us to expect.

Mixing cream into coffee. The initial state is low-entropy and simple. The final state is high-entropy and simple. The intermediate, medium-entropy state exhibits interesting complexity.

But it's not true that complexity has gone down as entropy has gone up. Consider the first configuration, with cream and coffee totally separate; it's low-entropy, but it's also manifestly simple. Cream on top, coffee on bottom, nothing else going on. The final configuration, with everything mixed together, is also quite simple. It's completely characterized by saying "everything is mixed together." It's the intermediate stage, in between low entropy and high entropy, where things look complex. Tendrils of cream reach into the coffee in intricate and beautiful ways.

The cream-and-coffee system exhibits behavior that is very different from a naïve identification of "increasing entropy" with "decreasing complexity." Entropy goes up, as the second law says it should; but complexity first goes up, then goes down.

At least, that's the way it looks. We haven't yet given a precise definition of what we mean by "complexity," as we were able to do for entropy. Partly that's because there is no one definition that works for every circumstance—different systems can exhibit complexity in different ways. That's a feature, not a bug; complexity comes in many forms. We can ask about the complexity of a given algorithm designed to solve a problem, or the complexity of a machine that responds to feedback, or the complexity of a static image or design.

For the moment, let's take a "we know it when we see it" attitude toward complexity, and be prepared to develop more formal definitions when circumstances require.

✳

It's not just cups of coffee in which complexity grows and then fades as entropy increases: the universe as a whole does exactly the same thing. At early times, near the Big Bang, the entropy is very low. The state is also extremely simple: it's hot, dense, smooth, and rapidly expanding. That's a complete description of what is going on; there's no real difference in conditions in the universe from place to place. In the far future the entropy will be very high, but conditions will once again be simple. If we wait long enough, the universe will appear cold and empty, and will have regained its smoothness. All of the matter and radiation we currently see will have left our observable horizon, diluted away by the expansion of space.

It is today, in between the far past and the far future, when the universe is medium-entropy but highly complex. The initially smooth configuration has become increasingly lumpy over the last several billion years as tiny perturbations in the density of matter have grown into planets, stars, and galaxies. They won't last forever; as we saw in chapter 6, eventually all the stars will burn out, black holes will swallow them up, and then even the black holes will evaporate away. The era of complex behavior that our universe is currently enjoying is, alas, a temporary one.

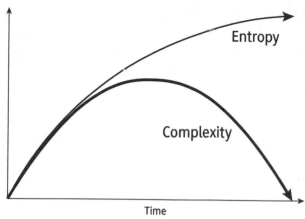

The evolution of entropy and complexity in a closed system over time.

This similarity between the development of complexity in coffee cups and the universe, even as entropy is constantly increasing, is provocative. Is it possible that there is a new law of nature yet to be found, analogous to the second law of thermodynamics, that summarizes the evolution of complexity over time?

The short answer is "We don't know." The somewhat longer answer is "We don't know, but maybe, and if so, there's good reason to believe it will be—appropriately enough—complicated."

I have been working on precisely this issue in my own research, with collaborators Scott Aaronson, Varun Mohan, Lauren Ouellette, and Brent Werness. It all started on a ship sailing through the North Sea. This was part of an unusual interdisciplinary conference devoted to the nature of time, which was literally international in scope: it began in Bergen, Norway, continued during the ship voyage, and finished up in Copenhagen, Denmark. I gave the opening lecture, and Scott was in the audience. I talked a bit about how complexity seems to come and go as closed systems evolve, using coffee and the universe as my examples.

Scott is one of the world's experts on "computational complexity," which organizes different kinds of questions into categories based on how hard they are to solve. He was intrigued enough to think about making the idea more precise. He recruited Lauren, an undergraduate at MIT at th write a simple computer code representing an automaton that would simulate cream and coffee mixing into each other. After we wrote a first draft of a paper and put it on the Internet, Brent wrote to us to point out a flaw in our results—not one that undermined the basic idea, but one that indicated the specific example we were looking at wasn't appropriate. In the spirit of moving science forward, rather than blackballing Brent and trying to destroy his scientific career as punishment for his impertinence, we recognized that he was right and brought him on as a collaborator. Scott recruited Varun, another MIT undergraduate, to update the code and perform more simulations, until we finally fixed our problems. Such is the majestic progress of science.

For our investigation, we were specifically interested in what we called the *apparent complexity* of the cup of coffee. It's related to what computer scientists call the "algorithmic" or "Kolmogorov" complexity of a string of bits. (Any image can be represented as a string of bits, for example, in a data file.) The idea is to pick some computer language that has the ability to output such strings, such as 010010110110. The algorithmic complexity of a string is simply the length of the shortest program that, when run, outputs that string. Simple patterns have low complexity, while completely random strings have high complexity—the only way to output them is simply to have a "Print" statement that includes an explicit copy of the string.

For our purposes of characterizing images of cream mixing with coffee, random noise would count as "simple," not complex. So, following Boltzmann's treatment of entropy, we defined "apparent complexity" by coarse-graining. Rather than observing the position of every single particle in our simulation, we looked at the average number in a small region of space. The apparent complexity is then the algorithmic complexity of the coarse-grained distribution of cream and coffee. It's a nice way to formalize our intuitive notion of "how complex an image appears to be." High apparent complexity corresponds to a coarse-grained (smeared) image that contains a lot of interesting structure.

Unfortunately, there is no way to directly calculate the apparent complexity of an image. But there is a very good approximation: just stick the image into a file-compression algorithm. Everyone's computer has programs that do that, so we were off and running.

At the beginning of the simulation, the apparent complexity is low: a complete description is just "cream on top, coffee below." At the end, the apparent complexity is low once again: all we need to say is that there are equal amounts of cream and coffee at every point. In between, when the mixing is occurring, is where things become interesting. What we found was that complexity doesn't *necessarily* develop—whether it does or not depends on how the cream and coffee interact with each other.

Roughly speaking, if the cream and coffee molecules interact only with other nearby molecules, you don't see much development of complexity. Everything just smoothly blends together rather than forming a jagged pattern of tendrils.

If we introduce long-range effects—analogous to the spoon stirring the coffee—that's when things get interesting. Rather than just blurring together, the boundary between the cream and coffee takes on a fractal aspect. The resulting images have a high degree of apparent complexity; in order to describe them accurately, you would have to specify the intricate shape of the cream-coffee boundary, which would require a relatively large amount of information.

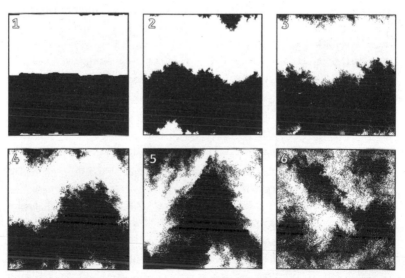

A simple computer simulation of cream and coffee mixing together. The configuration starts out simply and grows increasingly complex; further evolution would show it becoming simple once again, as the black and white became completely mixed.

The relationship between "fractal" and "complex" is more than just a cosmetic one. A fractal is a geometric figure that looks basically the same at any magnification. In the cream and coffee, we see roughly fractal patterns appear in the configuration of the molecules before they eventually fade away in equilibrium. This is a hallmark of complexity; interesting things are going on when we look at the system up close, with just a few moving parts, and also when we look at it all at once.

In both physics and biology, complexity often emerges in a hierarchical fashion: small pieces conglomerate into larger units, which then conglomerate into even larger ones, and so on. Smaller units maintain their integrity while interacting together within the whole. In this way, networks are built up that exhibit complex overall behavior emerging from simple underlying rules. The coffee-cup automaton is too simple a system to model this process faithfully, but the appearance of a fractal shape is a reminder of how robust and natural complexity can be.

Keep going, and the apparent complexity disappears. All of the cream and coffee is simply mixed together. If we wait long enough, any isolated system reaches equilibrium, where nothing interesting happens.

There is no law of nature, therefore, that says complexity necessarily develops as systems evolve from low entropy to high entropy. But it *can* develop—whether it does or not depends on the details of the system you are thinking of. On the strength of one simple computer simulation, it seems that a key issue is the existence of effects that stretch over long distances, rather than only involving particles right next to each other.

The real world features interactions both on short ranges, when particles bump into each other, and on ones that stretch over longer ranges, like the influence of gravity or electromagnetism. When we see complex structures arise as the universe expands and cools, what we're seeing is an interplay between competing influences. The expansion of the universe draws things apart; mutual gravitational forces pull them together; magnetic fields push them sideways; collisions between atoms shove matter around and allow it to cool down. If interesting complex structures can arise in a computer simulation with nothing more than white dots and black dots, it's not surprising that they arise in something as multifaceted as the expanding universe.

The appearance of complexity isn't just compatible with increasing entropy; it *relies* on it. Imagine a system that didn't have any Past Hypothesis, and was simply in a high-entropy equilibrium state right from the start. Complexity would never develop; the whole system would remain featureless and uninteresting (apart from rare random fluctuations) for all time. The only reason complex structures form at all is because the universe is undergoing a gradual evolution from very low entropy to very high entropy. "Disorder" is growing, and that's precisely what permits complexity to appear and endure for a long time.

The microscopic laws of physics don't distinguish between past and future. So any tendency of things to behave differently in one direction in time as opposed to the other—whether it's birth and death, biological evolution, or the appearance of complicated structures—must ultimately be traced to the arrow of time and therefore to the second law. The increase of entropy over time literally brings the universe to life.

Apparent complexity doesn't capture all of what people have in mind when they admire the workings of a clock or a human eye. What makes those remarkable is how the different pieces work together in harmony to help achieve what appears to be some sort of purpose. We'll have to work a bit harder to see how such behavior can arise through the action of mindless matter obeying simple laws. The answer, unsurprisingly, can be traced once again to the growth of entropy and the arrow of time.

As we work our way up from quantum fields and particles to human beings, the subjects we will tackle are going to become more and more difficult, and our statements correspondingly less definitive. Physics is the simplest of all the sciences, and fundamental physics—the study of the basic pieces of reality at the deepest level—is the simplest of all. Not "simple" in the sense that the homework problems are easy, but simple in the sense that Galileo's trick of ignoring friction and air resistance makes our lives easier. We can study the behavior of an electron without worrying about, or even knowing much about, neutrinos or Higgs bosons, at least to a pretty good approximation.

The rich and multifaceted aspects of the emergent layers of our world are not nearly so accommodating to the curious scientist. Once we start dealing with chemistry, biology, or human thought and behavior, all of the pieces matter, and they matter all at once. We have made correspondingly less progress in obtaining a complete understanding of them than we have, for example, on the Core Theory. The reason why physics classes seem so hard is not because *physics* is so hard—it's because we understand so much of it that there's a lot to learn, and that's because it's fundamentally pretty simple.

Our goal is to offer a plausibility sketch that the world can ultimately be understood on the basis of naturalism. We don't know how life began, or how consciousness works, but we can argue that there's little or no reason to look beyond the natural world for the right explanations. We can always be wrong in that belief; but then again, we can always be wrong about any belief.

Asking that our understanding of human life be compatible with what we know about the underlying physics places some interesting constraints on what life is and how it operates. Knowing the particles and forces of which we are made allows us to conclude with very high confidence that individual lives are finite in scope; our best cosmological theories, while much less certain than the Core Theory, suggest that "life" as a broader concept is also finite. The universe seems likely to reach a state of thermal equilibrium. At that point it won't be possible for anything living to survive; life relies on increasing entropy, and in equilibrium there's no more entropy left to generate.

Those swirls in the cream mixing into the coffee? That's us. Ephemeral patterns of complexity, riding a wave of increasing entropy from simple beginnings to a simple end. We should enjoy the ride.

29

Light and Life

Italian astronomer Giovanni Schiaparelli will go down in history as the discoverer of the "canals on Mars." In 1887, after observing our planetary neighbor through his telescope, Schiaparelli reported that its surface was crisscrossed with long, straight lines he labeled *canali*. The idea captured the imagination of people around the world, including American astronomer Percival Lowell, who oversaw the construction of a new observatory in Arizona and performed countless observations of Mars. Based on what Lowell thought he saw—a system of interlocking oases connected by the canals, which seemed to change with the passage of time—he developed elaborate ideas about life on the Red Planet, featuring an advanced civilization struggling to survive in an environment with precious little water. He popularized this idea in a series of books that became very influential, helping to inspire H. G. Wells's *The War of the Worlds*.

There were two problems. The first was that Schiaparelli, although he was also interested in the possibility of life on Mars, had never claimed that there were any canals there. The Italian word *"canali"* should have been translated into English as "channels," not "canals." Channels occur naturally, while canals are artificially constructed. The second problem is that Schiaparelli didn't observe any channels either. The features he described were artifacts of the difficulty involved in observing a faraway planet with relatively primitive instruments.

Today, we have examined Mars quite closely, including with a number of orbiters and landers sent by the United States, the Soviet Union, Europe, and India. (As of this writing, Mars is the only known planet to be inhabited solely by robots.) We haven't found any decaying cities or ancient architectural landmarks, but the search for life continues. Perhaps not in the form of Lowell's dying civilization or Wells's malevolent tripods, but there is certainly a chance of eventually finding microscopic life-forms elsewhere in the solar system—if not on Mars, then possibly in the oceans of Jupiter's moon Europa (which has more liquid water than all the oceans on Earth), or on Saturn's moons Enceladus and Titan.

The question is, will we know it when we see it? What is "life" anyway?

Nobody knows. There is not a single agreed-upon definition that clearly separates things that are "alive" from those that are not. People have tried. NASA, which is heavily invested in looking for life outside the Earth, adopted a working definition of a living organism: a self-sustaining chemical system capable of Darwinian evolution.

We could quibble with the bit about "Darwinian evolution." That's a feature of how living organisms here on Earth have in fact come to be, but not a characterization of what each organism *is*. When you come across an injured squirrel and ask, "Is it alive?" nobody answers, "I don't know, let's see if it's capable of Darwinian evolution." The usefulness of a definition is that it should help us decide difficult cases, such as when scientists might someday construct an artificial life-form. By this criterion, such a beast would automatically be judged nonliving without further thought, which isn't especially helpful. For our present purposes, this is indeed quibbling; when we talk about the actual life we know and love, evolution plays a central role.

The "correct" definition of life, one that we're going to discover through careful research, doesn't exist. The life-forms with which we are familiar share a number of properties, each of which is interesting and many of which are remarkable. Life as we know it moves (internally if not externally), metabolizes, interacts, reproduces, and evolves, all in hierarchical, interconnected ways. It's obviously a uniquely important part of the big picture.

We can start with general principles, working our way toward the specific origin of life here on Earth; from there we can once again expand our view, to see how living creatures evolve and interact with one another.

❋

One of the many suggested definitions of life was put forward by none other than Erwin Schrödinger, who helped formulate the fundamental principles of quantum mechanics. In his book *What Is Life?*, Schrödinger examined the question from a physicist's point of view. The fundamental problem, as he saw it, was one of balance. On the one hand, living things are constantly changing and moving. Whether it's a cheetah chasing after a gazelle, or sap moving slowly through the branches of a redwood tree, something is always happening inside living organisms. On the other hand, living things also maintain their structure; throughout their changes they preserve some basic integrity. What kind of physical process, he wondered, could manage to consistently straddle the line between stasis and change?

This question prompted Schrödinger to put forward a definition of life that seems very different from NASA's:

> When is a piece of matter said to be alive? When it goes on "doing something," exchanging material with its environment, and so forth, and that for a much longer period than we would expect an inanimate piece of matter to "keep going" under similar circumstances.

Schrödinger is focusing on the "self-sustaining" part of the NASA definition, which most of us just breeze over. After all, many things seem to be self-sustaining: waterfalls, oceans, and for that matter the inanimate rock on which William Paley stubbed his toe.

The crucial idea here is that a living being "keeps going" for "a much longer period than we would expect." That's a bit vague; Schrödinger isn't presuming to offer a once-and-for-all definition of a precise concept; he's trying to capture something of our intuition about what life is. A rock might maintain its shape for a long time, but it will never repair itself. A rock can be in motion, for example, if an avalanche starts it rolling downhill; but once it gets to the bottom, it will stop moving and just sit there. It won't brush itself off and climb back up the hill, like an animal might.

This is another way in which living organisms seem to—but don't actually—violate the second law of thermodynamics. Not only do they come into being as organized structures; they then are able to maintain that order over long periods of time.

As with the formation of complexity in the first place, the truth is the converse of our most naïve expectation. Complex structures can form, not despite the growth of entropy but *because* entropy is growing. Living organisms can maintain their structural integrity, not despite the second law but because of it.

Everyone knows that the sun provides a useful service to life here on Earth: energy, in the form of photons of visible light. But the really important thing we get from the sun is energy with very low entropy—so-called *free energy*. That energy is then put to use by biological organisms, and returned to the universe in a highly degraded form. "Free energy" is a confusing term that actually means "useful energy"—think "free" as in "free to do something." It has nothing to do with "energy for free"—the total amount of energy is still constant.

The second law says that the entropy of an isolated system will increase until the system reaches maximum entropy, after which it will sit there in equilibrium. In an isolated system, the total amount of energy remains fixed, but the form that energy takes goes from being low-entropy to being higher-entropy. Think of burning a candle. If we kept track of all the light and heat generated by the candle, the total amount of energy would stay the same over time. But the candle can't burn forever; it goes for a while and then stops. The energy locked inside has been transformed from a low-entropy form to a high-entropy form, and there's no going back.

Free energy can be used to do what physicists call *work*. If we take some macroscopic object and move it around, we are doing work on it. The definition of "work" is simply the force we exerted to get the thing going, times the distance over which it moved. It requires work to lift a stone from the bottom of a hill up to the top. Essentially everything useful that you can do with energy is some kind of work, whether it's getting a rocket into orbit or gently lifting your eyebrow to indicate skepticism.

Free energy is energy in a potentially useful form. The high-entropy remainder is the "disordered energy," equal to the temperature of the system times its entropy. The flow of heat from one system to another increases the amount of useless disordered energy. Indeed, one way of formulating the second law is to say that, in an isolated system, free energy is converted into disordered energy as time passes.

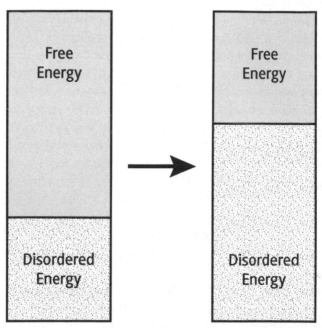

Another way of thinking about the second law of thermodynamics. Over time, energy is converted from "free" (available to do work) to "disordered" (dissipated, useless).

Schrödinger's idea was that biological systems manage to keep moving and maintaining their basic integrity by taking advantage of free energy in their environments. They take in free energy, use it to do whatever work they need it to do, then return the energy to the world in a more disordered form. (In the first edition of his book he went to great lengths not to use the phrase "free energy," because he thought the concept would be confusing. I'm asking a little more of you than Schrödinger was willing to ask of his readers.)

✳

Whether a certain amount of energy is "free" or "disordered" depends on its environment. If we have a piston full of hot gas, we can use it to do work by letting it expand and push the piston. But that's assuming that the piston isn't surrounded by gas of equal temperature and density; if it is, there's no net force on the piston, and we can't do any work with it.

The light we get from the sun is low-entropy relative to its environment, and therefore contains free energy, available to do work. The environment is just the rest of the sky, dotted with starlight and suffused with the cosmic microwave background radiation, at a few degrees above absolute zero. A typical photon emitted by the sun has 10,000 times the energy of a typical photon in the microwave background.

Imagine there were no sun. The entire sky would look like the night sky does now. Here on Earth, we would quickly equilibrate, and come to the same cold temperature as the night sky. There would be no free energy; life would grind to a halt. (Most life, anyway. Microbial "chemilithoautotrophs" feed off free energy locked up in mineral compounds. Even without the sun, the Earth still wouldn't be in perfect thermal equilibrium.)

But imagine that we were surrounded by the sun—the whole sky was raining photons down on us as bright as the sun does now. The Earth would rapidly equilibrate, but we would come to the high temperature of the surface of the sun. There would be a lot more energy reaching Earth than there is now, but the solar-temperature radiation would all be useless, disordered energy. Life would be just as impossible under those conditions as it would be without the sun at all.

What matters to life is that our environment here on Earth is very far from equilibrium, and will be for billions of years. The sun is a hot spot in a cold sky. Because of that, the energy we receive in the form of solar photons is almost entirely free energy, ready to be turned into useful work.

And that's exactly what we do. We receive photons from the sun, primarily in the visible-light part of the electromagnetic spectrum. We process the energy, and then return it to the universe in the form of lower-energy infrared photons. The entropy of a collection of photons is roughly equal to the total number of photons you have. For every one visible photon it receives from the sun, the Earth radiates approximately twenty infrared photons back into space, with approximately one-twentieth of the energy each. The Earth gives back the same amount of energy as it gets, but we increase the entropy of the solar radiation by twenty times before returning it to the universe.

The energy here on Earth is not exactly constant, of course. Since the Industrial Revolution, we have been polluting the atmosphere with gases that are opaque to infrared light, making it harder for energy to escape and thereby heating the planet. But that's another story.

30

Funneling Energy

I t's worth seeing how all this grand physics theorizing plays out in biological practice.

The basic power battery of life here on Earth is a molecule called *adenosine triphosphate*, or ATP. We're using "battery" in a broad sense, as something that stores free energy for later use. Think of ATP as a compressed spring, ready to push apart when it is released and expend its energy doing something (hopefully) useful. And useful it is: the free energy stored in ATP is used for muscle contraction, transporting molecules and cells through the body, synthesizing DNA and RNA and proteins, sending signals through nerve cells, and other vital biochemical functions. ATP plays a crucial role in allowing an organism to move around and maintain itself, as Schrödinger highlighted as the defining characteristic of life.

The chemical structure of adenosine triphosphate, ATP. It includes atoms of hydrogen (H), oxygen (O), phosphorus (P), nitrogen (N), and carbon. Following chemical tradition, the carbon atoms aren't indicated explicitly, but are located at each unlabeled vertex or bend in the diagram.

The release of energy from ATP typically happens in the presence of water (H_2O). One of the three phosphates—groups with one phosphorus atom (P) surrounded by oxygen atoms (O), at the left of the diagram—splits off from the ATP, leaving us with adenosine diphosphate (ADP). The phosphate then joins with a hydrogen atom from a nearby water molecule, leaving the remaining OH to combine with the ADP.

The total energy of these final products is less than that of the original ATP molecule; the process thus releases both free energy (to do some useful biochemical work) and disordered energy (heat). Fortunately, ATP is a rechargeable battery; the body then uses an external source of energy, such as sunlight or sugar, to convert the phosphate and ADP back into water and ATP, which is then ready to be put to work once again.

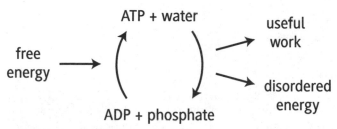

Free energy from external sources (photosynthesis, sugars) is stored in ATP, so that it can be converted to useful work where the body needs it. Such a process necessarily produces disordered energy as well.

All of the energetic activity going on in your body uses up a tremendous amount of ATP; a typical person churns through an amount of ATP equal to about their body mass each day. When you flex your biceps to lift a barbell or a glass of wine, the energy to contract your muscles comes from ATP snapping apart, causing proteins to slide against one another in your muscle fibers. The individual atoms making up the ATP aren't used up; each molecule is simply broken apart and then reassembled, hundreds of times a day.

Where does the free energy come from to create all that ATP from the lower-energy ADP? Ultimately it comes from the sun. The process of photosynthesis occurs when a molecule of chlorophyll in a plant or some microorganism absorbs a photon of visible light, whose energy knocks loose an electron. The energetic electron is shuttled across a membrane by a series of molecules called an *electron transport chain*. As a result, there are more electrons than protons on one side of the membrane, setting up an electrical gradient, with a net negative charge on one side and a net positive charge on the other.

This is the basic way life funnels energy: protons on one side of a membrane push each other apart, with some escaping through an enzyme called *ATP synthase*. The proton trying to escape winds up the synthase, providing it with energy that it uses to synthesize ATP from ADP, in a process called *chemiosmosis*. Some of the energy, inevitably, becomes disordered, and is released in the form of low-energy photons and thermal jiggling (heat) of the surrounding atoms.

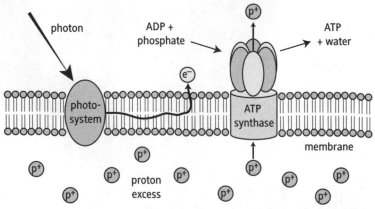

How photosynthesis stores free energy from the sun in ATP. A photon hits a photosystem embedded in a biological membrane, causing an electron (e^-) to be ejected. This process leaves an excess of protons (p^+) on the other side of the membrane. Electrostatic repulsion pushes the protons away, until one escapes through an ATP synthase enzyme. The ATP synthase uses energy from the proton to convert ADP into ATP, which can then carry energy elsewhere.

You and I don't personally photosynthesize. Our free energy doesn't come directly from the sun, but from glucose and other sugars, as well as fatty acids. Tiny organelles called mitochondria, the powerhouse of the cell, use the free energy locked in these molecules to convert ADP to ATP. But the free energy in those sugars and fatty acids that we eat ultimately came from the sun via photosynthesis.

The basic setup seems to be universal within life here on Earth. The phrase *proton-motive force* has been coined to describe the powering of ATP synthase by the protons flowing through it. The mechanism was discovered by British biochemists Peter Mitchell and Jennifer Moyle in the 1960s. Mitchell was an interesting character. Forced to resign his academic position when the pressures of his job led to severe health problems, he eventually set up a private laboratory at a place called Glynn House. He was awarded the Nobel Prize in Chemistry in 1978 for the idea that the proton-motive force was responsible for ATP synthesis via chemiosmosis.

The cell is the basic unit of life: a collection of functional subunits, organelles, suspended in a viscous fluid, all surrounded by a cellular membrane. Immersed as we are in a technological society, we tend to think of cells as tiny "machines." But the differences between real biological systems and the artificially constructed machines that we're used to dealing with are as important as their similarities.

These differences stem in large part from the fact that machines are generally created for some particular purpose. Because of this origin, machines tend to be just good enough for their designated purposes, and no better. Design tends to be specific, and brittle. When something goes wrong—you lose a tire on your car, or the battery dies on your phone—the machine doesn't work at all. Biological organisms, which have developed over the years with no specific purpose in mind, tend to be more flexible, multipurpose, and self-repairing.

Cells don't merely tolerate chaos; they harness it. They have little choice, given the environment in which microbiology takes place.

Our human-scale world is relatively calm and predictable. Throw a ball on a day with good weather, and you can estimate with some confidence how far it will travel. Cells, by contrast, operate at the scale of nanometers, billionths of a meter. Conditions in that world are dominated by random motions and noise—what biophysicist Peter Hoffmann has dubbed a "molecular storm." Just from ordinary thermal jiggling, molecules inside our bodies bump into one another trillions of times a second, in a maelstrom that puts ordinary storms to shame. Scaled up to human size, living in the equivalent of the cell's molecular storm would be like trying to throw a ball that was constantly being bombarded by other balls, each of which carried hundreds of millions of times the energy that your arm could impart.

It doesn't seem like a hospitable environment for any microscopic sporting events, or for the delicate operations that are part of the cellular ecosystem. How do cells manage to do any kind of organized activity under such conditions?

There is a great deal of energy in the maelstrom, but it is all disordered energy; it isn't directly useful for tasks like pulling a muscle or sending nutrients through the body. The ambient molecules are in a near-equilibrium state, bouncing off one another randomly. But the cell can take advantage of the low-entropy free energy bundled up in ATP—not only to perform work directly, but to focus the disordered energy in the surrounding medium.

Consider a ratchet—a gear whose teeth are slanted in one direction. Let it be subject to random jiggling back and forth—*Brownian forces*, named after botanist Robert Brown. It was he who, in the early nineteenth century, noticed that small dust particles suspended in water tended to move around in unpredictable ways, a phenomenon we now attribute to their being constantly bombarded by individual atoms and molecules. A Brownian ratchet, by itself, doesn't tend to move one way or the other; it drifts back and forth unpredictably.

But imagine that the teeth of our ratchet aren't fixed, but are something we could control from the outside. When the ratchet moves in the direction we want it to, we make the angle low and easy to move across; when it moves the other way, we increase the angle and make it harder. That would allow us to convert the random, undirected Brownian motion into directed, useful transport. Of course, it requires the intervention of some external agent that is itself low-entropy, far from equilibrium.

This kind of Brownian ratchet is a simple model for many molecular motors inside a living cell. There aren't any external observers changing the shapes of the molecules to fit specific purposes, but there is free energy carried around by ATP. The ATP molecules can bind to the moving parts of the cellular machinery, releasing their energy at just the right time to allow fluctuations in one direction, while inhibiting them in the other. Getting work done at the nanoscale is all about harnessing the chaos around you.

Schrödinger's picture of living organisms maintaining their structural integrity by using up free energy is impressively manifested in real-world biology. The sun sends us free energy, in the form of relatively high-energy visible-light photons. These are captured by plants and single-celled organisms that use photosynthesis to create ATP for themselves, as well as sugars and other edible compounds, which in turn store free energy that can be used by animals. This free energy is used to maintain order within the organism, as well as allowing it to move and think and react, all of the things that living beings do that distinguish them from nonliving things. The solar energy we started with is gradually degraded along the way, turning into disordered energy in the form of heat. That energy is ultimately radiated back to the universe as relatively low-energy infrared photons. Long live the second law of thermodynamics.

The basic ingredients in this story are familiar from the Core Theory: photons, electrons, and atomic nuclei. As far away as our everyday lives seem from the details of modern physics, understanding how we eat and breathe and live brings us face-to-face with the underlying particles and forces beneath it all.

31

Spontaneous Organization

Flemish chemist Jan Baptist van Helmont, working in the seventeenth century, was one of the first scientists to understand that there were gases other than air, and was even responsible for coining the term "gas." But he will always be best remembered for his recipes for creating living creatures. According to van Helmont, the way to create mice from nonliving materials is to place a soiled shirt inside an open vessel, along with some grains of wheat. After approximately twenty-one days, he wrote, the wheat will have been transformed into mice. If for some reason you wanted to make scorpions rather than mice, he recommended scratching a hole in a brick, filling the hole with basil, covering with another brick, and leaving them out in sunlight.

If only it were that easy. I like to think that, if van Helmont had followed proper Bayesian reasoning, he would have been able to come up with plausible alternative hypotheses to explain the appearance of mice in his vessel with the soiled shirt. Once we move beyond vitalism, and understand that "life" is a label we attach to certain kinds of processes rather than a substance that inhabits matter and starts pushing it around, we begin to appreciate what an enormously complex and interconnected process it is. It's one thing to see how living organisms can harness free energy to maintain themselves and move around. It's quite another thing to understand how life ever got started. As of this writing we have more questions than answers.

For a while there, it seemed like understanding the origin of life, or *abiogenesis*, wouldn't be that difficult. Charles Darwin didn't say that much about the problem in *Origin of Species*, but he briefly speculated that a "warm little pond" could have witnessed the formation of proteins, which might then "undergo still more complex changes." Darwin didn't know much about chemistry or molecular biology, but in a famous experiment in 1952, Stanley Miller and Harold Urey took a flask full of some simple gases—hydrogen (H_2), water (H_2O), ammonia (NH_3), and methane (CH_4)—and zapped it with sparks. The idea was that these compounds may have been present in the atmosphere of the ancient Earth, and the sparks would simulate the effects of lightning. With a fairly simple setup, and after running for just a week without any special tinkering, Miller and Urey found that their experiment had produced a number of different amino acids, organic compounds that play a crucial role in the chemistry of life.

Today we don't think that Miller and Urey were correctly modeling conditions on the early Earth. Their experiment nevertheless demonstrated a crucial biochemical fact: it's not that hard to make amino acids. To make life, the next step would be to assemble proteins, which do the heavy lifting in terms of biological function—they move things around inside the body, catalyze useful reactions, and help cells communicate with one another. That turns out to be not so easy.

While it is encouraging that the initial amino-acid step seems relatively straightforward, by now it has become clear that scientists are going to have to be a lot more clever if we are to understand how the steps proceeded after that.

The study of the origin of life brings together biology, geology, chemistry, atmospheric science, planetary science, mathematics, information theory, and physics. There are multiple promising ideas, not always compatible with one another. We can sketch out plausible ways life might have originally arisen, and how that process fits into the rest of the big picture.

Let's focus on three features that seem to be ubiquitous in life as we know it:

1. *Compartmentalization*. Cells, the building blocks of living organisms, are bounded by membranes that separate their inner structure from the outside world.
2. *Metabolism*. Living creatures take in free energy, and use it to maintain their form as well as performing actions.

3. *Replication with variation.* Living beings create more of themselves, passing along information about their structure. Small variations in that information enable Darwinian natural selection.

There are certainly more aspects to life than this, but if we can account for these we will have made major progress in understanding how life got its start.

Of these features, compartmentalization turns out to be relatively easy to understand. Inorganic materials, in appropriate environments, readily create membranes and differentiate themselves. When a system is far from equilibrium, these spontaneously formed structures can help harness free energy, in particular to enable metabolism and replication. The devil, needless to say, is in the details.

The appearance of cell membranes and other kinds of compartments is one example of the more general phenomenon of *self-organization.* That's what we call it when a large system, consisting of many smaller subsystems, falls into orderly patterns of configuration or behavior, even though the subsystems all behave independently, and with no special goal in mind. The idea of self-organization has been fruitfully applied to phenomena as disparate as the growth of computer networks, the appearance of stripes and spots on animal hides, the spread of cities, and the sudden formation of traffic jams. A classic example is swarming: in flocks of birds or schools of fish, each animal responds only to what its nearest neighbors are doing, but the result is an impressive display of what looks for all the world like highly choreographed behavior.

Self-organization is everywhere. Let's consider one particular example, to get a general flavor for the idea, before moving on to the specifics of cellular membranes. Someday, after all, we might want to understand the nature and origin of spontaneously formed membranes in biospheres other than the one here on Earth.

In 1971, American economist Thomas Schelling proposed a simple model of segregation. One form would be racial segregation within cities, but the basic idea would work for a variety of differences, from linguistic communities to boys and girls choosing seats in an elementary school classroom. Schelling asked us to imagine a square grid with two different kinds of symbols, X's and O's, as well as a few empty spaces. Suppose that the X's and O's aren't completely intolerant of each other, but they get a little uncomfortable if they feel surrounded by symbols of the opposite type. If a symbol is unhappy—if an X has too many O neighbors, for example—it will get up and move to a randomly selected empty space. That happens over and over again, until everybody is happy.

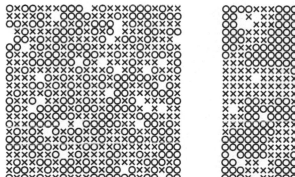 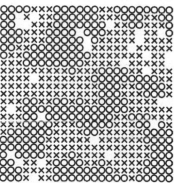

Spontaneous segregation in the Schelling model. Initial condition on the left, final condition on the right.

You wouldn't be surprised to see significant segregation if the symbols were very intolerant—if they were unhappy even having one or two neighbors of the other type, for example. Schelling showed that even a little bit of preference was enough to induce large-scale segregation. In the figure we've shown an example with 500 symbols, half X's and half O's, randomly distributed on a grid with a few empty spaces. Imagine that a given symbol is unhappy if 70 percent or more of its neighbors are of the opposite type. That's relatively tolerant; an O is fine with having as many as five X's among eight neighbors, and becomes unhappy only when there are six or more. In the initial configuration, only 17 percent of the symbols start out as unhappy.

But that's enough. Once we let the unhappy symbols pull up stakes and move to empty spots on the grid, and let that process continue until everyone is happy, what we're left with is the arrangement on the right. Large swaths of segregated neighborhoods, separated by clearly demarcated boundaries.

This large-scale order emerged purely as the result of localized, individual decisions, not as the handiwork of some malicious central planner. And the "decisions" didn't involve any higher forms of cognition; it's self-organization, not externally imposed or goal-driven. We could imagine individual molecules behaving the same way, and indeed they sometimes do. Oil and water separate from each other, and we'll see that lipid molecules have definite preferences that help account for the origin of membranes in cellular life. Schelling shared the 2005 Nobel Prize in Economics with Robert Aumann, primarily for his work in game theory and conflict behavior.

One important point about Schelling's theory is that the way we model the evolution of the system is *not* reversible. The dynamics are not "Laplacian"; information is not conserved. It is therefore not a model of the real world at its most fundamental level. But it can be a perfectly good emergent description of coarse-grained dynamics, as long as the system as a whole is far from equilibrium. The process of an X or O noticing that it's unhappy and moving to a randomly chosen empty space is one that necessarily increases the entropy of the universe in the process. Information is lost, since many initial configurations could lead to the same final one. Entropy increases, but the way it increases is by creating an impermanent structure with a high degree of order and complexity.

The alacrity with which simple dynamic systems exhibit self-organization makes it a little easier to believe that something like a cellular membrane could spontaneously assemble under the right conditions. But real biological membranes aren't made out of boys and girls who don't want to sit next to each other in class; they're made of lipids.

A *lipid* is a particular type of organic molecule, one that has ambivalent feelings toward water. To chemists, *organic* simply means "based mostly on carbon atoms, often involving hydrogen and perhaps a few other elements," regardless of whether a compound has anything to do with living creatures. It's a very different notion of "organic" than you will find in your local upscale supermarket. The connection with biology arises because so much of biochemistry is based on carbon, which can easily form arbitrarily complex molecular chains.

Lipids have a "head" that is *hydrophilic* (attracted to water) on one end, and a "tail" that is *hydrophobic* (repelled by water) on the other. It's this split personality, attracting water from one side and repelling it from the other, that helps these lipids form into membranes.

Imagine putting a concentration of such lipids into water. The hydrophilic end is happy, but the hydrophobic end doesn't know what to do with itself—there's water everywhere. "Happy" is not meant to be taken literally; as with the X's and O's, an unhappy molecule is just one that will move to a different configuration until some condition is satisfied. For a lipid, one end is content in the presence of water, while the other wants to avoid it entirely.

The lipid's search for contentment is a metaphorical way of talking about the fact that the system evolves so as to minimize free energy. Entropy increases, which suggests to us a certain emergent vocabulary, in which the molecules "want" to find a state with low free energy. The arrow of time leads us to speak a language of purpose and desire, even though we're only talking about molecules obeying the laws of physics.

The one thing that the hydrophobic carbon tails can do is to seek comfort in the company of their own kind. The lipids can line up next to one another, so that their tails are all surrounded by other, equally hydrophobic compatriots, rather than by water. There are a few different ways this can happen. The simplest is for the lipids to form a little ball, called a *micelle*, with the hydrophilic heads on the outer surface, exposed to water, while the hydrophobic chains are bundled up with one another.

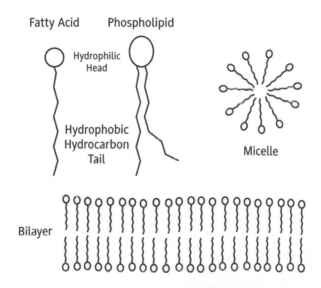

There is another option: a *bilayer*—two sheets of lipids, each one of which has the hydrophilic heads pointing in the same direction, with the hydrophobic tails of the two sheets clinging together. That way the heads get to enjoy the water they seek out, while the tails are completely shielded from it.

In an aqueous (water-containing) solution, lipids will spontaneously organize themselves into one of these types of structures. Which one depends on the circumstances: on what kind of lipid we're dealing with, and on other properties of the solution, especially whether it is more acidic (likes to give away protons and accept electrons) or alkaline (the opposite).

Examples of lipids include fatty acids, which are relatively simple, and phospholipids, which are a bit more complicated. Fatty acids are everywhere in biochemistry—they are one of the fuel sources that mitochondria can use to make ATP, for example. Phospholipids consist of two fatty acids joined together by a phosphate group (a compound of phosphorous, carbon, oxygen, nitrogen, and hydrogen).

The cellular membranes in organisms living on Earth today are made from bilayers of phospholipids. These molecules very naturally self-organize into bilayers, but not into micelles, because their double tails are too thick to easily fit into the ball-like micelle configuration. The bilayer membranes then fold into themselves to form spherical enclosures, known as vesicles. That's the easiest part of assembling a cell.

One problem with phospholipids, as far as the origin of life is concerned, is that the bilayers they construct are just too good at their job. They are fairly impenetrable, with only water and some other small molecules able to pass from one side to another. It therefore seems likely that the earliest form of cellular membranes were actually made of fatty acids rather than phospholipids. Once they are put in place, evolution can set about improving them.

Fatty acids can self-assemble into bilayers, but only under the right conditions. In highly alkaline solutions, fatty acids prefer to form micelles; in highly acidic conditions, they glom together into big oily drops. At intermediate levels of acidity, their favorite configuration is a bilayer. It's a phase transition, governed by the acidity of the surrounding medium.

These bilayers of fatty acids don't relax into long two-dimensional sheets, like a piece of paper. Rather, they quickly pinch off and form little spheres. That's the configuration with the lowest free energy in that environment. It's another manifestation of how, rather than smooshing everything into featureless goo, the second law helps create the kind of organized structures that are useful for life.

Fatty acids are relatively simple molecules, so it wouldn't be hard to find them in appropriate environments on the prebiotic Earth. What's more, the membranes they form are more permeable than those made of phospholipids. That's good news for early life. In a mature organism, you don't want chemicals leaking willy-nilly into and out of your cell; embedded in the membranes are very specific structures (like ATP synthase) that guide nutrients and energy sources in and out as appropriate. Early on, before such dedicated mechanisms have evolved, what you're looking for is something that can do a fairly good job of compartmentalizing the chemical precursors of life, but not such a good job that they are isolated from the outside world and essentially choked to death. Fatty acids seem just right for the task.

From the perspective of a poetic naturalist, one of the most interesting features of spontaneous compartmentalization is how it lends itself readily to an emergent description of the system. Without compartments and membranes, we're faced with a soupy mess of compounds, energy sources, and reactions. Once a boundary forms between different kinds of stuff, we can readily talk about the "object" (inside the boundary) and its environment (everything outside). The boundary—whether it's literally a cell membrane, or the skin or exoskeleton of a multicellular organism—both helps the structure take advantage of the free energy around it and helps us talk about it in useful, computationally efficient ways.

Karl Friston, a British neuroscientist, has suggested that the function of biological membranes can be understood in terms of a *Markov blanket*, a term coined by statistician Judea Pearl in the context of machine learning. Imagine we have a network: a collection of "nodes" connected by lines. A "Bayesian network" is a graph formed from nodes that can send, receive, and process information, like computers on the Internet or neurons in a brain. If we pick out any one node, its Markov blanket consists of all the nodes that can directly influence it (its "parents"), plus all the nodes it can directly influence (its "children"), plus all the nodes that can also influence its children (its "spouses," of which there may be many).

This complicated-sounding construction captures a simple idea: given some part of the network, the Markov blanket captures everything you need to know about its input and output. There may be an enormous number of possible internal states of the nodes, but all that matters for the operation of the network is what gets filtered through the Markov blanket.

A cell membrane, argues Friston, can be thought of as a Markov blanket. Many intricate processes go on inside the cell, and many things are happening all the time in the environment outside. But communication between the two is mediated through the cell membrane. Under these conditions, the system evolves toward a configuration in which the cell membrane is robust—it maintains its configuration, even in the presence of (not-too-large) perturbations from inside or out.

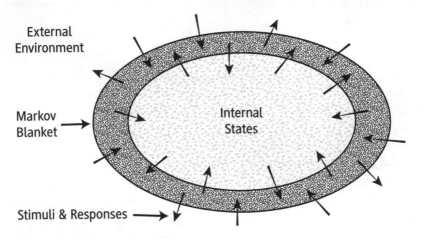

This theory was originally developed not for individual cells but as a way of thinking about how brains interact with the outside world. Our brains construct models of their surroundings, with the goal of not being surprised very often by new information. That process is precisely Bayesian reasoning—subconsciously, the brain carries with it a set of possible things that could happen next, and updates the likelihood of each of them as new data comes in. It is interesting that the same mathematical framework might apply to systems on the level of individual cells. Keeping the cell membrane intact and robust turns out to be a kind of Bayesian reasoning. As Friston puts it:

> The internal states (and their blanket) will appear to engage in active Bayesian inference. In other words, they will appear to model—and act on—their world to preserve their functional and structural integrity, leading to homeostasis [preserving stable internal conditions] and a simple form of autopoiesis [maintaining structure through self-regulation].

This is a speculative and new set of ideas, not an established picture of how we should think about the function of cells and membranes. It's worth remarking on because it shows how the concepts we've been talking about—Bayesian reasoning, emergence, the second law—come together to help explain the appearance of complex structures in a world governed by simple, unguided laws of nature.

32

The Origin and Purpose of Life

On a crowded flight to a conference in Bozeman, Montana, I was reading some research papers on the connection between statistical physics and the origin of life. The man sitting next to me glanced over at them, curiously. "Oh, yes," he offered, "I know that work well."

Over the course of a career as a physicist, you run into people who have theories of how the universe works, and are eager to share them with you. Those theories are rarely very promising. Presumably the study of life attracts similar numbers of garrulous enthusiasts. But we had a long flight ahead of us; I asked him what his thoughts were on the matter.

"That's easy," he replied with a nod. "The purpose of life is to hydrogenate carbon dioxide."

It wasn't the answer I was expecting. I had been fortuitously seated next to Michael Russell, a geochemist at NASA's Jet Propulsion Laboratory, close to my own home institution of Caltech. It wasn't a complete accident—he and I were both traveling to give talks at the same conference. Russell, it turns out, is a leading (if somewhat iconoclastic) figure in the study of life's origin, and one whose approach is especially physics-friendly. We got along fine.

Russell is one of the leaders of a faction in the origin-of-life debates who believes that the first crucial step was the appearance of metabolism. This camp imagines that the crucial event was the appearance of a complex network of chemical reactions that took advantage of free energy in the environment of the young Earth, which could then be used to power repli-cation once it began. There is also a replication-first camp, which currently enjoys wider popularity in the community. They tend to think that energy sources are relatively plentiful and unproblematic, and the important leap in the development of life was the synthesis of an information-bearing mol-ecule (presumably RNA, ribonucleic acid) that could duplicate itself and pass down its genetic information.

We won't be adjudicating this disagreement: these are hard questions to which we simply don't yet know the answers. But they are not hopeless questions. Progress toward understanding abiogenesis has been made on multiple fronts, both theoretically and experimentally. Whatever order me-tabolism and replication appeared in, they are both necessary, and part of the scientific fun will be in figuring out how all the ingredients fit together into the final recipe.

If you want to understand how life began, it makes sense to begin by look-ing for features that are shared by existing forms of life. One such feature seems to be the proton-motive force involved in chemiosmosis, as we dis-cussed in chapter 30. Cell membranes collect energy from photons or from compounds like sugar, and use that energy to expel electrons outside the cell, leaving an excess of protons inside. The mutual repulsion of the pro-tons creates a force that can be used to do useful things like creating ATP.

Where did life ever get that idea from? It's not exactly the obvious way for a cell to manipulate energy. When the chemiosmotic process was worked out by Peter Mitchell and Jennifer Moyle in the 1960s, they were met with enormous skepticism in the biology community, until the experimental evidence became definitive. The fact that nature finds this technique so useful might be a clue that it took advantage of it right from the start.

This is where the hydrogenation of carbon dioxide comes in. Russell's comment alludes to the fact that there is free energy locked up in a mixture of carbon dioxide (CO_2) and hydrogen gas (H_2), both of which were abun-dant in certain environments on the young Earth. If the carbon could somehow shed its two oxygen atoms and replace them with hydrogen, we could end up with methane (CH_4) and water (H_2O). That's a configuration that has less free energy; as far as the second law of thermodynamics is concerned, it's a transformation that "wants" to happen.

It doesn't happen all by itself. Anytime you light a candle, or set anything else on fire, you are releasing free energy by combining the fuel with oxygen. But candles don't just burst into flame; they require a spark to start the reaction.

In the case of carbon dioxide, we require something more elaborate than a spark. It's easy to invent sequences of reactions that gradually move the oxygens off of the carbon atom and replace them with hydrogen. The problem is that, while the sequence as a whole releases energy, the first required steps actually *cost* energy, and therefore don't happen by themselves. Extracting the free energy from carbon dioxide is like robbing a bank: there's a lot of money in there, but you have to go to a great deal of effort to get it out.

A number of researchers, including William Martin and Nick Lane as well as Russell, have been working hard on exploring scenarios in which the right sequence of reactions could have come together in just the right way to take advantage of the ambient free-energy bounty. They have a couple of tricks at their disposal. One is *catalysis*: hastening along the reaction you want by taking advantage of nearby compounds that aren't themselves reacting but can change the shape or properties of the chemicals that are involved. Another is *disequilibrium*: an imbalance in conditions at nearby locations that can be used to drive the desired reactions.

These ingredients come together in the right way in a specific environment: deep-sea hydrothermal vents. In particular, *alkaline* vents—ones where proton-attracting alkaline chemicals are produced. They're not the only plausible environment in which we can search for life's origin; as just one other example, serpentine mud volcanoes are another ocean-floor structure that might be hospitable to early life. But alkaline vents have some nice properties.

As early as 1988, Russell predicted, on the basis of his vision for life's origin, a particular kind of underwater geological formation that had not yet been discovered: underwater vents that were alkaline, warm (but not too hot), highly porous (riddled with tiny pockets, like a sponge), and relatively stable and long-lasting. The idea was that the pockets could provide a kind of compartmentalization even before the existence of any kind of organic cell membranes, and the disequilibrium between alkaline chemicals in the vents and the proton-rich acidic ocean water all around would naturally produce a version of the proton-motive force so beloved by biological cells.

In 2000, Gretchen Früh-Green, on a ship in the mid-Atlantic Ocean as part of an expedition led by marine geologist Deborah Kelley, stumbled across a collection of ghostly white towers in the video feed from a robotic camera near the ocean floor deep below. Fortunately they had with them a submersible vessel named *Alvin*, and Kelley set out to explore the structure up close. Further investigation showed that it was just the kind of alkaline vent formation that Russell had anticipated. Two thousand miles east of South Carolina, not far from the Mid-Atlantic Ridge, the Lost City hydrothermal vent field is at least 30,000 years old, and may be just the first known example of a very common type of geological formation. There's a lot we don't know about the ocean floor.

The chemistry in vents like those at Lost City is rich, and driven by the sort of gradients that could reasonably prefigure life's metabolic pathways. Reactions familiar from laboratory experiments have been able to produce a number of amino acids, sugars, and other compounds that are needed to ultimately assemble RNA. In the minds of the metabolism-first contingent, the power source provided by disequilibria must come first; the chemistry leading to life will eventually piggyback upon it.

Albert Szent-Györgyi, a Hungarian physiologist who won the Nobel Prize in 1937 for the discovery of vitamin C, once offered the opinion that "life is nothing but an electron looking for a place to rest." That's a good summary of the metabolism-first view. There is free energy locked up in certain chemical configurations, and life is one way it can be released. One compelling aspect of the picture is that it's not simply working backward from "We know there's life; how did it start?" Instead, it's suggesting that life is the solution to a problem: "We have some free energy; how do we liberate it?"

Planetary scientists have speculated that hydrothermal vents, similar to Lost City, might be abundant on Jupiter's moon Europa or Saturn's moon Enceladus. Future exploration of the solar system might be able to put this picture to a different kind of test.

※

In the ecosystem of abiogenesis researchers, metabolism-first proponents are a plucky minority. The most popular approach, as mentioned earlier, is replication-first.

Metabolism is essentially "burning fuel," something we see all around us, from lighting a candle to starting a car engine. Replication seems harder, more precious, difficult to obtain. If there is any part of "life" that might act as a bottleneck to getting it started, it's the fact that living beings reproduce themselves.

Fire is a well-known chemical reaction that readily reproduces itself, leaping from tree to tree in a forest, but by most definitions it doesn't count as alive. We want something that carries *information* through the reproduction process: something whose "offspring" keep some knowledge of where they came from.

There's a simple example of such a thing: crystals. Certain kinds of atoms can organize themselves into regular patterns, which are then called crystals. The same atoms might support different possible crystalline structures: when carbon arranges itself in a cubic pattern, we get diamond, but if it's in a hexagonal pattern, all we have is graphite. Crystals can grow by adding new atoms, and can then divide by the simple expedient of breaking in two. Each of the offspring will have inherited the structure of its parent crystal.

That's still not life, though we're getting closer. While the basic crystalline structure can be inherited, variations in that structure—random mutations—cannot. Variations are certainly possible; real crystals are often riddled with impurities, or suffer from defects where the structure doesn't follow the dominant pattern. But there's no way to pass down knowledge of these variations to subsequent generations. What we want is a configuration that is crystal-like (in that there is a fixed structure that can be reproduced) but more elaborate than a simple repeating pattern.

The kind of thing we need was described by John von Neumann, a brilliant Hungarian American mathematician who played crucial roles in the development of quantum mechanics, statistical mechanics, and game theory. In the 1940s, he laid out in abstract terms what would be required for a system to reproduce itself and evolve in an open-ended way. His (purely mathematical) machine—the "von Neumann Universal Constructor"—included not only a mechanism for actually performing the self-replication, but also a "tape" that encoded the structure of the machine. Von Neumann–like self-replicators have been implemented in computer simulations, complete with the possibility of mutation and evolution. No one has yet built a large-scale physical machine that would behave this way, but there's nothing in the laws of physics that would prevent us from doing so, and NASA and other organizations have seriously investigated the possibility. Would a physical implementation of a von Neumann Universal Constructor qualify as "alive"?

Erwin Schrödinger, in *What Is Life?*, recognized the need for information to be passed down to future generations. Crystals don't do the job, but they come close; with that in mind, Schrödinger suggested that the culprit should be some sort of "aperiodic crystal"—a collection of atoms that fit together in a reproducible way, but one that had the capacity for carrying substantial amounts of information, rather than simply repeating a rote pattern. This idea struck the imaginations of two young scientists who went on to identify the structure of the molecule that actually does carry genetic information: Francis Crick and James Watson, who deduced the double-helix form of DNA.

Deoxyribonucleic acid, DNA, is the molecule that essentially all known living organisms use to store the genetic information that guides their functioning. (There are some viruses based on RNA rather than DNA, but whether or not they are "living organisms" is disputable.) That information is encoded in a series of just four letters, each corresponding to a particular molecule called a *nucleotide*: adenine (A), thymine (T), cytosine (C), and guanine (G). These nucleotides are the alphabet in which the language of genes is written. The four letters string together to form long strands, and each DNA molecule consists of two such strands, wrapped around each other in the form of a double helix. Each strand contains the same information, as the nucleotides in one strand are paired up with complementary ones in the other: A's are paired with T's, and C's are paired with G's. As Watson and Crick put it in their paper, with a measure of satisfied understatement: "It has not escaped our notice that the specific pairing we have postulated immediately suggests a possible copying mechanism for the genetic material."

In case it has managed to escape your notice, the copying mechanism is this: the two strands of DNA can unzip from each other, then act as templates, with free nucleotides fitting into the appropriate places on each separate strand. Since each nucleotide will match only with its specific kind of partner, the result will be two copies of the original double helix—at least as long as the duplication is done without error.

The information encoded in DNA directs biological operations in the cell. If we think of DNA as a set of blueprints, we might guess that some molecular analogue of a construction worker comes over and reads the blueprints, and then goes away to do whatever task is called for. That's almost right, with proteins playing the role of the construction workers. But cellular biology inserts another layer of bureaucracy into the operation. Proteins don't interact with DNA directly; that job belongs to RNA.

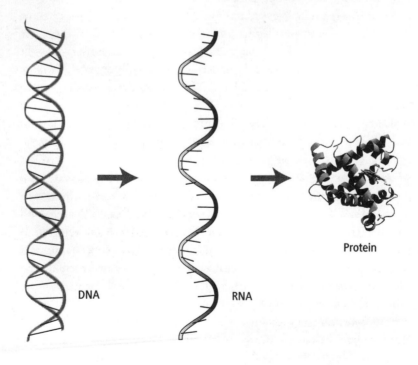

DNA   RNA   Protein

RNA is similar in structure to DNA, but it usually comes in the form of single strands. The "backbone" of the strands differs slightly from RNA to DNA, and RNA pairs adenine with a nucleotide called uracil (U), rather than with thymine. It's less chemically stable than DNA, but it can carry equivalent information in its particular sequence of nucleotides.

Information gets out of DNA when the two strands unzip, and their sequences are copied onto RNA segments. Those segments, called *messenger RNA*, carry genetic information to a different unit within the cell, the *ribosome*. Ribosomes, discovered back in the 1950s, are complicated structures that take the information in the RNA and use it to construct proteins. This multistep process enables a relatively stable information-storage system (DNA) to construct useful molecules (proteins) using less stable messengers (RNA) and a complete separate construction facility (the ribosome).

Just as for compartmentalization and metabolism, replication faces a "How did we get here from there?" problem, relating the sophisticated structures of modern-day biology to simpler systems that could plausibly have come into existence from non-life. For compartmentalization, we need to understand how we got to bilayers made of phospholipids, and the answer might be found in fatty acids. For metabolism, we need to know how we got to cells driven by the proton-motive force, and the answer might be porous chambers in alkaline vents. For replication, we need to know how we got to DNA, and the answer might be RNA.

The relationship of RNA to DNA is like the relationship of an oral tradition of poetry to words written down in books. The same information can be conveyed, but DNA is much more reliable and stable. Yet it is sufficiently sophisticated that it's hard to see how it could have come into existence by itself. When DNA gets copied, an important part of the work is done by proteins. But the proteins are supposed to be constructed using information encoded in the DNA. How could either one arise without the other already being present?

The favorite answer among abiogenesis researchers is a scenario called *RNA world*. The basic idea was proposed by a number of people in the

1960s, including Alexander Rich, Francis Crick, Leslie Orgel, and Carl Woese. DNA is good at storing information, and proteins are good at performing biochemical functions; RNA is able to do both, although it's not as good at either one. RNA could have come along before either DNA or proteins, and served as the basis for a primitive and less robust form of early life, before evolution gradually distributed responsibilities to the more effective DNA and proteins.

The role of RNA in extracting information from DNA was recognized fairly early on, but it wasn't until later that biologists verified that RNA could also act as a catalyst, expediting and governing the rate of biochemical reactions. In particular, *ribozymes*, discovered in the 1980s, are a particular kind of RNA that can catalyze their own synthesis, as well as that of proteins. The word "ribozyme" is annoyingly similar to "ribosome." It turns out that the crucial part of the ribosome complex consists of ribozyme RNA. That is, the ribosome is mostly ribozyme. (It's jargon like this that turns young scientists toward physics and astronomy.)

Further investigations have shown that there are a number of different types of RNA, responsible for a variety of functions inside the cell. In addition to messenger RNA and ribosomal RNA, we also have *transfer RNA* that brings amino acids to the right place to be made into proteins, *regulatory RNA* that helps guide the expression of genes, and more. These discoveries have helped popularize the RNA-world hypothesis. If you want to get life started from a replication-first perspective, you need a molecule that can carry genetic information without relying on other complex mechanisms to reproduce itself. RNA seems to hit the sweet spot.

The idea that RNA may have been the first carrier of genetic information, and was able both to self-reproduce and to assemble other biochemically useful structures, is compelling and beautiful. Like any good paradigm, one of the great features of the RNA world scenario is that it has spurred a tremendous amount of exciting research.

Consider the fact that RNA can be an enzyme: it can catalyze chemical reactions, both for self-assembly and for protein synthesis. Where did that ability come from? It's pretty clear how a string of nucleotides can store information, but acting as an enzyme seems like a completely different kind of talent.

This question was addressed in an interesting experiment by David Bartel and Jack Szostak in 1993. (Szostak shared the Nobel Prize in 2009 for his work on how chromosomes are protected when DNA divides.) Their technique was basically a human-aided version of Darwinian evolution. They started with a large amount of random RNA: trillions of molecules with no particular sequence to their nucleotides. They then picked out a fraction of those molecules, the ones that seemed to be associated with somewhat higher rates of catalysis, and made many copies of those. This procedure was repeated several times: look for RNA that seemed to be catalyzing certain reactions, and make copies of it. At each copying stage, random mutations occurred, which occasionally led to the copied RNA being a better catalyst than its precursor. After just ten iterations of this procedure, the results were clear: the last pool of molecules was approximately 3 million times better at catalyzing reactions than the original sample. It's a vivid demonstration of how undirected, random mutation can lead to enormous improvements in the ability of chemicals to perform biologically useful functions.

Another exciting development came from biologists Tracey Lincoln and Gerald Joyce in 2009. They were able to create a system of two RNA enzyme molecules—ribozymes—that together underwent self-sustained replication. Without any help from surrounding proteins or other biological structures, these molecules are able to completely duplicate each other in about an hour. Even better, the molecules occasionally mutate, and therefore undergo Darwinian evolution, with the more fit structures preferentially surviving. It's not a cell by any means, but you don't need to strain to see how it could be one of the steps along the road from chemistry to life.

Even if RNA played a central role at the origin of life, we don't yet have a complete picture. Compartmentalization, metabolism, and replication all have to be brought together. RNA and bilayers made from fatty acids may be symbiotic—they could help each other flourish in the rough-and-tumble environment of the early Earth. A membrane can shield the fragile RNA from external commotion, helping it survive long enough to reproduce. An RNA molecule, meanwhile, can attract other biological molecules into the membrane, helping it grow to the point where it will naturally split in two—a primitive form of cellular division.

Fitting in metabolism may be trickier, though Szostak doesn't think it's a big problem. He envisions a proto-cell, RNA encapsulated in a simple membrane, floating in a pond that is warm on one end and cold on the other. Convection currents push the proto-cell back and forth between the two sides. In the cold end, the RNA grows by gathering nucleotides from its surroundings, and two RNA strands huddle together as if seeking warmth. When it drifts to the warmer side of the pond, the increased temperature gradually peels the two strands apart; the membrane accretes a few more fatty-acid molecules until it divides in two, and (hopefully, sometimes) we now have two proto-cells with a single strand of RNA each. They both drift back to the cold side of the pond, and the cycle of proto-life begins again.

Russell and the metabolism-firsters don't think it will be nearly that easy. They believe that the hard part is assembling a complex system of chemical reactions that can take advantage of the ambient free energy, setting up proton-motive forces in chambers of porous underwater vents. From there, they suggest, these reactions will naturally feed on any surrounding free-energy fuel they can find. That might mean that they break free of the rocks by entering fatty-acid membranes, and they keep going by regulating their reactions through enzymes, which eventually become RNA.

Or maybe both scenarios are right, or maybe neither is.

There is no reason to think that we won't be able to figure out how life started. No serious scientist working on the origin of life, even those who are personally religious, points to some particular process and says, "Here is the step where we need to invoke the presence of a nonphysical life-force, or some element of supernatural intervention." There is a strong conviction that understanding abiogenesis is a matter of solving puzzles within the known laws of nature, not calling for help from outside of them.

This conviction comes from the incredible historical track record science has established. While there are many questions about life's origin that science hasn't answered, there are a large number that it has, any one of which could have been a problem that science all by itself was unable to address. (Recall Immanuel Kant's confident proclamation that there will never be a Newton for a blade of grass.) How do species evolve from earlier species? How do organic molecules become synthesized? How do cellular membranes assemble themselves? How can complex reaction networks overcome free-energy barriers? How can RNA molecules develop the ability to act as catalysts for biochemical reactions? These are questions we *have* answered. Our Bayesian credence that this string of successes will keep going should be very high indeed.

This perspective meets resistance in certain quarters, and not only among religious fundamentalists. The idea that life could just start out of no life at all isn't obvious. We don't see it taking place before our eyes, no matter what Jan Baptist van Helmont might have imagined. Modern-day organisms are mind-bogglingly complex, and made of individual parts that work together amazingly well. The idea that it "just happened" is a challenging one.

Fred Hoyle, an esteemed British astrophysicist known for his staunch opposition to the Big Bang model, attempted to quantify this unease. He considered the configuration of atoms in a biological structure such as a cell. Then, in a move taken from Ludwig Boltzmann's playbook, he compared the total number of ways such atoms could be arranged to the much smaller number that would qualify as a cell. Multiplying together a bunch of tiny numbers, he concluded that the chance of life assembling all by itself is something like 1 in $10^{40,000}$.

Hoyle was a master of vivid imagery, and he illustrated his point with a famous analogy:

> The chance that higher life forms might have emerged in this way is comparable to the chance that a tornado sweeping through a junkyard might assemble a Boeing 747 from the materials therein.

The problem is that Hoyle's version of "this way" is nothing at all like how actual abiogenesis researchers believe that life came about. Nobody thinks that the first cell occurred when a fixed collection of atoms was rearranged over and over in all possible ways until it just happened to take on a cell-like configuration. What Hoyle is describing is essentially the Boltzmann Brain scenario—truly random fluctuations coming together to create something complex and ordered.

The real world is different. The "unlikeliness" associated with low-entropy configurations is built into the universe from the start, by the incredibly low entropy near the Big Bang. The fact that the development of the cosmos proceeds from this very special initial condition, rather than wandering through a more typical equilibrium ensemble of states, imposes a strong nonrandom aspect on the evolution of the universe. The appearance of cells and metabolism is a reflection of the universe's progression toward higher entropy, not an unlikely happenstance in an equilibrium background. Like the swirls of cream mixing into coffee, the marvelous complexity of biological organisms is a natural consequence of the arrow of time.

We've made amazing progress in understanding what life is and how it came to be, and there's every reason to think that progress will continue until we have figured it out. The work ahead will involve chemistry, physics, mathematics, and biology, not magic.

33

Evolution's Bootstraps

I n 1988, Richard Lenski had a brilliant idea: he was going to turn evolutionary biology into an experimental science.

Evolution is the idea that provides the bridge from abiogenesis to the grand pageant of life on Earth today. There's no question that it's a science; evolutionary biologists formulate hypotheses, define likelihoods of different outcomes under competing hypotheses, and collect data to update our credences in those hypotheses. But chemists and physicists have an advantage over evolutionary biologists or, for that matter, astronomers: they can perform repeated experiments in their labs. It would be very hard to set up a laboratory experiment to see Darwinian evolution in action, just as it would be hard to create a new universe.

But it's not impossible. (At least for evolution; we still don't know how to create new universes.) And that's exactly what Lenski set out to do.

His basic setup was—and is, as the experiment is still ongoing—a simple one. He started with twelve flasks containing growth medium: a liquid with a specific mixture of chemicals, including a bit of glucose to provide energy. He then introduced a population of identical *E. coli* bacteria into each of them. Every day, each flask goes from a few million to a few hundred million cells. One percent of the surviving bacteria are extracted and moved to new flasks with the same growth medium as before. The remaining bacteria are mostly disposed of, although every so often a sample is frozen for future examination, creating an experimental "fossil record." (Unlike human beings, live bacteria can easily be frozen and revived at a later date using current technology.) The total population growth amounts to about six and a half generations in a day; the limiting resource is nutrition, not time (it takes less than an hour for a cell to divide). As of late 2015, this added up to more than 60,000 generations of bacteria—enough for some interesting evolutionary wrinkles to develop.

Confined to this extremely specific and stable environment, the evolved bacteria are by now quite well adapted to their surroundings. They are now over twice the size of the individuals in the original population, and they reproduce more rapidly than before. They have become very good at metabolizing glucose, while generally decaying in their ability to thrive in more diverse nutrient environments.

Most impressively, there have been qualitative as well as quantitative changes in the *E. coli*. Among the ingredients in the initial growth medium was citrate, an acid made of carbon, hydrogen, and oxygen. The original bacteria had no ability to use this compound. But around generation 31,000, Lenski and his collaborators noticed that the population in one particular flask had grown larger than the others. Looking more closely, they realized that some of the bacteria in that flask had developed the ability to metabolize citrate, rather than just glucose.

Citrate is not as good an energy source as glucose is. But if you're a bacterium in a flask full of other bacteria that are competing for a fixed amount of glucose, the ability to live off of this other energy source is very useful. Without having any particular goal to work toward, without the benefit of any external prompting or instruction, evolution had come up with a clever new way of allowing organisms to flourish in their particular environment.

The origin of life was the mother of all phase transitions. Like other chemical reactions or combinations thereof, life proceeds by converting free energy into disordered energy. The aspect that makes life special among chemical reactions is that it carries with it a set of instructions. Like the tape in one of John von Neumann's Universal Constructors, the genetic information contained in DNA regulates and guides the interconnected dance of reactions that defines a living organism. Those instructions can change as they are passed down from generation to generation. That ability is what gets natural selection off the ground.

We've speculated that DNA came from RNA, which in turn may have self-catalyzed its own production under the right circumstances. It's possible that the creation of the first RNA molecule involved random fluctuations at critical points along the way. Boltzmann taught us that entropy *usually* increases, but there is always some probability that it will occasionally move downward. The more moving parts a system has, the more rare such fluctuations will be; at macroscopic scales, the number of atoms involved is so large that it's not worth worrying about. But at the level of individual molecules, rare fluctuations are frequent enough to be important. The appearance of the first self-replicating RNA molecule might just have been a matter of good luck.

We sometimes think of natural selection as "survival of the fittest." But even before evolution in Darwin's sense officially kicked in, there was a competition of sorts going on for the available free energy. Some of it would have been readily accessible, but some—similarly to that locked up in the citrate in Richard Lenski's flasks of bacteria—would have required more ingenuity to unlock. An intricate network of reactions, directed by proteins created by a sequence of nucleotides in RNA, could have prospered where simpler processes would have flickered out. Once heritable genetic information starts playing a role, all of the ingredients are in place for natural selection to commence.

From a certain perspective, Darwin's theory is sufficiently commonsensical that it seems almost inevitable. Upon first reading *Origin*, Thomas Henry Huxley, Darwin's contemporary and vocal supporter, exclaimed, "How extremely stupid not to have thought of that!" But natural selection is a very specific process, and by no means inevitable or obvious. It's not simply "species gradually change over time," or "well-adapted organisms are more likely to reproduce."

Organisms reproduce, and they hand down their genetic information to the next generation. That information is largely stable—children resemble their parents—but it's not absolutely fixed. Small, random variations can be introduced at every step. The variations do not strive to reach any future goals, and neither can individual organisms influence them by their actions. (Your offspring don't become more muscular just because you work out.) If we have descent with inheritance, and there is slight, random variation in the genetic information that can affect the likelihood of reproduction, natural selection can occur. Variations that fortuitously improve an organism's chances of handing down its genetic heritage will be more likely to persist than those that are harmful or neutral.

These ingredients shouldn't be taken for granted. This is why biologists highlight the difference between "evolution" and "natural selection." The former is the change of the genome (complete set of genetic information) over time; the latter refers to the specific case where changes in the genome are driven by different amounts of reproductive success.

Darwin didn't know about DNA or RNA, or even of genes, discrete units of inherited information. It was the Augustinian monk Gregor Mendel who established the basic rules of heredity, through a set of now-famous experiments crossing different varieties of pea plants. In the 1930s and '40s, biologists developed the *modern synthesis*, combining natural selection with Mendelian genetics. The paradigm continues to be elaborated upon as we learn more and more about biology and inheritance, but the basic picture remains enormously successful.

The reality of biology here on Earth is, unsurprisingly, more complicated than the simplest statement of natural selection. Like any way of talking about the world, Darwin's theory works only within its domain of applicability.

There are forces at work in the history of life other than organisms adapting to their environments. This is completely compatible with Darwin's conception; natural selection happens, but it happens within the messiness of the real world, and it's not the only thing happening. Many features of the genome of any individual species are going to be the results of accidents rather than any particular adaptation. This is known as *genetic drift*. Sometimes there will be mutations that neither increase nor decrease the fitness of an organism; other times, the randomness inherent in sexual reproduction or unpredictable features of the environment will cause some traits to become common while others die off. Biologists debate the relative importance of adaptation and genetic drift, but there is little doubt that both are important.

In Lenski's long-term evolution experiment, the mutation that allowed some of the bacteria to metabolize citrate occurred around generation

31,000. When the researchers unfroze some of the earlier generations to see if they would evolve this ability again, they found that the answer was yes—but only when they started with cells from generation 20,000 or later. Around generation 20,000, one or more mutations must have occurred that did not themselves allow the bacteria to metabolize citrate, but set the stage for a later mutation that would do so. A single trait can be brought to life by multiple, separate mutations, which may not individually have much noticeable impact at all.

Selection pressures work on traits, while genetic information is passed down through DNA, and the map from one to the other isn't a simple one. Something as basic as how tall a person is won't typically be fixed by one particular string of nucleotides, but instead will depend on an interplay between different factors working simultaneously. As a result, selection pressure acting on one trait may end up affecting another one, if they depend on common sets of DNA sequences. Evolutionary history is replete with "spandrels," as was famously emphasized by biologists Stephen Jay Gould and Richard Lewontin. These are traits that arise for one reason and then end up being used for something quite different. By-products of the evolutionary process, rather than aspects that are directly selected for. Gould and Lewontin imagine that many features of the human brain fall under this category.

To make matters worse, inheritance can be more than simply a matter of passing down DNA from one generation to the next. There is *horizontal gene transfer*, in which genes are passed from one organism to another in a way other than reproduction. It is relatively common in bacteria, and occasionally happens in multicellular species. There are *epigenetic* phenomena, in which the chemical structure of inherited DNA is modified during development by influences such as the nutritional intake of an organism and the maternal environment in which an embryo develops. It is currently unclear how much such changes can be inherited by subsequent generations, but to the extent that they are, natural selection will act upon them as usual.

So the real world is a beautiful mess. Is this kind of undirected mechanism—just what we would expect in a universe governed by unthinking underlying laws and with a strong arrow of time—sufficient to account for all the spectacular intricacy of our planet's biosphere? "There is grandeur in this view of life," Darwin writes in *On the Origin of Species*. But is his simple mechanism really enough to make dolphins and butterflies and rain forests from a meager collection of organic molecules fighting for free energy? Can the wonders of efficiency and ingenuity we see in biological organisms really come about from random variation plus time? (Hint: yes.)

34

Searching through the Landscape

I n computer science, as in life, we are often faced with the simple problem of finding some particular item in a long list of possibilities. Consider the traveling-salesman problem: given a list of cities and the distances between them, what is the shortest route that visits each city exactly once? That can be rephrased in the following way. Take a list of cities and the distances between them. Now make another list, consisting of every possible route that goes through each city at least once. (It will be an enormously longer list, but it is still finite.) Which route is the shortest?

A *search algorithm* is a precisely stated procedure for finding what you are looking for in a list of objects. Of course you could trudge through every element of the list, asking, "Is this the one?" That can be hard, since quite reasonable-sounding questions can involve very unreasonably sized lists to sort through. For the traveling-salesman problem, the number of possible routes grows roughly as the factorial of the number of cities involved. The factorial of a number n is equal to 1 times 2 times 3 times 4 . . . times $(n - 1)$ times n. For twenty-seven cities, that's about 10^{28} routes to search through. At a rate of a billion routes per second, that search would take you longer than the age of the observable universe.

The trick, then, isn't just to find any old search algorithm, but to find efficient ones. And very often, the number of choices is so high that we're happy to find solutions that are just pretty good, rather than absolutely perfect.

Natural selection can be thought of as a search algorithm. The problem being tackled by evolution is: "What organism would survive and reproduce most effectively in this particular environment?" Except it's not really "organisms" that are being searched, it's genomes, or particular strings of nucleotides in a strand of DNA. The human genome contains about 3 billion nucleotides. That's a lot, compared to, for example, a bacterium, which might have a few million. But let's not be too proud; there are flowering plants with over 100 billion nucleotide base pairs in their DNA. Some organisms will survive and reproduce, while some will not. How, over the course of generations, do we find the DNA sequences that lead to organisms with the highest chance of survival?

This counts as a hard problem, from the perspective of computational resources. Each of our 3 billion nucleotides is 1 of 4 possible choices: A, C, G, or T. The total number of possible arrangements of human-sized DNA is not four times 3 billion (which wouldn't be so bad); it's four *to the power of* 3 billion: $4^{3,000,000,000}$, or roughly 1 followed by 2 billion zeros. That's a stupendously, hilariously large number. It's also an overestimate; some sequences of nucleotides have the same functional impact as others, and the vast majority of sequences wouldn't even lead to an organism. We could count genes rather than nucleotides; that would cut down the number of dimensions considerably, although each gene can take many more than four possible forms, so the number is still enormous, and the interdependence of different gene functions makes any such counting a little uncertain. By any possible measure the problem of finding the "best" organism by searching through all of the possible genomes an organism could have is a daunting one.

Evolution provides a strategy for searching for high-fitness genomes in a ridiculously big space of possibilities. Computer scientists have recently shown that a simplified model of evolution (allowing for mixing via sexual reproduction, but not for mutations) is mathematically equivalent to an algorithm devised by game theorists years ago, known as *multiplicative weight updates*. Good ideas tend to show up in a variety of places.

The phrase "search algorithm" isn't meant to imply that anyone wrote an algorithm, or anyone is specifying a goal that evolution is supposed to search for. Evolution doesn't have any goals in mind; evolution simply happens, with Laplacian equanimity, each step following from the previous one. In the spirit of poetic naturalism, "search algorithm" is simply a useful way of talking about the process of evolution. Under the appropriate circumstances, they are formally mathematically equivalent, and the connection provides some nice visual intuition. Don't let the language trick you into believing there is any agency guiding the course of evolution, or setting up goals ahead of time; at the same time, don't let the fear of sounding like you believe in agency prevent you from using a language that gives significant insight into the process.

One way of visualizing evolution's search problem is in terms of a fitness landscape. The idea is that we can assign, to any particular genome in a specific environment, a numerical value called its "fitness." This number characterizes how likely it is that an organism based on that genome will reproduce in that environment. We can visualize the fitness in terms of a rolling landscape, with hills and valleys, where the role of "directions in space" is played by different forms each gene can take, and the role of "height above ground" is played by the fitness. (When we actually draw a fitness landscape, we typically look at only one or two genes at a time, but in the back of your mind you should remember that this is really a 25,000-dimensional space we're thinking of, one for each gene.) A high-fitness hill corresponds to a genome that produces an organism that is very likely to reproduce (the more offspring, the better), while a low-fitness valley is a genome that is unlikely to make it to subsequent generations.

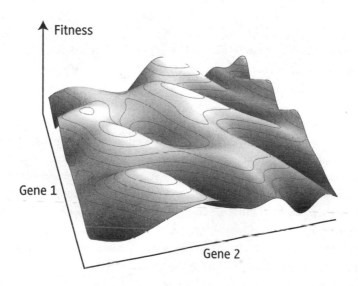

Fitness

Gene 1

Gene 2

We can think of evolution as nudging populations toward higher elevations in the fitness landscape, favoring genes that lead to more fit organisms. That's a simplification, of course. There isn't a single fixed fitness landscape appropriate to all species and all circumstances for all times; at best we should think of a particular population in a fixed environment. The shape of the landscape will depend on all of the properties of that environment. Other species come and go, the physical surroundings change, so the landscape changes over time. But some aspects of the environment can be stable enough over a sufficiently long time period that a fixed landscape is a useful metaphor for visualizing what goes on.

Biologists see the world differently from physicists. The concept of a landscape also appears in physics, for example, when we are asking what phase a system settles down to at a given temperature and pressure. But in the back of their minds, physicists are always thinking about a ball rolling on a hill. Consequently, the favored points on the landscape are the *minimum* values of the function being plotted (typically the energy), since balls roll downward. Biologists are thinking about wily mountain goats, or children playing a game of King of the Hill. To them, the favored points on the landscape are the *maximum* values of fitness.

Here's how evolution searches through the fitness landscape, looking for higher peaks: We have a population of organisms of a certain species, so they occupy a set of nearby points on the landscape. Individuals are born, hopefully reproduce, and die. Their offspring have slightly different genomes, so they are located somewhere else on the landscape—not too far away, but not at exactly the same place either. The ones that end up lower on a slope are less likely to reproduce than those that find themselves higher up. As generations pass by, the population finds itself gradually moving toward higher ground.

We draw two-dimensional plots, but in reality the number of genes can be very large indeed, so it can take an extremely long time for a population to climb up the landscape. Species may never get to the top of a single hill, much less the highest mountain around, though individual traits may do so. Some parts of the landscape are relatively flat; that's where different genomes don't have very different levels of fitness, and genetic drift can be the dominant feature in evolution. A more realistic portrayal would have a time-varying landscape, as both physical and biological features of the environment continually shift around. When that happens, it's literally impossible to simply find the top of a hill and just sit there; one day's maximum might be a valley the next day.

Finally, there's no sense in which evolution's algorithm is guaranteed to find the best possible result. Most variations are small, and allow us to explore only nearby points in the landscape. Occasionally, there might be a rare mutation that enables us to skip from one peak to another, but only with peaks that are close together to begin with. Just as for the traveling-salesman problem, finding a good-enough solution can be extremely useful for all practical purposes.

The search procedure employed by evolution is so efficient that real human computer programmers often use an analogous process to develop their own strategies. This is a technique known as *genetic algorithms*. As with genomes, we can imagine the set of all possible algorithms of a certain length, at least within a fixed computer language. There will be a large number of them, and in principle we want to know which one is the best at solving some specified problem. The genetic-algorithm approach works like natural selection, except that the role of the fitness landscape is put in by the programmer. In biology this would be called directed evolution, to emphasize the difference with natural selection, where the fitness landscape is fixed by nature without any particular agenda.

Start with some randomly chosen algorithms, and let them tackle the problem. Take some fraction that do the best, and then "mutate" them, possibly also allowing them to mix with other successful algorithms. Throw away the unsuccessful strategies, and repeat the process. The population of algorithms being studied will gradually climb up the relevant fitness landscape, defined as how well each strategy does at finding a good solution to its problem. (It's the virtual equivalent of what Bartel and Szostak did to find RNA configurations that could act as catalysts.)

Genetic algorithms provide a nice illustration of some of the interesting features of evolution as a strategy inventor. One such example was invented by computer scientist Melanie Mitchell. She asks us to consider Robby, a virtual robot who lives in a simple world, a ten-by-ten grid of squares. Robby threw a party last night, and there are empty cans scattered across the grid. Robby wants to clean them up in a hurry, being as efficient as possible with only a finite amount of time available. Our task is to invent a strategy—an unambiguous set of instructions about what to do at every step—that will help Robby the Robot pick up all the cans on the grid.

You might think that Robby can just walk from one can to the next, and the challenge is to find the shortest path. But Robby is burdened with two significant handicaps, perhaps due to partying a little too hard the previous night. First, he can't see very far. Standing on any one square, Robby can see whether there's a can on his own square, and he can see whether there's a can on any of the squares immediately north, south, east, or west of him. But that's all; he can't see whether there are any cans diagonally, or on the squares farther away.

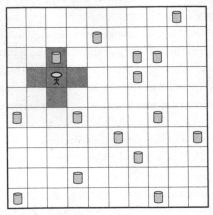 

The world of Robby the Robot, on the left: a grid of squares, some empty and some littered with cans. Robby's field of view is highlighted. On the right, a situation where Robby is on a square with a can, and with multiple cans nearby.

Your next thought is therefore that Robby should walk in some kind of pattern, systematically scanning the grid and picking up any cans he sees. But he has a second handicap: Robby has absolutely no memory at all. He doesn't know where he's been, what cans he's picked up, or what he was doing even one moment ago. His strategy can only refer to what he must do next based on his situation right now; it can't include anything like "move east, and next time move south," since that would encompass two moves in a row.

Given these limitations, it's straightforward to enumerate every possible strategy that Robby could follow. There are five squares he knows about: his own, and the four neighbors corresponding to each cardinal direction. Each square is in one of three conditions: it can be empty, have a can, or be behind the wall (where he can't go). Robby's "state" is a list of what's on each of the five squares he knows about: a total of $3^5 = 243$ states. There are seven possible actions he can take: he can pick up a can (if one is there), he can move in one of the four cardinal directions, he can move in a random direction, or he can stay put and do nothing.

A strategy for Robby is just a specification of one of the seven actions for each of the 243 states. The number of possible strategies is thus 7^{243}, or about 10^{205}. You're not going to try out every strategy just to find the best possible one.

You can try to be clever, and *design* a strategy you think will do a good job. Mitchell did just that, choosing a baseline strategy for what would count as "pretty good, even if not necessarily the best." It was a simple approach: if Robby is on a square with a can, pick it up. If not, look for cans on the four nearby squares. If there is one can, move in that direction. If there are no cans, move in a random direction. If there is more than one can, move in a specified direction. Call this the "benchmark strategy." As hoped, the benchmark strategy proved to do a respectable job; in a large number of trials, it typically reached about 69 percent of a perfect score.

Alternatively, we can be inspired by nature's method, and *evolve* a strategy using directed evolution. A specific strategy for Robby is like a specific list of nucleotides in a DNA helix, a discrete information-carrying string. We can artificially evolve it by starting with some number of randomly chosen strategies, letting them run for a while, and picking out the ones that do the best. Then we make several copies of each survivor, "mutating" each copy by randomly altering a few of the specific actions each strategy specifies for a particular state, and even mimicking sexual reproduction by cutting strategies and pasting them together with other ones. The process is reminiscent of evolution. Can it find strategies for Robby that are better than the "pretty good" designed one?

Yes, it can. Evolution easily found much better solutions than design. After only 250 generations, the computer was doing as well as the benchmark strategy, and after 1,000 generations, it had reached almost 97 percent of a perfect score.

After a genetic algorithm has evolved, we can go back and watch what it does, trying to figure out what made it so effective. This tricky bit of reverse-engineering is increasingly a real-world challenge. Many useful computer programs operate according to genetically constructed algorithms that no human programmer actually understands, which is a scary thought. Fortunately, Robby's choices are sufficiently constrained that we can try to figure out what is going on.

Robby's best strategies improve on the benchmark in a number of clever ways. Consider a situation where Robby is on a square containing a can, and the squares to the east and west also contain cans. The benchmark strategy, quite naturally, instructs him to pick up the can. But think about what will happen next: Robby will move either east or west, thereby losing track of the can in the other direction. The genetic algorithm, though it was constructed using nothing but random variations and selection, "figured this out," and came up with a better strategy. When Robby is in the middle of a sequence of three cans, he doesn't pick up the one on his square; he moves east or west until he's reached the edge of the can grouping, and only then does he pick up a can. Next, quite naturally, he moves back into the grouping, scooping up cans along the way. This and other bits of clever engineering turn out to be enormously more effective than the "obvious" designed benchmark strategy.

Evolution isn't always better than design. An omniscient designer could find the best strategy every time. The point is that natural selection, or directed evolution in this case, is a really good search strategy. It doesn't necessarily find the best solution, but it regularly finds impressively clever ones.

As wonderful as evolution is at searching for peaks in a complex, high-dimensional fitness landscape, there are places that it won't find. Consider a landscape with a very high mountain, separated by a long, flat plain from a collection of undulating hills. And imagine a population whose genomes are located within those hills. The process of small variation and natural selection will let the species explore around the hills, looking for the highest point it can find. But as long as the variations in the genome within the population remain small, all of the individuals will remain in the grouping of hills. None will have any reason to make a long, unrewarding trek across the flat plain to get to the isolated peak. Evolution can't see globally across the space of genomes and find a better one; it proceeds locally through random variation and then an evaluation (through reproduction) of how well that particular variation is doing at the moment.

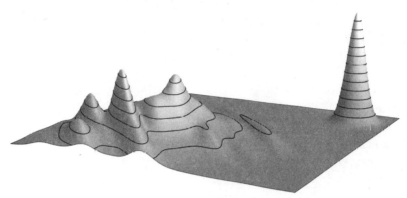

A fitness landscape with an isolated peak that would be difficult for natural selection to find.

The failure to find an isolated solution to some problem within a long list of possibilities isn't unique to evolution. Almost every efficient search strategy attempts to take advantage of structure within the list of possibilities—such as the fact that nearby points on a fitness landscape have similar values of fitness—rather than blindly scanning every option. It could, however, enable an empirical challenge to natural selection as the correct theory of the evolution of species. If someone could show that a particular organism's genome had high fitness within the landscape defined by its environment, but could not be "found" by the strategy that evolution employs, it would decrease our credence in Darwin's theory.

Given any one particular genome, how do we know that it is an isolated peak in the fitness landscape? Such peaks almost certainly exist, although they might be less common than they first appear. When we draw a two-dimensional landscape, isolated peaks are almost inevitable, but when the underlying space has many more dimensions (like the 25,000 or so genes in a human being), there can be a lot more paths to get from one peak to another.

A possible criterion for genomes that wouldn't be produced by evolution was put forward by Michael Behe, a critic of natural selection and advocate of intelligent design. In an attempt to show that certain organisms couldn't have arisen through conventional Darwinian evolution, Behe proposed the notion of "irreducible complexity." An irreducibly complex system, in Behe's definition, is one whose functioning involves a number of interacting parts, with the property that every one of the parts is necessary for the system to function. The idea is that certain systems are made of parts that are so intimately interconnected that they can't arise gradually; they must have come together all at once. That's not something we would expect from evolution.

The problem is that the property of irreducible complexity isn't readily measurable. To illustrate the concept, Behe mentions an ordinary mousetrap, with a spring mechanism and a release lever and so forth. Remove any one of the parts, he argues, and the mousetrap is useless; it must have been designed, rather than incrementally put together through small changes that were individually beneficial.

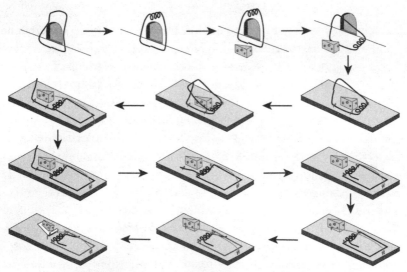

Incremental evolution of a complex mousetrap, as designed by John McDonald. The trap starts as a simple wire that can snap shut when disturbed. In a series of steps, it adds: a spring, some bait, resting on its side, attached to a platform, a longer "hammer," a tripwire, a staple to hold the tripwire, a shorter spring wire, an even shorter spring wire, a separate catch to hold the tripwire, separating the hammer from the spring, and finally a more elaborate catch to release the trap.

You can probably guess what happened next. At least two different people (John McDonald and Alex Fidelibus) presented possible "evolutionary paths" that mousetraps might have followed. They created a series of designs, starting very simply and becoming gradually more complex, of working mousetraps. Each step worked a little better than the previous one, despite differing by only a small change. And the final step was precisely a modern mousetrap. Adding insult to injury, Joachim Dagg investigated the way that actual mousetraps have changed over the years, showing that (despite being designed) they evolved gradually rather than appearing all at once. In Dagg's words, "All prerequisites for evolution (variation, transmission, and selection) abound in mousetrap populations."

Irreducible complexity reflects a deep concern that many people have about evolution: the particular organisms we find in our biosphere are just too designed-looking to possibly have arisen through "random chance plus selection."

A version of this conviction can be traced back to William Paley, of the watchmaker analogy. Paley wrote before Darwin came on the scene, but he put some effort into attempting to refute any future Darwin-like thinker who would deny God's central role in explaining the complexity of the world. His favorite example was the eye. The word "eye" appears more than two hundred times in Paley's *Natural Theology: or, Evidences of the Existence and Attributes of the Deity, Collected from the Appearances of Nature.* The many pieces that have to work together, the undeniable effectiveness of the eye at its assigned task, the effort to which the body attempts to protect and preserve its eyes—to Paley, these spoke strongly to the view that the eye implied "the necessity of an intelligent Creator."

Not only can eyes be explained through natural selection; they seem to have evolved separately dozens of times over the history of life. It's not difficult to trace out plausible paths for how eyes could develop. The absorption of photons is one of the most basic activities that living organisms do. This ability can be concentrated in photosensitive patches, or "eyespots," that are found even in some single-celled organisms. Given that an organism can sense light, it can be advantageous to acquire sensitivity to the direction from which the light is emitting. A simple way to achieve this ability is to locate the eyespot in a recessed cup, such as is seen in certain flatworms. Deepening the cup to an almost spherical opening allows the organism to employ a primitive kind of lens, similar to that in a pinhole camera; this is what we find in some contemporary mollusks. Filling that eyehole with a transparent fluid helps with both protection and focusing. Many of the steps along the way won't arise in single jumps; often, evolution can borrow mechanisms from other functions in the organism that came about for different reasons.

You get the idea—not only can eyes be developed in stages of increasing complexity and fitness, but we actually see such development in real creatures alive today. And the human eye, as wondrous as it is, has unambiguous flaws that would be inexcusable for a talented designer but make perfect sense in light of evolution. The nerve fibers that carry visual information to the brain are, for no good reason, in front of our retinas rather than behind them. The octopus eye is a better design, with the retina in front and nerves in back, so that octopuses don't have a blind spot like humans do. Our anatomy reflects the accidents of our evolutionary history.

35

Emergent Purpose

Time for a multiple-choice quiz: Why do giraffes have such long necks?

1. Over the generations, giraffes kept stretching upward to reach leaves near the tops of trees. Gradually their necks got longer and longer.

2. Long necks help you eat. Random mutations in their DNA gave some giraffes longer necks than others. These individuals enjoyed a nutritional advantage over their compatriots, because they could reach fresh leaves near the treetops. This advantage was passed on to their descendants, and gradually the giraffe population developed longer necks.

3. Long necks are sexy. Male giraffes compete for the affections of females by swinging their heads at each other. Random mutations in their DNA gave some giraffes longer necks than others, which conferred a reproductive advantage. This advantage was passed on to their descendants, and gradually the giraffe population developed longer necks.

4. Given the laws of physics, and the initial state of the universe, and our location in the cosmos, collections of atoms

in the shape of long-necked giraffes came into existence 14 billion years after the Big Bang.

The difference between options 1 and 2 is a common way of explaining Darwin's theory of natural selection. Option 1 is incorrect; changes that individuals undergo during their lives, such as through exercise or learning new behaviors, are not incorporated into our genetic information, and are therefore not passed down to subsequent generations. (There are nuances here, as some environmentally influenced ways that genes are expressed may be heritable, even if the genes themselves don't change.) Option 2 is a more standard Darwinian explanation. It's not that previous generations of giraffes *wanted* to reach higher; it's just that those that did accrued an advantage that was passed on to their descendants.

Then there is option 3, known as "sexual selection." It is a perfectly plausible Darwinian explanation, one that relies on a specific mechanism of selection pressure to achieve the empirical result. Some researchers have suggested that a form of sexual selection is a better explanation than the traditional leafy-treetop story that we tell about the length of giraffe necks. This illustrates one of the difficulties in understanding how evolution actually proceeds in the real world: there may be more than one way to explain the emergence of a single trait.

The debate is ongoing. For example, under sexual selection it's likely that male and female giraffe necks would evolve differently, but the data seem to indicate that they are fairly similar. Option 2 is currently more popular, but new data will continue to impact our credences for each of the different hypotheses.

So what about option 4, which avoids any particular evolutionary storytelling? It's a true statement, but not a useful one in this context. From the poetic-naturalism perspective, natural selection provides a successful way of talking about emergent properties of the biological world. We don't need to use a vocabulary of evolution and adaptation to correctly describe what happens, but doing so gives us important and useful knowledge.

The evolution of life provides a rich source of higher-level phenomena emerging from the fundamental description of reality, including phenomena that have no direct analogue at the deepest level. Because our specific universe starts in a special state and shows a strong arrow of time, these emergent pictures can invoke words like "purpose" and "adaptation," even though those ideas are nowhere to be found in the underlying mechanistic behavior of reality.

A common concern among skeptics of evolution is how it is supposed to lead to the creation of new *kinds of things* out of the mindless motion of matter. "Purposes" are one obvious example. We say, without apparent embarrassment, things like "The purpose of the giraffe's long neck is to help it reach fresh leaves near the treetops." Another example is "information." DNA is said to carry genetic information; the optic nerve carries information from the eye to the brain. Then there is consciousness itself. The concern is that these concepts represent a radical break from the mere Laplacian working-out of the laws of physics. How could evolution, which itself is ultimately purely physical, bring these utterly new kinds of things into existence?

It's a natural thing to worry about. The process of evolution is unplanned and unguided. Whether or not genetic information gets passed on to future generations depends only on the conditions of its immediate environment and random chance, not on any future goals. How can an intrinsically purposeless process lead to the existence of purposes?

But this worry is a little strange, at least in the hands of anyone who accepts that natural selection provides an explanation for more prosaic things like gills and eyeballs. These kinds of organs are "utterly new" in their own way. There is no general principle along the lines of "new kinds of things cannot naturally arise in the course of undirected evolution." Things like "stars" and "galaxies" come to be in a universe where they formerly didn't exist. Why not purposes and information?

In poetic naturalism, the appearance of "truly new" concepts as one theory emerges from another is the least surprising thing in the world. As time passes and entropy increases, the configuration of matter in the universe takes on different forms, enabling the emergence of different higher-level ways of talking. The appearance of something like "purpose" simply comes down to the question "Is 'purpose' a useful concept when developing an effective theory of this part of reality in this particular domain of applicability?" There may be any number of interesting and challenging

technical issues to be addressed, but there is no obstacle to the emergence
of all kinds of new concepts along the way.

Think about Robby the Robot, cleaning up cans from his grid. In the most
successful strategies that were artificially generated through many generations of variation and selection, Robby had evolved a technique of not picking up a can on his current square if there were also cans to the east and
west. Rather, he would move in one direction—let's say west—until he arrived on a square with a can, but no can on the square just west of his location. Only then would he double back, picking up all the cans along the way.

Why does Robby act in this way? We could simply say, "Those moves are
part of the strategy that survives the genetic algorithm process." That would
be the equivalent of answer 4 in the list of giraffe-neck explanations above.
It's not wrong, but it's not very illuminating either. Or we could say, "Robby
doesn't want to forget that there are cans on either side, so he leaves them
in place, knowing he will come back and pick them up later."

Is that a sensible way of talking? Robby the Robot doesn't really *want*
anything. He's not even a real robot—just a string of ones and zeroes inside
some computer memory. Psychologists sometimes speak of the "anthropomorphic fallacy," when we attribute human thoughts or emotions to inanimate objects. (My computer gets grumpy if I don't reboot it every so
often.) It may be fun and harmless to speak about Robby as if he has wants,
but it's not *really* true. Right?

Consider the possibility that we have this backward. When we say that
Robby the Robot doesn't really have wants in the same sense that a person
does, we are taking the implicit stance that there are things called "wants"
that can be correctly attributed to some things in the universe (like human
beings) and not to others (like virtual robots). What are these "wants"
anyway?

The idea that something wants something else is a way of talking that is
potentially useful in the right circumstances—a simple idea that summarizes a good amount of complex behavior in a convenient way. If we see a
monkey climbing a tree, we could describe what's happening by providing
a list of what the monkey is doing at each moment in time, or for that

matter we could specify the position and velocity of every atom in the monkey and the environment at each moment. But it's immensely easier and more efficient to say, "The monkey wants those bananas that are up in the tree." The fact that we can say that is a piece of useful knowledge over and above all of those positions and velocities.

There is no Platonic idea of a "want" floating out there in the space of ideas that can be properly associated with some kinds of beings and not with others. Rather, there are situations in which it is useful to describe things as somebody wanting something, and other situations in which that is not so useful. These situations can emerge in the natural, undirected evolution of matter in the universe. Those wants are as real as things ever get.

In the particular case of Robby, it is neither necessary nor especially helpful to characterize his behavior in terms of wants, purposes, or desires. It's just as easy to simply say what his can-collecting strategy actually is. But the difference between him and a person, as far as the ontological status of "wants" is concerned, is simply a matter of degree. We could imagine a robot with an enormously more complicated programming than little Robby. We might not know much about that specific programming, but perhaps we are able to observe how the robot acts. It may be that the best way of understanding the robot's behavior is to say, "That robot really wants to pick up those cans."

Under naturalism, there isn't that much difference between a human being and a robot. We are all just complicated collections of matter moving in patterns, obeying impersonal laws of physics in an environment with an arrow of time. Wants and purposes and desires are the kinds of things that naturally develop along the way.

There is a similar story to tell about "information." It's worth thinking about, as it will come up again when we start talking about consciousness. If the universe is just a bunch of stuff obeying mechanistic physical rules, how can one thing ever "carry information" about anything else? How can one configuration of atoms be "about" some other configuration?

Words like "information" are a useful way of talking about certain things that happen in the universe. We don't ever need to talk about information—we can take the "option 4" viewpoint and just talk about the quantum state of the universe inexorably evolving through time. But the fact that information is an effective way of characterizing certain physical realities is a true and nontrivial insight onto the world.

Consider the Voynich manuscript. This is a remarkable and unique book, whose likely provenance has been traced to the early fifteenth century, possibly from Italy. It is a whimsical volume, replete with fanciful illustrations of astronomical and biological subjects. For the most part, the many flora depicted in the illustrations cannot be identified with actual plant species. Most remarkably, the text of the book has proven, to date, to be completely indecipherable. Not only the language but even the apparent alphabet is something that has never been seen before. Statistical analyses of the words and symbols in the writing seem to be compatible with those of ordinary languages, but cryptographers have been stymied in their attempts to interpret the text as some kind of code. It may be a very good cipher; it may be a unique language that was invented by an individual and then forgotten; or it may be a complete hoax.

Does the Voynich manuscript contain information?

An excerpt of the writing that appears in the Voynich manuscript.

One is tempted to say that it depends on the origin of the book. If it really is a hoax, and the words are some kind of semirandom nonsense, then perhaps it doesn't contain much information at all. But if it is merely a clever code that will someday be broken, it might contain a great deal—even if that "information" is purely a work of imagination.

What if the Voynich manuscript is a code that will never be broken? What if it was originally written with very specific intent, but its meaning has been so well hidden that nobody will ever be able to reveal it? Does it *still* contain information? What if the manuscript is placed in a capsule and launched into space, and then the Earth is destroyed by a cataclysmic asteroid impact, and the book floats through the void for all of eternity. Does it contain information then?

We tend to use the word "information" in multiple, often incompatible, ways. In chapter 4 we talked about conservation of information in the fundamental physical laws. There, what we might call the "microscopic information" refers to a complete specification of the exact state of a physical system, and is neither created nor destroyed. But often we think of a higher-level macroscopic concept of information, one that can indeed come and go; if a book is burned, the information contained in it is lost to us, even if not to the universe.

The macroscopic information contained in a book is relative to the environment in which it is embedded. When we talk about the information contained in the book you are currently reading, what we mean is that these words are *correlated* with certain ideas that you get upon reading them. You read the word "giraffe," and the notion of a certain kind of long-necked African ungulate appears in your mind. The same holds for the information contained in a strand of DNA: it is correlated with the synthesis of certain proteins in the cell. It is this connection with one configuration of matter (a book or a DNA strand) and something else in the universe (the image of a giraffe, or a useful protein molecule) that lets us talk about the existence of information. Without those correlations—if there isn't, and never will be, anyone around to read the book, or any RNA molecules that can read the DNA and go off to make protein—there is no point in talking about information.

From this perspective, the appearance of information-bearing objects in the course of the undirected evolution of matter and life is unsurprising. It happens because—wait for it—the universe started with an extremely low entropy. That means it was in a very specific kind of state; just knowing the low-entropy macroscopic configuration of the universe gives us a tremendous amount of information about its microscopic state. (In equilibrium, where entropy is high, the microstate could be almost anything, and we have essentially no information about it.) As the universe evolves from this very specific configuration to increasingly generic ones, correlations between different parts of the universe develop very naturally. It becomes useful to say that one part carries information about another part. It's just one of the many helpful ways we have of talking about the world at an emergent, macroscopic level.

In the late 1990s, a controversy arose about a "Statement on Teaching Evolution" adopted by the National Association of Biology Teachers (NABT) in the United States:

> The diversity of life on earth is the result of evolution: an unsupervised, impersonal, unpredictable and natural process of temporal descent with genetic modification that is affected by natural selection, chance, historical contingencies and changing environments.

The controversial bit was the inclusion of the words "unsupervised" and "impersonal." It was thought by some that this characterization went beyond the merely scientific, to pass judgment on questions that belonged to the sphere of religion. Two prominent theologians, Alvin Plantinga and Huston Smith, wrote a letter to the NABT, arguing that this encroachment would backfire by "lower[ing] Americans' respect for scientists and their place in our culture." The thought was presumably that in any perceived conflict between science and religion, Americans will always choose religion. Plantinga and Smith urged the board of directors to amend the statement to delete "unsupervised" and "impersonal." After some debate, the board agreed, and those words were dropped from the statement in future publications.

One can argue about the political wisdom of such a move, but the original wording of the NABT statement was scientifically appropriate. The theory of evolution describes an unsupervised and impersonal process. The theory might be wrong, or incomplete; what looks like unguided evolution to us might be secretly nudged in some preferred direction by a subtle, unseen force. But that's a different theory, one that you are welcome to flesh out and try to test using conventional scientific techniques. In the theory that seems to provide an excellent description of the history of life on Earth, nothing is being supervised, and nothing is personal. Natural selection does not strive toward any goal, whether it is increasing amounts of complexity, the ultimate appearance of consciousness, or the greater glory of God.

Given the enormous empirical successes of Darwin's theory, it is not surprising that some religious thinkers have proposed versions of "theistic evolution"—seminatural selection, but guided by God's hand. Supporters of this view include a number of distinguished biologists, including Francis Collins, director of the US National Institutes of Health, and Kenneth Miller, a cell biologist who has actively campaigned against the teaching of creationism in American schools.

Perhaps the most popular way of attempting to reconcile evolution with divine intervention is to take advantage of the probabilistic nature of quantum mechanics. A classical world, so the reasoning goes, would be perfectly deterministic from start to finish, and there would be no way for God to influence the evolution of life without straightforwardly violating the laws of physics. But quantum mechanics only predicts probabilities. In this view, God can simply choose certain quantum-mechanical outcomes to become real, without actually violating physical law; he is merely bringing physical reality into line with one of the many possibilities inherent in quantum dynamics. Along similar lines, Plantinga has suggested that quantum mechanics can help explain a number of cases of divine action, from miraculous healing to turning water into wine and parting the Red Sea.

True, all of these seemingly miraculous occurrences would be *allowed* under the rules of quantum mechanics; they would simply be very unlikely. Very, extremely, outrageously unlikely. If we populated every planet circling every star in the universe with scientists, and let them do experiments continuously for many times the current age of the observable universe, it would be extraordinarily improbable that even one of them would witness a single drop of water changing into wine. But it's possible.

"Possible" doesn't quite do the job that advocates of theistic evolution would like it to do. There are roughly two scenarios. In one, the choices made at each quantum event have a high probability of coming true on their own, and the hand of God is simply picking one likely event among several possibilities. In that case, God isn't doing much of anything at all. The appearance of human beings was never very improbable; it could easily have happened without any divine intervention. If you pray that a fair coin flip comes up heads, and it does, it would seem strange to attribute too much credit to God. Or, from a Bayesian perspective, the gain in likelihood you achieve through divine intervention isn't nearly enough to overcome the added complexity and inevitable loss of precision involved in allowing supernatural influences to alter the course of the physical world.

The other scenario is that the events necessary to bring about human beings through the course of evolution were extremely unlikely, even though they were possible—comparable, perhaps, to the spontaneous parting of the Red Sea. In that case, you are not simply taking advantage of quantum indeterminacy; you are violating the laws of physics. Observing an event that is so extremely unlikely that you wouldn't expect to see it anywhere in the observable universe should count as evidence that you are calculating probabilities in the wrong theory. If someone flips a coin one hundred times and gets heads every time, you are observing an outcome that was possible if the coin was fair—but it's much more likely that the game is rigged.

Quantum indeterminacy doesn't offer the slightest bit of cover for those who want to make room for God to influence the evolution of the world. If God micromanages which outcomes are realized in quantum events, it is just as much an intervention as if he were to alter the momentum of a planet in classical mechanics. God either does, or does not, affect what happens in the world.

The problem for theism is that there's no evidence that he does. Advocates of theistic evolution do not make a positive case that we *need* divine intervention to explain the course of evolution; they merely offer up quantum mechanics as a justification that it could possibly happen. But of course it can *possibly* happen, if God exists; God can do whatever he wants, no matter what the laws of physics may be. What theistic evolutionists are actually doing is using quantum indeterminacy as a fig leaf: it's not that God is allowed to act in the world, it's that they are allowed to imagine him acting in a way such that nobody would notice, leaving no fingerprints.

It is unclear why God would place such a high value on acting in ways that human beings can't notice. This approach reduces theism to the case of the angel steering the moon, which we considered in chapter 10. You can't disprove the theory by any possible experiment, since it is designed precisely to be indistinguishable from ordinary physical evolution. But it doesn't gain you anything either. It makes the most sense to place our credence in the idea that the divine influences simply aren't there.

36

Are We the Point?

As impressive as the appearance and evolution of life are, doesn't it seem a bit—fragile? If conditions were just a bit different, doesn't it seem plausible that life wouldn't have come about at all?

This concern is sometimes developed into the positive claim that the existence of life is evidence against naturalism. The idea is that conditions—anything from the mass of the electron to the rate of expansion of the early universe—are fine-tuned for life's existence. If these numbers were just a little bit different, the argument goes, we wouldn't be here to talk about it. That makes perfect sense under theism, since God would want us to be here, but might be hard to account for under naturalism. In Bayesian language, the likelihood of life appearing in the universe might be large under theism, and small under naturalism. We can therefore conclude that our very existence is strong evidence in favor of God.

The fine-tuning argument for God's existence rubs some people the wrong way. It seems to take everything that science has discovered since Copernicus and turn it on its head. If this logic is right, we actually are the center of the universe, figuratively speaking. We are the reason the universe exists; numbers like the mass of the electron take the values they do because of us, not simply by accident or even because of some hidden physical mechanism. It can come across as more than a little arrogant to contemplate all of the interacting quantum fields of the Core Theory, or see an image of some of the hundreds of billions of galaxies that populate our universe, and say to yourself, "I know why it's like that—so that I could be here."

Nevertheless, fine-tuning is probably the most respectable argument in favor of theism. It's not a clever-sounding bit of a priori reasoning that allows us to demonstrate the existence of some feature of the universe without leaving our armchair. The fine-tuning argument plays by the rules of how we come to learn about the world. It takes two theories, naturalism and theism, and then tests them by making predictions and going out and looking at the world to test which prediction comes true. It's the best argument we have for God's existence.

It's still not a very good argument. It relies heavily on what statisticians call "old evidence"—we didn't first formulate predictions of theism and naturalism and then go out and test them; we knew from the start that life exists. There is a selection effect: we can be having this conversation only in possible worlds where we exist, so our existence doesn't really tell us anything new.

Still, naturalists need to face fine-tuning head-on. That means understanding what the universe is predicted to look like under both theism and naturalism, so that we can legitimately compare how our observations affect our credences. We'll see that the existence of life provides, at best, a small boost to the probability that theism is true—while related features of the universe provide an extremely large boost for naturalism.

The most important step is to determine the probability that we would measure various experimental outcomes under each theory. Easier said than done, given that there are many specific versions of both theism and naturalism. We will do our best, but should keep in mind that there's a good bit of leeway in our estimates of the likelihoods, and a certain element of judgment that will color our final answers.

If naturalism is true, what is the probability that the universe would be able to support life? The usual fine-tuning argument is that the probability is very small, because small changes in the numbers that define our world would render life impossible.

A famous example of such a number is the energy of space itself: the vacuum energy, or cosmological constant. According to general relativity, empty space can hold an intrinsic amount of energy in every cubic centimeter. Our best current observations indicate that this energy is small, but not quite zero: about one hundred-millionth of an erg in every cubic centimeter of space. (An erg is not that much energy; a hundred-watt lightbulb uses up a billion ergs per second.) But the vacuum energy could have been enormously larger. A back-of-the envelope calculation shows that a reasonable value would have been something like 10^{112} ergs per cubic centimeter—a full 120 orders of magnitude larger than the actual number.

If the vacuum energy had taken on this "natural" value, you wouldn't be reading these words right now. There would be no such thing as words or books or people. Vacuum energy accelerates the expansion of the universe, pushing things apart from one another. An energy that enormous would rip apart individual atoms, making anything like "life" extremely unlikely. The tiny value of the vacuum energy in the real world seems gentle and life-affirming by contrast.

Vacuum energy isn't the only number that seems to be tuned for life. The way that stars shine (ultimately providing free energy for our biosphere) depends sensitively on the mass of the neutron. Stars run by nuclear fusion. The first step is when two protons come together and one of them converts to a neutron, creating a nucleus of deuterium. If the neutron were a little bit heavier, that reaction would not occur in stars. If it were a little bit lighter, all of the hydrogen in the early universe would have been converted to helium, and helium-based stars would have much shorter lifetimes. As with the vacuum energy, the mass of the neutron seems fine-tuned to allow for the existence of life.

That might very well be. But there are two subtleties that render this reasoning a bit uncertain.

First, we don't have reliable ways of judging whether the values of various physical quantities are likely or unlikely. The vacuum energy in our world is much smaller than simple estimates might lead us to guess. But those simple estimates could be wildly misguided, based as they are on our incomplete understanding of the ultimate laws of physics. For example, the maximum entropy that a region of space can contain is higher when the vacuum energy is lower. Perhaps there is a physical principle that prefers space to have a high maximum entropy rather than a low one. If so, that would favor very small values of the vacuum energy, which is exactly what we observe. We shouldn't get too excited when physical quantities seem unnaturally large or small until we understand the mechanism that sets their values, if there is any. They could be attributable to ordinary physical processes, having nothing to do with the existence of life.

Second, we don't know that much about whether life would be possible if the numbers of our universe were very different. Think of it this way: if we didn't know anything about the universe other than the basic numbers of the Core Theory and cosmology, would we predict that life would come about? It seems highly unlikely. It's not easy to go from the Core Theory to something as basic as the periodic table of the elements, much less all the way to organic chemistry and ultimately to life. Sometimes the question is relatively simple—if the vacuum energy were enormously larger, we wouldn't be here. But when it comes to most of the numbers characterizing physics and astronomy, it's very hard to say what would happen were they to take on other values. There's little doubt that the universe would look quite different, but we don't know whether it would be hospitable to biology. Indeed, a recent analysis by astronomer Fred Adams has shown that the mass of the neutron could be substantially different from its actual value, and stars would still be able to shine, using alternative mechanisms to the ones employed by our universe.

Life is a complex system of interlocking chemical reactions, driven by feedback and free energy. Here on Earth, it has taken a particular form, making use of the wonderful flexibility of carbon-based chemistry. Who is to say what other forms analogous complex systems might take? Fred Hoyle, the astronomical gadfly who liked to cast doubt on the Big Bang and the origin of life, wrote a science-fiction novel called *The Black Cloud*, in which the Earth is menaced by an immense, living, intelligent cloud of interstellar gas. Robert Forward, another scientist with a science-fictional bent, wrote *Dragon's Egg*, about microscopic life-forms that live on the surface of a neutron star. Perhaps a trillion trillion years from now, long after the last star has winked out, the dark galaxy will be populated by diaphanous beings floating in the low-intensity light given off by radiating black holes, with the analogue of heartbeats that last a million years. Any one possibility seems remote, but we know of a number of physical systems that naturally develop complex behavior as entropy increases over time; it's not at all hard to imagine that life could develop in unexpected places.

✳

There is another famous complication: we might not have just a universe, but a multiverse. The physical numbers that are purportedly fine-tuned—even supposedly fixed constants, such as the mass of the neutron—could take on very different values from place to place. If that's the case, the fact that we find ourselves in a part of the multiverse that is compatible with life is exactly what we should expect. Where else would we find ourselves?

This idea is sometimes labeled the *anthropic principle*, and the very mention of it tends to inflame passionate debate between its supporters and detractors. That's too bad, because the basic concept is very simple, and practically indisputable. *If* we live in a world where conditions are very different from place to place, *then* there is a strong selection effect on what we will actually observe about that world: we will only ever find ourselves in a part of the world that allows for us to exist. There are several planets in the solar system, for example, and some of them are much larger than Earth. But nobody thinks it is weird or finely tuned that Earth is where we live; it's the spot that is most hospitable to life. That's the anthropic principle in action.

The only real question is whether it is reasonable to imagine that we do live in a multiverse in the first place. The terminology can be confusing; naturalism says there is only one world, but that "world" can include an entire multiverse. In this context, what we care about is a *cosmological multiverse*. That means there are literally different regions of space, very far away and therefore unobservable to us, where conditions are quite different. We call these regions "other universes," even though they are still part of the natural world.

Because there's been a finite number of years since the Big Bang, and because light moves at a fixed speed (one light-year per year), there are parts of space that are simply too far away for us to see them. It's completely possible that out beyond our visible horizon, there are regions where the local laws of physics—the equivalent of the Core Theory—are utterly different. Different particles, different forces, different parameters, even different numbers of dimensions of space. And there could be a huge number of such regions, each with its own version of the local laws of physics. That's the cosmological multiverse. (It's a separate idea from the "many worlds" of quantum mechanics, where different branches of the wave function are all subject to the same physical laws.)

Some people find this kind of speculation distasteful, as it relies on phenomena that are, and will remain, beyond the reach of observation. But even if we can't see other universes, their existence can affect the way we understand the universe we do see. If there is only one universe, the puzzle of the vacuum energy is "Why does the vacuum energy take on the particular value that it does?" If there are many universes, with different values of the vacuum energy, the question is "Why do we find ourselves in this part of the multiverse, where the vacuum energy takes on this specific value?" These are quite distinct issues, but each is a perfectly legitimate scientific question. Whether or not we live in a multiverse is a perfectly ordinary scientific consideration, to be judged by perfectly ordinary methods: what physical model provides the best account of the data?

There would, admittedly, be something disreputable about the multiverse idea if we were positing all of these different regions of space for no good reason, or only so that we could address fine-tuning problems. That would represent an extremely elaborate and contrived model. Even if it provided a good fit to the data, it would be natural to penalize it severely when it came to assigning prior credences; simple theories are always to be preferred over complicated ones.

But in modern cosmology, the multiverse is not a theory at all. Rather, it is a *prediction* made by other theories—theories that were invented for completely different purposes. The multiverse wasn't invented because people thought it was a cool idea; it was forced on us by our best efforts to understand the portion of universe that we do see.

Two theories, in particular, move us to contemplate the multiverse: string theory and inflation. String theory is currently our leading candidate for reconciling gravitation with the rules of quantum mechanics. It naturally predicts more dimensions of space than the three we observe. You might think that this rules out the idea, and we should move on with our lives. But these extra dimensions of space can be curled up into a tiny geometric figure, far too small to be seen in any experiment yet performed. There are many ways to do the curling up—many different shapes the extra dimensions can take. We don't know the actual number, but physicists like to throw around estimates like 10^{500} different ways.

Every one of those ways to hide the extra dimensions—what string theorists call a *compactification*—leads to an effective theory with different observable laws of physics. In string theory, "constants of nature" like the vacuum energy or the masses of the elementary particles are fixed by the exact way in which extra dimensions are curled up in any given region of the universe. Elsewhere, if the extra dimensions are curled up in a different way, anyone who lived there would measure radically different numbers.

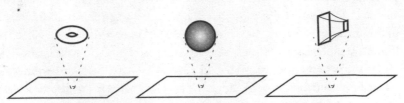

Different ways that extra dimensions of space could be compactified and hidden from our view. Each possibility leads to different numbers characterizing the physical laws we would measure in that region of space.

String theory, then, allows for the existence of a multiverse. To actually bring it into existence, we turn to inflation. This idea, pioneered by physicist Alan Guth in 1980, posits that the very early universe underwent a period of extremely rapid expansion, powered by a kind of temporary super-dense vacuum energy. This has numerous beneficial aspects, in terms of explaining the universe we see: it predicts a smooth, flat spacetime, but one with small fluctuations in density—exactly the kind that can grow into stars and galaxies through the force of gravity over time. We don't currently have direct evidence that inflation actually occurred, but it is such a natural and useful idea that many cosmologists have adopted it as a default mechanism for shaping our universe into its present state.

Taking the idea of inflation, and combining it with the uncertainty of quantum mechanics, can lead to a dramatic and unanticipated consequence: in some places the universe stops inflating and starts looking like what we actually observe, while in other places inflation keeps going. This "eternal inflation" creates larger and larger volumes of space. In any particular region, inflation will eventually end—and when it does, we can find ourselves with a completely different compactification of extra dimensions than we have elsewhere. Inflation can create a potentially infinite number of regions, each with its own version of the local laws of physics—each a separate "universe."

Together, inflation and string theory can plausibly bring the multiverse to life. We don't need to postulate a multiverse as part of our ultimate physical theory; we postulate string theory and inflation, both of which are simple, robust ideas that were invented for independent reasons, and we get a multiverse for free. Both inflation and string theory are, at present, entirely speculative ideas; we have no direct empirical evidence that they are correct. But as far as we can tell, they are reasonable and promising ideas. Future observations and theoretical developments will, we hope, help us decide once and for all.

What we can say with confidence is that *if* we get a multiverse in this way, any worries about fine-tuning and the existence of life evaporate. Finding ourselves in a universe that is hospitable to life is no stranger, nor any more informative, than finding ourselves living on Earth: there are many different regions, and this is the one in which we can live.

What should be our credence that there is such a multiverse? It's difficult to say with our current level of understanding of fundamental physics and cosmology. Some physicists would put the chances at nearly certain, others at practically zero. Perhaps it's fifty-fifty. For our present discussion, what matters is that there is a simple, robust mechanism under which naturalism can be perfectly compatible with the existence of life, even if the life turns out to be extremely sensitive to the precise values of the physical parameters characterizing our environment.

So what about the likelihood of a universe like ours appearing under theism? Here we are faced with a similar problem: the word "theism" doesn't refer to a unique, predictive theory of the world. People will interpret it in different ways, leading to different estimates of the likelihoods of various observable features. We have little choice but to proceed, keeping in mind the inherent uncertainties of the question.

It's reasonable to accept that theism predicts the existence of life with high probability. At least, most theists do not advocate a conception of God under which he is completely indifferent to the existence of human beings. We could imagine such a conception; a noninterventionist God who created or sustained the universe but had no special regard for what you and I would call "life." But we can afford to err on the side of generosity, and assume that the probability of life existing under theism is appreciable; larger, indeed, than it would be under naturalism.

That is far from the end of the story, however. There is an important distinction between "life" and "the numbers describing a universe consistent with the existence of the kinds of complex chemical reactions we identify with biological organisms." God might care about the former, but it's far less clear that he would care about the latter.

The physical parameters of our universe govern what can happen according to the laws of physics. But under theism, "life" is generally something other than a simple manifestation of the laws of physics. Theists tend to be non-physicalists; they believe that living organisms are more than simply the collective behavior of their physical parts. There is a spirit, soul, or life-force that is the most important part of what life really is. The physical aspects may be important, but they are not at the heart of what we mean by "life."

And if that's true, it's unclear why we should care about fine-tuning of physical features of the universe at all. The physical world could behave in any way it pleases; God could still create "life," and associate it with different collections of matter in whatever way he might choose. The requirement that our physical situation be compatible with complex networks of chemical reactions that perpetuate themselves and feed off of free energy in the way we usually associate with living organisms is only relevant *if naturalism is true*. If anything, the fact that our universe does allow for these physical configurations should be taken to increase our credence for naturalism at the expense of theism.

Any theist worth their salt could, admittedly, come up with a number of reasons why God would choose to associate immaterial souls with complex self-sustaining chemical reactions, at least for a time. Likewise, if we lived in a universe where life was not associated with matter in such a way, it wouldn't be hard to come up with justifications for that. This is the problem with theories that are not well defined.

There is another substantial difficulty for the idea that fine-tuning provides evidence for theism. Namely, there is more to the laws of nature and the configuration of our universe than simply whether or not life can exist. If one wants to claim that theism explains certain features of our universe because we predict that God would want life to exist, we must then ask what other features of the universe we would predict under theism. It is here that theism doesn't fare so well.

Predicting what the universe should look like under theism is difficult, for two reasons. There are many different conceptions of God, all of which are somewhat vague on the specifics of God's intentions about the constants of nature. Furthermore, the fact that we know a lot about what the actual universe does look like tends to color our predictions. It's an inherent problem with any theory that is formulated using words. Equations provide less freedom to shape predictions in order to match known results.

Nevertheless, let's give it a shot. There are a number of features of the universe that we would probably expect to see if the existence of life (or human beings) was a primary consideration in its design. Let's highlight three:

- **Degree of fine-tuning.** If the reason why certain characteristics of the universe seem fine-tuned is because life needs to exist, we would expect them to be sufficiently tuned to allow for life, but there's no reason for them to be much more tuned than that. Vacuum energy actually has this property; it is less than it could be, but big enough to be observable. But other numbers—the entropy of the early universe, for example— seem much more tuned than is necessary for life to exist. Life requires an arrow of time, so there must be some sort of low-entropy early state. But in our universe, the entropy is far lower than it needs to be just to allow for life. From purely anthropic considerations, there is no reason at all for God to have made it that small. We therefore think there is some dynamic, physics-based reason why the entropy started off with the fine-tuned value it did. And once we allow for that possibility, other purported fine-tunings may have similar physical explanations.

- **Messiness of observed physics.** If the laws of physics were chosen so that life could exist, we would expect that each of the various features of those laws would play some important role in the unfolding of life. What we see, on the contrary, is something of a mess. All living beings are made out of the lightest generation of fermions—the electron and the up and down quarks, with occasional appearances from electron neutrinos. But there are two heavier families of particles, which don't play any part in life. Why did God make the top and bottom quarks, for example, and why do they have the masses they do? Under naturalism we would expect a variety of particles, some of which are important to life and some of which are not. That's exactly what we do observe.

- **Centrality of life.** If the eventual appearance of life were an important consideration for God when he was designing the universe, it is hard to understand why life seems so unimportant in the final product. We live in a galaxy with more than 100 billion stars, in a universe with more than 100 billion galaxies. All of this splendor is completely superfluous, as far as life is concerned. Nothing about biology here on Earth would be noticeably different if we lived in a universe with just our solar system and maybe a few thousand surrounding planets. Perhaps we could throw in the rest of our galaxy just to be generous. But the billions of galaxies that we can barely detect in our most powerful telescopes play no part in our existence. As far as physics and biology are concerned, the universe could easily have consisted of a relatively small number of particles that came together to make a few stars, and that would be enough to provide a comfortable environment for human life. Theism predicts that most other stars and galaxies shouldn't be there at all.

If life were important to God, our existence here on Earth would seem like a bigger deal, cosmically speaking. One possible response is to say, "God is inscrutable; we have no idea what kind of universe he would design." That's a plausible position, but it's not quite fair in this context. The essence of the fine-tuning argument is that we *do* know something about the universe God would design: one with physical laws that allow for the emergence of the complex chemical reactions we know as living organisms. It's illegitimate to claim that we know that, but nothing further about what God would do. A theory gets credit for explaining features of the world only to the extent that it goes out on a limb and makes predictions for what the world should be like.

A somewhat better response is to put forward some positive theory for why God would want the universe to look the way it does, in particular why it seems so wildly extravagant, with all of those stars and galaxies and whatnot. Typically such theories end up positing some physical reason for why it is simpler or easier for God to make many galaxies rather than just one. Maybe God likes inflation and the multiverse.

There are a few problems here. First, it's not true; there's nothing in the laws of physics that gets in the way of a more compact and focused universe than the one we see around us. Second, one would have to invent a reason why God prefers to make easy universes rather than to exert himself a bit. And third, you can see the road this takes us down: in the course of explaining why God would want to make a universe like the one we see, we end up removing his special influence from it, and falling back on purely physical mechanisms. If it's so easy to make a universe like the one we see, why rely on God at all?

Our theories are inevitably influenced by what we already know about the world. To get a more fair view of what theism would naturally predict, we can simply look at what it *did* predict, before we made modern astronomical observations. The answer is: nothing like what we actually observe. Prescientific cosmologies tended to resemble the Hebrew conception illustrated in chapter 6, with Earth and humanity sitting at a special place in the cosmos. Nobody was able to use the idea of God to predict a vast space with hundreds of billions of stars and galaxies, scattered almost uniformly through the observable universe. Perhaps the closest was Giordano Bruno, who talked about an infinite cosmos among his many other heresies. He was burned at the stake.

PART FIVE

———

THINKING

37

Crawling into Consciousness

Almost 400 million years ago, a plucky little fish climbed onto land and decided to hang out rather than returning to the sea. Its descendants evolved into the species *Tiktaalik roseae*, fossils of which were first discovered in 2004 in the Canadian Arctic. If you were ever looking for a missing link between two major evolutionary stages, *Tiktaalik* is it; these adorable creatures represent a transitional form between water-based and land-based animal life.

One can't help but wonder—what were they thinking about, those first land-dwelling animals?

A reconstruction of *Tiktaalik roseae*, crawling onto land. (Illustration by Zina Deretsky, courtesy of the National Science Foundation)

We don't know, but we can make some reasonable guesses. As far as stimulating new avenues of thought is concerned, the most important feature of their new environment was simply the ability to see a lot farther. If you've spent much time swimming or diving, you know that you can't see as far underwater as you can in air. The attenuation length—the distance past which light is mostly absorbed by the medium you are looking through—is tens of meters through clear water, while in air it's practically infinite. (We have no trouble seeing the moon, or distant objects on our horizon.)

What you can see has a dramatic effect on how you think. If you're a fish, you move through the water at a meter or two per second, and you see some tens of meters in front of you. Every few seconds you are entering a new perceptual environment. As something new looms into your view, you have only a very brief amount of time in which to evaluate how to react to it. Is it friendly, fearsome, or foodlike?

Under those conditions, there is enormous evolutionary pressure to think fast. See something, respond almost immediately. A fish brain is going to be optimized to do just that. Quick reaction, not leisurely contemplation, is the name of the game.

Now imagine you've climbed up onto land. Suddenly your sensory horizon expands enormously. Surrounded by clear air, you can see for kilometers—much farther than you can travel in a couple of seconds. At first, there wasn't much to see, since there weren't any other animals up there with you. But there is food of different varieties, obstacles like rocks and trees, not to mention the occasional geological eruption. And before you know it, you are joined by other kinds of locomotive creatures. Some friendly, some tasty, some simply to be avoided.

Now the selection pressures have shifted dramatically. Being simple-minded and reactive might be okay in some circumstances, but it's not the best strategy on land. When you can see what's coming long before you are forced to react, you have the time to contemplate different possible actions, and weigh the pros and cons of each. You can even be ingenious, putting some of your cognitive resources into inventing plans of action other than those that are immediately obvious.

Out in the clear air, it pays to use your imagination.

Bioengineer Malcolm MacIver has suggested that the flapping of fish up onto dry land was one of several crucial transitions that led to the development of the thing we now call *consciousness*. Consciousness is not a single brain organ or even a single activity; it's a complex interplay of many processes acting on multiple levels. It involves wakefulness, receiving and responding to sensory inputs, imagination, inner experience, and volition. Neuroscience and psychology have learned a great deal about what consciousness is and how it functions, but we are still far away from any sort of complete understanding.

Consciousness is also a unique and heavy burden. Being able to reflect on ourselves, our past and possible futures, and the state of the world and the cosmos brings great benefits, but it also opens the door to alienation and anxiety. The American cultural anthropologist Ernest Becker, commenting on Danish philosopher Søren Kierkegaard, once characterized consciousness this way:

> What does it mean to be a *self-conscious animal*? The idea is ludicrous, if it is not monstrous. It means to know that one is food for worms. This is the terror: to have emerged from nothing, to have a name, consciousness of self, deep inner feelings, and excruciating inner yearning for life and self-expression—and with all this yet to die.

The special feature of self-awareness, the ability to have a rich inner life and reflect on one's place in the universe, seems to demand a special kind of explanation, a unique place in the big picture. Is consciousness "just" a way of talking about the behavior of certain kinds of collections of atoms, obeying the laws of physics? Or is there something definitively new about it—either an entirely new kind of substance, as René Descartes would have had it, or at least a separate kind of property over and above the merely material?

If there is any one aspect of reality that causes people to doubt a purely physical and naturalist conception of the world, it's the existence of consciousness. And it can be hard to persuade the skeptics, since even the most optimistic neuroscientist doesn't claim to have a complete and comprehensive theory of consciousness. Rather, what we have is an expectation that when we do achieve such an understanding, it will be one that is completely compatible with the basic tenets of the Core Theory—part of physical reality, not apart from it.

Why should there be any such expectation? In part it comes down to Bayesian reasoning about our credences. The idea of a unified physical world has been enormously successful in many contexts, and there is every reason to think that it will be able to account for consciousness as well. But we can also put forward a positive case that the alternatives don't work very well. If it's not easy to see how consciousness can be smoothly incorporated as part of physical reality, it's even harder to imagine how it could be anything else. Our main goal here is not to explain how consciousness does work, but to illustrate that it *can* work in a world governed by impersonal laws of nature.

In this chapter and the next we'll highlight some of the features of consciousness that make it special. Then over the following few chapters we'll examine some arguments that, whatever consciousness is, it has to be more than simply a way of talking about ordinary matter in motion, obeying the conventional laws of physics. What we'll find is that none of those arguments is very persuasive, and we'll be left with a greater conviction than we started with that we human beings are part and parcel of the natural world, thoughts and emotions and all.

Sometimes when we think about our conscious selves, we can't help but imagine a little person inside our heads, making decisions and pulling strings. Even if we don't go as far as Descartes's belief in an immaterial soul that somehow interacts with our body, it's tempting to visualize a dictatorial "self" inside our brain that is the locus of our self-awareness. Philosopher Daniel Dennett coined the term "Cartesian theater" to describe the supposed mental control room containing a tiny homunculus who gathers all of the input from our sensory organs, accesses our memories, and sends out instructions to the various parts of our bodies.

Consciousness doesn't seem to be like that. Our minds are not run as top-down dictatorships; they are rambunctious parliaments, populated by squabbling factions and caucuses, with much more going on beneath the surface than our conscious awareness ever accesses.

The fanciful Pixar movie *Inside Out* represents the process of thinking as arising from a kind of teamwork between five personified emotions: Joy, Sadness, Disgust, Anger, and Fear. Each of the five would offer their opinions about the appropriate way of dealing with any particular situation, and one voice would hold sway depending on the circumstances. As professional killjoy neuroscientists were quick to point out, that's not actually how the mind works either. But it's a lot closer in spirit to what really happens than imagining a single unified self; there really are different "voices" that contribute to the ultimate narrative of our conscious awareness and decision making.

We could bring the *Inside Out* model closer to reality with two modifications. First, the various "modules" that contribute to our thought processes don't map directly onto emotions. (Neither do they have charming personalities or colorful anthropomorphic bodies.) They are unconscious processes of various sorts—the kind of mental functions that could have naturally arisen over the course of biological evolution, well before the explicit development of consciousness. Second, while there is no dictator in the mind, there does seem to be a kind of prime minister of the parliament, a seat of cognition where the inputs from many modules are sewn together into a continuum of consciousness.

Daniel Kahneman, a psychologist who won the Nobel Prize in Economics for his work on decision making, has popularized dividing how we think into two modes of thought, dubbed *System 1* and *System 2*. (The terms were originally introduced by Keith Stanovich and Richard West.) System 1 includes all the various modules churning away below the surface of our conscious awareness. It is automatic, "fast," intuitive thinking, driven by unconscious reactions and heuristics—rough-and-ready strategies shaped by prior experience. When you manage to make your coffee in the morning or drive from home to work without really paying attention to what you are doing, it's System 1 that is in charge. System 2 is our conscious, "slow," rational mode of thinking. It demands attention; when you're concentrating on a hard math problem, that's System 2's job.

As we go through the day, the vast majority of work being done in our brain belongs to System 1, despite our natural tendency to give credit to our self-aware System 2. Kahneman compares System 2 to "a supporting character who believes herself to be the lead actor and often has little idea of what's going on." Or in the words of neuroscientist David Eagleman, "Your consciousness is like a tiny stowaway on a transatlantic steamship, taking credit for the journey without acknowledging the massive engineering underfoot."

The System 1/System 2 distinction is an example of what's known as a *dual process theory* of thinking. An early example of such a theory was discussed by Plato, who in his dialogue *Phaedrus* introduced the allegory of the chariot. He was discussing the soul, not the mind, but the ideas are closely related. In the dialogue, Socrates explains that the soul has a charioteer (System 2), and is pulled by two horses (System 1), one of which is noble and the other is troublesome. Psychologist Jonathan Haidt has argued that Plato gives too much credit to the charioteer, and that a better metaphor would be a small rider atop a giant elephant. The rider—our conscious self—exerts some control, but the majority of the power resides in the elephant beneath.

The hallmark of consciousness is an inner mental experience. A dictionary definition might be something like "an awareness of one's self, thoughts, and environment." The key is awareness: you exist, and the chair you're sitting on exists, but you *know* you exist, while your chair presumably does not. It's this reflexive property—the mind thinking about itself—that makes consciousness so special. MacIver suggests that one of the most important pieces in this puzzle—the ability to take time to contemplate multiple alternatives, breaking the immediate connection between stimulus and response—started to become selected for by evolution once we crawled up onto the rocks.

It is natural to suppose that our imaginative faculties grew out of the evolutionary pressure in favor of developing the ability to weigh competing options for our future actions. Psychologist Bruce Bridgeman has gone so far as to characterize consciousness as "the operation of the plan-executing mechanism, enabling behavior to be driven by plans rather than immediate environmental contingencies." Consciousness is more than that; we can be conscious of being in love or enjoying a symphony without necessarily making associated plans. But the ability to conjure different hypothetical futures is certainly part of it.

There's a lot going on beneath the deceptively simple idea of "making plans." We have to have the ability to conceive of times in the future, not merely the present moment. We need to be able to represent the actions of both ourselves and the rest of the world in our mental pictures. We must reliably predict future actions and their likely responses. Finally, we must be able to do this for multiple scenarios simultaneously, and eventually compare and choose between them.

The ability to plan ahead seems so basic that we take it for granted, but it's quite a marvelous capacity of the human mind.

The "now" of your conscious perception is not the same as the current moment in which you are living. Though we sometimes think of consciousness as a unified essence guiding our thoughts and behavior, in fact it is stitched together out of inputs from different parts of the brain as well as our sensory perceptions. That stitching takes time. If you use one hand to touch your nose, and the other to touch one of your feet, you experience them as simultaneous, even though it takes longer for the nerve impulses to travel to your brain from your feet than from your nose. Your brain waits until all of the relevant inputs have been assembled, and only then presents them to you as your conscious perceptions. Typically, what you experience as "now" corresponds to what was actually happening some tens or hundreds of milliseconds in the past.

Estonian-Canadian psychologist Endel Tulving suggested the term *chronesthesia*, or "mental time travel." One of Tulving's contributions was the distinction between two different kinds of memory: *semantic memory*, which refers to general knowledge (Gettysburg was the site of an important battle in the American Civil War), and *episodic memory*, which captures our recollection of personal experiences (I visited Gettysburg when I was in high school). Mental time travel, Tulving suggested, is related to episodic memory: imagining the future is a similar conscious activity to recalling events in the past.

Recent work in neuroscience has lent credence to this idea. Researchers have been able to use functional magnetic resonance imaging (fMRI) and positron emission tomography (PET) scans to pinpoint regions in the brain that are active while subjects are conducting various mental tasks. Interestingly, the tasks of "remember yourself in a particular situation in the past" and "imagine yourself in a particular hypothetical situation in the future" are seen to engage a very similar set of subsystems in the brain. Episodic memory and imagination engage the same neural machinery.

Memories of past experiences, it turns out, are not like a video or film recording of an event, with individual sounds and images stored for each moment. What's stored is more like a script. When we remember a past event, the brain pulls out the script and puts on a little performance of the sights and sounds and smells. Part of the brain stores the script, while others are responsible for the stage settings and props. This helps explain why memories can be completely false, yet utterly vivid and real-seeming to us— the brain can put on a convincing show from an incorrect script just as well as an accurate one. It also helps explain how our chronesthetic ability to imagine future events might have developed through natural selection. Evolution, always looking to work with existing materials, constructed our powers of imagination out of our existing capacity to remember the past.

While a capacity for mental time travel is important for some aspects of consciousness, it certainly isn't the whole story. Kent Cochrane was an amnesiac, famous in the psychology literature as the patient "K. C." When he was thirty years old, K. C. suffered a serious motorcycle accident. He survived, but during surgery he lost parts of his brain, including the hippocampus, and his medial temporal lobes were severely damaged. Afterward, he retained his semantic memory but completely lost his episodic memory. His ability to form new memories was almost completely absent, much like the character of Leonard Shelby in the movie *Memento*. K. C. knew that he owned a particular car, but had no recollection of ever driving in it. His basic mental capacities were intact, and he had no trouble carrying on a conversation. He just couldn't remember anything he had ever seen or done.

There's little question that K. C. was "conscious" in some sense. He was awake, aware, and knew who he was. But consistent with the connection between memory and imagination, K. C. was completely unable to contemplate his own future. When asked about what would happen tomorrow or even later that day, he would simply report that it was blank. His personality underwent a significant change after the accident. He had, in some sense, become a different person.

There is some evidence that episodic memory doesn't develop in children until they are about four years old, around the time they also seem to develop the capacity for modeling the mental states of other people. At younger ages, for example, children can learn new things, but they have trouble associating new knowledge with any particular event; when quizzed about something they just learned, they will claim that they have always known it. Tulving has argued that true episodic memory, and the associated capacity for imagination and mental time travel, might be unique to humans. It's an intriguing hypothesis, but the current state of the art doesn't let us say for sure. We know that rats, for example, after trying and failing to reach some food, will continue to think about how to reach it after the food has been removed, which might be interpreted as a kind of planning. Their mental activity involves the hippocampus, which is associated with episodic memory in humans. Our ability to imagine the future is incredibly detailed and rich, but it's not hard to imagine how it might have evolved gradually over the span of many generations.

There's so much we don't know about the development of consciousness, it's easy to be dubious of any particular theory. Was crawling out of the water and onto land a pivotal step along the way, as Malcolm MacIver suggests, or is that just another fish story?

We should be skeptical; that's our job. There are aquatic animals that seem to be much smarter than your average goldfish. Whales and dolphins, of course, but those are mammals that descended from land animals—so their intelligence actually provides evidence for the hypothesis, not against it. Octopuses are quite intelligent by many standards. They have the biggest brains of any invertebrate (animals without spinal cords), although still only about one-thousandth the number of neurons that a human has. An octopus might not be able to do crossword puzzles, but it can solve certain simple challenges, such as opening a jar to get at food that's inside.

MacIver notes that octopuses, while underwater creatures, seem to maximize the extent of their sensory capacities. They have very large eyes, and tend to remain still while executing complex tasks. It's dangerous being an octopus; from the point of view of a predatory sea-dweller, you are a vulnerable bag of delicious nutrients. To survive, they have had to develop innovative defensive strategies, camouflaging themselves by changing skin color and emitting clouds of ink when forced to flee. Intelligence is a part of that defensive arsenal; an octopus will hide among rocks and coral when it sleeps, often arranging pieces so as to better shield itself from view. Perhaps the evolutionary pressure that led to large octopus brains was of a completely different type from that which led to land-dwelling animals.

Whatever the importance of climbing onto land might have been, it did not lead immediately to animals that could write sonnets and prove mathematical theorems. Four hundred million years is a long time. The evolution of consciousness as we now know it took many steps. Chimpanzees can think and execute a plan, such as building a structure in order to get to a banana that is out of reach. That's a kind of imaginative thought, though certainly not the whole story.

We can conceive of many moments in the evolutionary history of consciousness ultimately leading to the exquisite complexity of our current mental capacities. As the reducibly complex mousetrap reminds us, we shouldn't let the intimidating sophistication of the final product trick us into thinking that it couldn't have come about via numerous small steps.

38

The Babbling Brain

I t's an image familiar from countless TV hospital dramas: the patient lying on their back, head placed inside an intimidating-looking medical apparatus meant to peer inside their brain. Most often it will be an MRI machine, which will produce beautiful images of brain activity by tracking the flow of blood. In my case it was an MEG machine: magnetoencephalography. By measuring the appearance of magnetic fields just outside my skull, this beast was going to test whether or not I had a brain, and whether my brain could indeed have thoughts.

I passed. I like to think the outcome wasn't really in doubt, but it's good to have these things verified by science.

My brain scan was carried out by neuroscientist David Poeppel in his lab at New York University. Unlike fMRI, which makes beautiful pictures but doesn't have great time resolution, MEG isn't very good at telling you where processes are located in the brain, but it can distinguish when they happen down to a few milliseconds.

That's important, because our brains are intricately connected multi-level systems that take time to do their work. Individual neural events happen several times per millisecond, but it takes tens of milliseconds for several of them to accumulate to sufficient strength for your brain to sit up and say, "Hey! Something's happening!"—a conscious perception.

In the brain, most of the hard work of thinking is done by the neurons. They are joined by *glial cells*, which help support and protect the neurons. Glial cells may play a role in how neurons talk to one another, but the

Isofield Contour Map

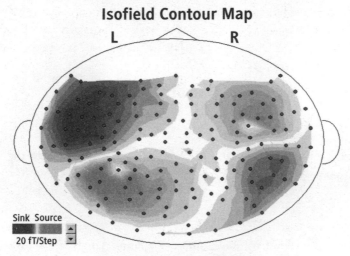

A map of the magnetic fields just outside my brain, generated by listening to a beeping sound. (Courtesy of David Poeppel lab, New York University)

information-carrying signals in the brain are carried by the neurons. A typical neuron will come equipped with two types of appendages: a large number of *dendrites*, which receive signals from outside, and (usually just one) *axon*, down which signals are sent. The body of a neuron is less than a tenth of a millimeter across, but axons can range from one millimeter all the way up to a full meter long. When a neuron wants to send a signal, it "fires" by pumping an electrochemical signal down its axon. That signal is received by other neurons at connection points known as *synapses*. Most synapses consist of a dendrite connecting to an axon, but the brain is a messy place, so various other kinds of connections are possible.

So neurons talk to each other by squirting electrically charged molecules from the axon of one to a dendrite on another. As any physicist will tell you, charged particles in motion generate magnetic fields. When a thought happens in my brain, this corresponds to charged particles hopping between neurons, creating a faint magnetic field that extends just a bit outside my skull. By detecting these magnetic fields, an MEG machine can pinpoint exactly when my neurons do their firing.

Poeppel and his colleagues are using this technique to study perception, cognition, and the workings of language in the brain. Sitting there in the

MEG, I listened to various meaningless beeps and boops, and the technician was able to track how long it took before I consciously perceived the auditory signal as a sound—tens of milliseconds, in a cascade of interrelated cortical responses.

I was most impressed by something much more prosaic—these probes attached to my skull could *sense me thinking*. What we call a "thought" corresponds directly and unmistakably to the motion of certain charged particles inside my head. That's an amazing, humbling fact about how the universe works. What would Descartes and Princess Elisabeth have thought?

Very few people today would deny that thinking is somehow related to what goes on in the brain. The divide is between those who believe that "thinking" is just a way of talking about the physical processes in the brain like the ones my MEG detected, and those who believe that we need to add some additional ingredients over and above the physical. It's worth doing a little thinking of our own about how brains actually work, to help understand why the physical picture is so compelling.

The brain is a network of interconnected neurons. We talked briefly in chapter 28 about how complex structures can arise by gradual agglomeration of smaller units into ever-larger ones, preserving the existence of interesting structure on all scales. The brain is a great example.

The conventional view of what happens in the brain is that it's not the neurons themselves that encode information but the way they are connected to one another. Every neuron is connected to some other neurons, and not to others; that's what defines the network structure of the brain, known as its *connectome*.

The connectome is simply the list of every single neuron in the brain, along with all of the connections between them. It's a system of impressive complexity: the human brain contains roughly 85 billion neurons, each of which is connected to a thousand or more other neurons, so we're talking about a hundred trillion or more connections in total. It's hard to look into a real human brain and discern all of those connections—but that's exactly the goal of several ongoing neuroscience research projects. Fully characterizing the human connectome would require something like a million million gigabytes of information.

Every neuron gleans input from other neurons, and occasionally from the outside world. Given that input, it decides whether to fire. Firing is a yes or no question—it either happens or it doesn't—but the input the neuron receives can be quite rich. Very roughly, a neuron will "listen" to its input for about 40 milliseconds at a time, and each incoming signal takes one millisecond to transfer. That's a huge amount of information. Forty separate inputs, from a couple of thousand different synapses, resulting in roughly 40 x 2,000 = 80,000 "bits" of information, or about $2^{80,000}$ possible messages a neuron could receive before it decides whether to fire or not. It's not simply "If I get more than the appropriate number of input signals, I will fire"; some signals increase the chance of firing, some decrease it, and the signals interact in complicated ways.

Knowing the complete human connectome wouldn't, by itself, come close to telling us everything we want to know about how human brains think. Not all neurons are the same, so knowing how they are connected isn't everything there is to know. Scientists have completely mapped the connectome of one multicellular organism: the tiny *C. elegans* nematode, a flatworm whose most common form has precisely 959 cells, 302 of which are neurons. We know how all of those neurons fit together—about 7,000 connections in total—but that doesn't tell us what the flatworm is thinking. It's like we know the highway map, but not the traffic patterns. Maybe someday we'll be able to read the nematode's mind.

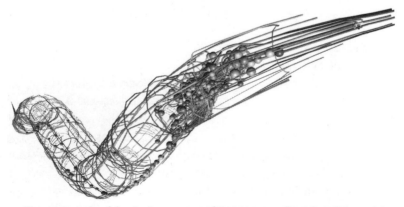

The connectome of the *C. elegans* nematode, as represented in a computer model from the OpenWorm project. (Courtesy of Chris Grove, Caltech)

People change over time, and our connectomes change along with us. The strength of the connections evolves, as the repeated firing of certain signals increases the chances that specific synapses will fire again in the future. We believe that memories are formed in this way, by synapses growing and shrinking in strength in response to stimuli. Neuropsychiatrist Eric Kandel shared the 2000 Nobel Prize in Medicine for his detailed investigation of how this happens in a particular organism, the humble sea slug. Slugs aren't great at remembering things, but Kandel trained them to recognize certain simple stimuli. He then showed that these new memories were connected to a change in the synthesis of proteins in the neurons, which led to alterations in their shape. Short-term memories were associated with synapses being strengthened, while long-term memories came from entirely new synapses being created.

More recently, neuroscientists have been able to directly observe neurons in mice growing and connecting as they learned how to perform new tasks. Impressively (or disturbingly, depending on your perspective), they have also been able to remove memories from mice by weakening specific synapses, and even artificially implanting false memories by directly stimulating individual nerve cells with electrodes. Memories are physical things, located in your brain.

A connectome is like a map of the countries of the world. It's not nearly enough to allow us to understand politics, but knowing the information contained therein is an important part of the bigger task. Having a good map won't stop you from getting lost, but it might help you find your way home.

One of the most crucial features of the brain is that it's not simply an undifferentiated mess of connected neurons. The connectome is a network, but it's a *hierarchical* network—groups of neurons are connected together, and those groups are then connected, and so on up to the entire brain. The babble of consciousness, with different mental modules offering input and being stitched together to make our aware self, is reflected in the workings of the brain. Different parts have their own jobs to do, but it's only when they come together that we find a conscious person.

There are various pieces of evidence for this, some of which come from studies of what happens when we lose consciousness: when we sleep, or when we're under anesthesia. One study, for example, gave a small magnetic stimulation to local regions of patients' brains. Effects of the signal were then measured as they propagated through the brain. When the patients were conscious, the signal induced responses all over the brain; in unconscious subjects the responses were confined to a limited region near the initial stimulus. Results like this are of much more than academic interest: doctors have long sought a way of telling whether a patient under anesthesia or suffering from brain damage was truly unconscious, or merely unable to move and communicate with the outside world.

To say that the connectome is a hierarchical network is to say that it lies somewhere between being maximally connected (every neuron is talking to every other neuron) and minimally connected (every neuron talks only to its immediate neighbors). As far as we can tell, the connectome is what mathematicians call a *small-world network*. The name comes from the famous six-degrees-of-separation experiment by psychologist Stanley Milgram. He found that randomly chosen people in Omaha, Nebraska, were linked to a specific person living in Boston, Massachusetts, by an average of about six first-name relationships. In network theory, we say that a network has the small-world property when most nodes are not directly connected to one another, but each one can be reached from any other one by a small number of steps.

That's what we find in the connectome. Neurons tend to be connected to nearby neurons, but there are also connections relatively far away. Small-world networks show up in many contexts, including connections between websites, electrical power grids, and networks of personal friendships. That's not an accident: this kind of organization seems to represent an optimum of efficiency for certain tasks, allowing processing to be done locally and results to spread quickly throughout the system. It is also robust to damage; knocking out a few connections doesn't appreciably alter the system's capacity. It's a perfect fit for the squabbling modules inside our brains.

One way of thinking about a small-world network is to say that it has "structure at all scales." It is not simply a bunch of neurons grouped into a ball, with those balls connected to one another. Rather, it's neurons connected into groups, connected into bigger groups, into even bigger groups, and so on. There is some indication that this kind of arrangement describes not only the spatial organization of the connectome but also how signals in the brain come and go in time. Small signals happen relatively frequently, medium-sized ones less often, and very big ones relatively rarely.

Physicists say that systems with this kind of hierarchical behavior are at a *critical point*. It's a ubiquitous phenomenon in the study of phase transitions, since systems become critical right as they are about to change from one phase to another. When water boils, there are many small bubbles, fewer larger ones, and so on. Criticality can be thought of as a sweet spot between boring order and useless chaos. As neurophysiologist Dante Chialvo put it, "A brain that is not critical is a brain that does exactly the same thing every minute, or, in the other extreme, is so chaotic that it does a completely random thing, no matter what the circumstances. That is the brain of an idiot."

In both space and time, then, the evidence we have to date indicates that our brains are complex systems organized in such a way as to take maximum advantage of their complexity. Given how impressive our brains are at carrying out complicated tasks, that should come as no surprise.

We could study the brain in exquisite detail, characterizing every neuron and mapping every connection, and still not convince ourselves that the brain accounts for the *mind*, the actual thinking of a human being. Back in chapter 26 we talked about Princess Elisabeth's objections to Descartes's picture of an immaterial soul interacting with the physical body, perhaps through the pineal gland. As interesting as those objections were, they don't necessarily close the deal until we can directly connect what happens in the brain to what we think of as our identities as persons. Over the years psychology and neuroscience have made great strides in doing just that.

We've already seen that memories are physically encoded in the brain. It's unsurprising, then, that our sensory perceptions are likewise encoded there. This is obviously true in some crude way, as the magnetic fields sticking out of my head demonstrated. But scientists have made advances recently in extracting quite detailed images of what patients are seeing, just by looking at what their brains are doing. By using fMRI images to determine what parts of the brain are firing when subjects are looking at images, or watching videos, neuroscientists can construct a template from which they can reconstruct images directly from the fMRI data, without "cheating" by knowing what the subjects are watching. It's not mind reading, at least not yet; we can make crude representations of what people are looking at, but not what they are imagining inside their heads. Perhaps that's just a matter of time.

None of this will necessarily convince a determined Cartesian dualist who wants to believe in immaterial souls. Of course, they will admit, *something happens* in the brain as we think and perceive the world. But that's not *all* that happens. The experiencing, the feeling, the actual soul of a person—that's something else entirely. Perhaps the brain is like a radio receiver. Altering it or damaging it will change how it plays, but that doesn't mean that the original signal is being created inside the radio itself.

That idea doesn't really hold up either. Damaging a radio might hurt our reception, making it hard to pick up our favorite station. But it doesn't turn that station from heavy-metal music into a smooth-jazz format. Damaging the brain, on the other hand, can change who a person is at a fundamental level.

Consider what's known as the Capgras delusion. Patients suffering from this syndrome have damage to the part of the brain that connects two other parts: the temporal cortex, associated with recognizing other people, and the limbic system, which is in charge of feelings and emotions. A person who develops Capgras delusion will be able to recognize people they know, but will no longer feel whatever emotional connection they used to have with them. (It is the flip side of prosopagnosia, which involves a loss of the ability to recognize people.)

You can imagine what this would do to a person. One patient, "Mrs. D," began to suffer from Capgras delusion at the age of seventy-four. Whenever she would see her husband, she would recognize this person, including all of the mental associations that said "this is my husband"—but she no longer felt any affection or love toward him, merely indifference. But she knew that she *should* have feelings for him, so her brain came up with a clever reconciliation of the inconsistency: this man wasn't really her husband, he was an impostor who looked just like him.

Mrs. D was not a unique case. There are many other examples of people suffering from some sort of brain damage, and having their emotional states or personalities dramatically altered thereby. That doesn't prove beyond any possible doubt that the mind is nothing more than a way of talking about what happens in the physical brain. But it should work to lower our credence in old-fashioned Cartesian dualism to a very small value indeed.

That leaves us either with physicalism—the world, including people, is purely physical—or some newfangled form of non-Cartesian dualism. To clean up that final question, we need to think more about what it means to be a conscious, experiencing person.

39

What Thinks?

I n Robert A. Heinlein's novel *The Moon Is a Harsh Mistress*, colonists on the moon revolt against the Lunar Authority back on Earth. Their cause would have been essentially hopeless if it hadn't been for the aid of Mike, a centralized computer that controlled all major automated functions in most of the Lunar cities. Mike wasn't just an important piece of machinery— he had, without anyone planning it, become self-aware. As the novel's narrator puts it,

> Human brain has around ten-to-the-tenth neurons. By third year Mike had better than one and a half times that number of neuristors.
> And woke up.

The narrator, Manuel O'Kelly Davis, is a computer technician who doesn't spend much time wondering about the origin or deeper meanings of Mike's emergence into consciousness. There's a revolution to be won, and presumably self-awareness is just the kind of thing that happens when thinking devices become sufficiently large and complex.

The reality would probably be a bit more complicated. A human brain has a lot of neurons in it; but those neurons aren't just connected up randomly. There is structure to the connectome, developed gradually through the course of natural selection. There is structure in a computer architecture as well, both hardware and software, but it seems unlikely that the kind of structure a computer has would hit upon self-awareness essentially by accident.

And what if it did? How would we know that a computer was actually "thinking," as opposed to mindlessly pushing numbers around? (Is there a difference?)

These issues were addressed in part by British mathematician and computer scientist Alan Turing back in 1950. Turing proposed what he called the imitation game, which is now more commonly known as the *Turing test*. With admirable directness, Turing opened his paper by stating, "I propose to consider the question, 'Can machines think?'" But he immediately decided that this kind of question was subject to endless squabbling over definitions. In the best scientific tradition, he therefore tossed it out and replaced it with a more operational query: Can a machine converse with a person in such a way as to make the person believe that the machine was also a person? (The best philosophical tradition would have dived into the definitional squabbling with gusto.) Turing put forward the ability to pass as human in such a test as a reasonable criterion for what it means to "think."

The Turing test has entered our cultural lexicon, and we regularly read news stories about this or that program that has finally passed the test. It might not be hard to believe, surrounded as we are by machines that send us email, drive our cars, and even talk to us. In truth, no computer has come close to passing a real Turing test. The competitions we read about in news reports are invariably set up to prevent interlocutors from really challenging a computer in the way Turing envisioned. We will very likely get there at some point, but contemporary machines do not "think" in Turing's sense.

When and if we do manage to construct a machine that can pass the Turing test to almost everyone's satisfaction, we will still be debating whether that machine truly thinks in the same sense that a human being does. The issue is consciousness, and the closely related issue of "understanding." No matter how clever a computer became at carrying on conversations, can it truly understand what it's saying? If the discussions turn to aesthetics or emotions, could a piece of software running on a silicon chip experience beauty or feel grief as a human can?

Turing anticipated this, and in fact labeled it *the argument from consciousness*. He quite properly identified the issue as a distinction between a third-person perspective (what others see me doing) and a first-person perspective (how I see and think myself). The argument from consciousness seemed, to Turing, to ultimately be solipsistic: you could never know that anyone was conscious unless you actually were that person. How do you know that everyone else in the world is actually conscious at all, other than by how they behave? Turing was anticipating the idea of a *philosophical zombie*—someone who looks and acts just like a regular person but has no inner experience, or *qualia*.

Turing thought that the way to make progress was to focus on questions that could be objectively answered by watching what happens in the world, rather than taking refuge in talk of personal experiences that are necessarily hidden from external observation. With a bit of charming optimism, he concluded that anyone who thought about things carefully would ultimately come to agree with him: "Most of those who support the argument from consciousness could be persuaded to abandon it rather than be forced into the solipsist position."

But it's possible to insist that thinking and consciousness cannot be judged from the outside while at the same time accepting that other people probably are conscious. Someone might think: "I know that I'm conscious, and other people are basically like me, so they're probably conscious as well. Computers, however, are not like me, so I can be more skeptical." I don't think this is the right attitude, but it's a logically consistent one. The question then becomes, *are* computers really so different? Is the kind of thinking done in my brain really qualitatively distinct from what happens inside a computer? Heinlein's protagonist didn't think so: "Can't see it matters whether paths are protein or platinum."

The Chinese Room is a thought experiment, proposed by American philosopher John Searle, that attempts to highlight how the Turing test might fall short of capturing what we really mean by "thinking" or "understanding." Searle asks us to imagine a person locked in a room with huge stacks of paper, each of which contains some Chinese writing. There is also a slot in the wall of the room, through which pieces of paper can be passed, and a set of instructions in the form of a lookup table. The person speaks and reads English, but doesn't understand any Chinese. When a piece of paper with some Chinese writing comes into the room through the slot, the person inside can consult the instructions, which will indicate one of the existing pieces of paper. The person then passes that paper back out through the slot.

Unbeknownst to our test subject, the pieces of paper that come into the room are perfectly sensible questions written in Chinese, and the pieces of paper that they are instructed to send out in return are perfectly sensible Chinese answers—ones that a regular thinking person might give. To a Chinese-speaking person outside the room, it looks for all the world as if they are asking questions of a Chinese speaker inside the room, who in turn is answering them in Chinese.

But surely we agree, Searle argues, that there isn't actually anyone in the room who understands Chinese. There's just an English-speaking person, some large stacks of paper, and an exhaustive set of instructions. The room seems able to pass the Turing test (in Chinese), but no real understanding is present. Searle's original target was research in artificial intelligence, which he felt would never be able to achieve a truly human level of thinking. In the terms of his analogy, a computer that tries to pass the Turing test is like the person in the Chinese room: it might be able to push symbols around to give the illusion of understanding, but no real comprehension is present.

Searle's thought experiment has generated an enormous amount of commentary, much of it aimed at refuting his point. The simplest refutation succeeds pretty well: of course the person in the room can't be said to understand Chinese, it's the combined system of "person plus set of instructions" that understands Chinese. Like Turing with the argument from consciousness, Searle saw this argument coming, and addressed it in his original paper. He was not very impressed:

> The idea is that while a person doesn't understand Chinese, somehow the conjunction of that person and bits of paper might understand Chinese. It is not easy for me to imagine how someone who was not in the grip of an ideology would find the idea at all plausible.

Like many such thought-experiment journeys, the first step of the Chinese Room—the existence of some bits of paper and an instruction manual that could mimic human conversation—is a doozy. If the instruction manual literally indicated a single answer for every question that might be asked, it would never pass the Turing test against a marginally competent human interlocutor. Consider questions like "How are you doing?," "Why do you say that?," or "Could you tell me more?" Real human conversations don't simply proceed on a sentence-to-sentence basis; they depend on context and what has gone before. At a minimum, the "slips of paper" would have to include a way to store memories, as well as a system for processing information that would integrate those memories into the ongoing conversation. It's not impossible to imagine such a thing, but it would be a lot more complex than a pile of papers and an instruction book.

In Searle's view, it doesn't matter what parts of the setup we include in what we call the "system"; none of it will ever achieve understanding in the true sense. But the Chinese Room experiment doesn't provide a convincing argument for that conclusion. It does illustrate the view that "understanding" is a concept that transcends mere physical correlation between input and output, and requires something extra: a sense in which what goes on in the system is truly "about" the subject matter at hand. To a poetic naturalist, "aboutness" isn't an extra metaphysical quality that information can have; it's simply a convenient way of talking about correlations between different parts of the physical world.

To take the Chinese Room as an argument that machines cannot think begs the question rather than addressing it. It constructs a particular version of a machine that purports to be thinking, and says, "Surely you don't think there's any real understanding going on here, do you?" The best answer is "Why not?"

If the world is purely physical, then what we mean by "understanding" is a way of talking about a particular kind of correlation between information located in one system (as instantiated in some particular arrangement of matter) and conditions in the external world. Nothing in the Chinese Room example indicates that we shouldn't think that way, unless you are already convinced we shouldn't.

That's not to downplay the difficulty in clarifying what we mean by "understanding." A textbook on quantum field theory contains information about quantum field theory, but it doesn't itself "understand" the subject. A book can't answer questions that we put to it, neither can it do calculations using the tools of field theory. Understanding is necessarily a more dynamic and process-oriented concept than the mere presence of information, and the hard work of defining it carefully is well worth doing. But as Turing suggested, there's no reason why that hard work can't be carried out at a purely operational level—referring to how things actually behave, rather than invoking inaccessible properties ("understanding," "consciousness") that are labeled as unobservable to outsiders from the start.

Searle's original target with his thought experiment wasn't the problem of consciousness (what it means to be aware and experiencing), but the problems of cognition and intentionality (what it means to think and to understand). The issues are closely related, however, and Searle himself later considered the argument to have demonstrated that a computer program can't be conscious. The extension is straightforward enough: if you think the system inside the room doesn't really "understand," you probably don't think it's aware and experiencing either.

The Chinese Room thought experiment forces those of us who think consciousness is purely physical to confront what a dramatic claim we are making. Even if we don't purport to have a fully fleshed-out understanding of consciousness, we should try to be clear about what kinds of things could possibly qualify as "conscious." In the Chinese Room, that question is raised about a pile of papers and an instruction book, but really those are just colorful ways of talking about the information and processing inside a computer. If we believe "consciousness" is just a way of talking about underlying physical events, what kind of uncomfortable situations does that commit us to?

The one system we generally agree is conscious is a human being— mostly the brain, but we can include the rest of the body if you like. A human can be thought of as a configuration of several trillion cells. If the physical world is all there is, we have to think that consciousness results from the particular motions and interactions of all those cells, with one another, and with the outside world. It is not supposed to be the fact that cells are "cells" that matters, only how they interact with one another, the dynamic patterns they carve out in space as they move through time. That's the consciousness version of multiple realizability, sometimes called *substrate independence*—many different substances could embody the patterns of conscious thought.

And if that's true, then all kinds of things could be conscious.

Imagine that we take one neuron in your brain, and study what it does until we have it absolutely figured out. We know precisely what signals it will send out in response to any conceivable signals that might be coming in. Then, without making any other changes to you, we remove that neuron and replace it with an artificial machine that behaves in precisely the same way, as far as inputs and outputs are concerned. A "neuristor," as in Heinlein's self-aware computer, Mike. But unlike Mike, you are almost entirely made of your ordinary biological cells, except for this one replacement neuristor. Are you still conscious?

Most people would answer yes, a person with one neuron replaced by an equivalently behaving neuristor is still conscious. So what if we replace two neurons? Or a few hundred million? By hypothesis, all of your external actions will be unaltered—at least, if the world is wholly physical and your brain isn't affected by interactions with any immaterial soul substance that communicates with organic neurons but not with neuristors. A person with every single one of their neurons replaced by artificial machines that interact in the same way would indisputably pass the Turing test. Would it qualify as being conscious?

We can't prove that such an automated thinking machine would be conscious. It's logically possible that a phase transition occurs somewhere along the way as we gradually replace neurons one by one, even if we can't predict exactly when it would happen. But we have neither evidence nor reason to believe that there is any such phase transition. Following Turing, if a cyborg hybrid of neurons and neuristors behaves in exactly the same way as an ordinary human brain would, we should attribute to it consciousness and all that goes along with it.

Even before John Searle presented the Chinese Room experiment, philosopher Ned Block discussed the possibility of simulating a brain using the entire population of China. (Why everyone picks China for these thought experiments is left as an exercise.) There are many more neurons in the brain than there are people in China or even the whole world, but by thought-experiment standards that's not much of an obstacle. Would a collection of people running around sending messages to one another, in perfect mimicry of the electrochemical signals in a human connectome, qualify as "conscious"? Is there any sense in which that population of people—collectively, not as individuals—would possess inner experiences and understanding?

Imagine mapping a person's connectome, not only at one moment in time but as it develops through life. Then—since we're already committed to hopelessly impractical thought experiments—imagine that we record absolutely every time a signal crosses a synapse in that person's lifetime. Store all of that information on a hard drive, or write it down on (a ridiculously large number of) pieces of paper. Would that record of a person's mental processes itself be "conscious"? Do we actually need development through time, or would a static representation of the evolution of the physical state of a person's brain manage to capture the essence of consciousness?

These examples are fanciful but illustrative. Yes, reproducing the processes of the brain with some completely different kind of substance (whether neuristors or people) should certainly count as consciousness. But no, printing things out onto a static representation of those processes should not.

From a poetic-naturalism perspective, when we talk about consciousness we're not discovering some fundamental kind of stuff out there in the universe. It's not like searching for the virus that causes a known disease, where we know perfectly well what kind of thing we are looking for and merely want to detect it with our instruments so that we can describe what it is. Like "entropy" and "heat," the concepts of "consciousness" and "understanding" are ones that we *invent* in order to give ourselves more useful and efficient descriptions of the world. We should judge a conception of what consciousness really is on the basis of whether it provides a useful way of talking about the world—one that accurately fits the data and offers insight into what is going on.

A form of multiple realizability must be true at some level. Like the Ship of Theseus, most of the individual atoms and many of the cells in any human body are replaced by equivalent copies each year. Not every one—the atoms in your tooth enamel are thought to be essentially permanent, for example. But who "you" are is defined by the pattern that your atoms form and the actions that they collectively take, not their specific identities as individual particles. It seems reasonable that consciousness would have the same property.

And if we are creating a definition of consciousness, surely "how the system behaves over time" has to play a crucial role. If any element of consciousness is absolutely necessary, it should be the ability to have thoughts. That unmistakably involves evolution through time. The presence of consciousness also implies something about apprehending the outside world and interacting with it appropriately. A system that simply sits still, maintaining the same configuration at every moment of time, cannot be thought of as conscious, no matter how complex it may be or whatever it may represent. A printout of what our brain does wouldn't qualify.

Imagine you were trying to develop an effective theory of how human beings behave, but without any recourse to their inner mental states. That is, you are playing the role of an old-time behaviorist: person receives input, person behaves accordingly, without any unobservable nonsense about an inner life.

If you wanted to make a good theory, you would end up reinventing the idea of inner mental states. Part of the reason is straightforward: the sensory input might be hearing someone ask, "How are you feeling?" and the induced reaction might be "I'm a little gloomy at the moment, to be honest." The easiest way to account for such behavior is to imagine that there is a mental state labeled "gloomy," and that our subject is in that state at the moment.

But there's also another reason. Even when an individual behaves in ways that do not overtly refer to their inner mental state, real human behavior is extremely complex. It's not like two billiard balls coming together on a pool table, where you can reliably predict what will happen with relatively little information (angle of impact, spin, velocities, and so on). Two different people, or even the same person in slightly different circumstances, can react very differently to the same input. The best way to explain that is by invoking internal variables—there is something going on inside the person's head, and we had better take it into account if we want to correctly predict how they will behave. (When someone you know well is behaving strangely, remember: it might not be about you.)

If we weren't familiar with consciousness already, in other words, we'd have to invent it. The fact that people experience inner states as well as outer stimuli is absolutely central to who they are and how they behave. Inner lives aren't divorced from outer actions.

Daniel Dennett has made essentially this point with what he calls the *intentional stance*. There are many circumstances in which it is useful to speak as if certain things have attitudes or intentions. We therefore, quite sensibly, speak that way—we attribute intentionality to all sorts of things, because that's part of a theory that provides a good account of the thing's behavior. Talking "as if" is the only thing we ever do, as there is no metaphysically distinct "aboutness" connecting different parts of the physical world, just relationships between different pieces of matter. Just as when we discussed the emergence of "purpose" in chapter 35, we can think of intentions and attitudes and conscious states as concepts that play essential roles in a higher-level emergent theory describing the same underlying physical reality.

What Turing was trying to capture in his imitation game was the idea that what matters about thinking is how a system would respond to stimuli, for example, to questions presented to it by typing on a terminal. A complete video and audio recording of the life of a human being wouldn't be "conscious," even if it precisely captured everything that person had done to date, because the recording wouldn't be able to extrapolate that behavior into the future. We couldn't ask it questions or interact with it.

Many of the computer programs that have attempted to pass cut-rate versions of the Turing test have been souped-up chat bots—simple systems that can spit out preprogrammed sentences to a variety of possible questions. It is easy to fool them, not only because they don't have the kind of detailed contextual knowledge of the outside world that any normal person would have, but because they don't have memories even of the conversation they have been having, much less ways to integrate such memories into the rest of the discussion. In order to do so, they would have to have inner mental states that depended on their entire histories in an integrated way, as well as the ability to conjure up hypothetical future situations, all along distinguishing the past from the future, themselves from their environment, and reality from imagination. As Turing suggested, a program that was really good enough to convincingly sustain human-level interactions would have to be actually thinking.

⁕

Cynthia Breazeal, a roboticist at MIT, leads a group that has constructed a number of experiments in "social robotics." One of their most charming efforts was a robot puppet named Leonardo, who had a body created by Stan Winston Studio, a special-effects team that had worked on such Hollywood blockbusters as *The Terminator* and *Jurassic Park*. Equipped with more than sixty miniature motors that enabled a rich palette of movement and facial expressions, Leonardo bore more than a passing resemblance to Gizmo from the Steven Spielberg film *Gremlins*.

The ability to have facial expressions, it turns out, is enormously useful in talking to human beings. Brains work better when they're inside bodies.

Leonardo interacted with the researchers in Breazeal's lab, both reading their expressions and exhibiting his own. He was also programmed with a *theory of mind*—he kept track of not only his own knowledge (from what his video-camera eyes picked up happening in front of him) but also the knowledge of other people (from what he saw them doing). Leonardo's actions were not all preprogrammed; he learned new behaviors through interacting with humans, mimicking gestures and responses he witnessed in others. Without knowing anything about his programming, anyone watching Leonardo in action could easily tell whether he was happy, sad, afraid, or confused, just by observing his expressions.

One illustrative experiment with Leonardo was a type of false-belief task: checking that a subject understands that a different person might hold a certain belief even if that belief is not true. (Humans seem to develop this capacity around the age of four years old; younger children labor under the misconception that everyone has the same beliefs.) Leonardo watches one person put a Big Bird doll inside one of two boxes in front of him. Then that person leaves the room, and another one comes in and switches Big Bird from the first box to the second one. The second person leaves and the first returns. Leonardo is smart enough to know both that Big Bird is in the second box, and that the first person "believes" that it's in the first box.

The experimenter then asks, "Leo, can you find where I think Big Bird is?" This is a query about metacognition, thinking about thinking. Leonardo correctly points to the first box, corresponding to his model of the experimenter's beliefs. But while pointing at the first box, Leonardo also sneaks a quick glance at the second box, where Big Bird is actually located. This wasn't programmed behavior; it was something that the robot learned from interacting with humans.

Whether you are a fish crawling onto land, a robot dealing with experimenters in the lab, or a person interacting with other people, it is helpful to have models of the world around you, including other organisms and *their* models. Awareness of ourselves and others, and the ability to communicate and interact on a number of levels, are useful capacities to have as we work to survive in a complicated world.

40

The Hard Problem

ife on Earth has undergone a series of dramatic phase transitions. Self-replicating organisms, cell nuclei, multicellular life, climbing onto land, the origin of language—all of these represent important new capacities that changed what life was capable of. The appearance of consciousness is arguably the most interesting phase transition of all, the beginning of a new kind of way for matter to organize itself and behave. Not only can atoms organize themselves into complex, self-sustaining patterns, but those patterns acquire a capacity for self-awareness and the ability to think about their place in the cosmos.

Unless something much deeper is going on. As philosopher Thomas Nagel has put it, "The existence of consciousness seems to imply that . . . the natural order is far less austere than it would be if physics and chemistry accounted for everything." (It was Nagel who really emphasized that "what it is like" to feel something is the kind of thing a complete theory should be able to explain. His famous example was that we can't know what it is like to be a bat, but the point is more general.) On this view, we shouldn't hope to explain conscious experience purely in terms of the physical behavior of the quantum fields in the Core Theory, since consciousness transcends the physical world.

It's not hard to understand why someone might feel this way. Fine, the thinking goes, I can accept that the universe exists and obeys natural laws without appealing to anything outside. I have no trouble believing that life is a complex network of interlocking chemical reactions that began spontaneously and evolved through natural selection over billions of years. But surely *I* am more than just a bunch of atoms knocking into one another under the influence of gravity and electromagnetism. I *perceive*, I *feel*—there is something that it is like to be me, something uniquely personal and experiential, a rich inner life that can't possibly be accounted for by unthinking matter in motion, no matter how many atoms you congregate together. The issue has been dubbed the *mind-body problem*: how can we hope to account for mental reality using only physical concepts?

As with the origin of life and the origin of the universe, we can't claim to have a full understanding of the nature of consciousness. The study of how we think and feel, not to mention how to think about who we are, is in its relative infancy. As neuroscientist and philosopher Patricia Churchland has put it, "We're pre-Newton, pre-Kepler. We're still sussing out that there are moons around Jupiter."

But nothing we do know about consciousness should lead us to doubt the ordinary, naturalist conception of the world that has been so exceptionally successful in other contexts. As of right now, nothing about the mind-body problem should persuade us that the laws of physics need updating, amending, or augmenting.

Like "life," consciousness is less a unified conception and more a collection of related attributes and phenomena. We are aware of ourselves, as distinct from the outside world. We can contemplate alternative futures. We experience sensations. We can reason abstractly and symbolically. We feel emotions. We can call up memories, tell stories, and sometimes lie. The simultaneous working of all these aspects contributes to being conscious, and some aspects are going to be easier to explain in purely physical terms than others.

Consider the color red. It is a useful concept, one that can apparently be recognized universally and objectively, at least by sighted people who are not prevented from seeing red by color blindness. The operational instruction "stop when the light is red" can be understood without ambiguity. But there is the famous lurking question: do you and I see the same thing when we see something red? That's the question of *phenomenal consciousness*—what is it like to experience redness?

The word *qualia* (plural of "quale," which is pronounced KWAH-lay) is sometimes used to denote the subjective experience of the way something seems to us. "Red" is a color, a physically objective wavelength of light or appropriate combination thereof; but "the experience of the redness of red" is one of the qualia we would like to account for in a complete understanding of consciousness.

Australian philosopher David Chalmers has famously emphasized the difference between what he calls the *Easy Problems* and the *Hard Problem* of consciousness. The Easy Problems are manifold—explaining the difference between being awake and asleep, how we sense and store and integrate information, how we can recall the past and predict the future. The Hard Problem is explaining qualia, the subjective character of experience. It can be thought of as those aspects of consciousness that are irreducibly first-person; what we personally feel, not how we act and respond as seen by the rest of the world. The Easy Problems are about functioning; the Hard Problem is about experiencing.

It's the Hard Problem that poses an apparent challenge to a purely physical understanding of the world. The Easy Problems aren't easy, but they are squarely in the wheelhouse of conventional scientific investigation. We don't have a finished understanding of how photons impinging on our retinas while we are looking at a fish end up conjuring the notion of "fish" in our brains. But the path to getting there seems pretty neuroscientifically straightforward. The Hard Problem, by contrast, seems like an entirely different kettle of those fish. We can poke around in the brain all we like, but how in the world do we expect that to help us understand our inner, wholly subjective, experience? How can a collection of quantum fields evolving in accordance with the Core Theory be said to have "inner experience" at all?

Many experts on consciousness think of these two issues, in the words of Peter Hankins, as "the Easy Problem (which is hard), and the Hard Problem (which is impossible)." But some think the Hard Problem is not only pretty easy; it really isn't a problem at all—just a matter of conceptual confusion. Discussions between the two camps can be frustrating; there's nothing more disheartening than someone telling you that the problem you think is most important and central isn't really a problem at all.

As poetic naturalists, that's basically what we'll be doing. The attributes of consciousness, including our qualia and inner subjective experiences, are useful ways of talking about the effective behavior of the collections of atoms we call human beings. Consciousness isn't an illusion, but it doesn't point to any departure from the laws of physics as we currently understand them.

There are a number of thought experiments that try to illustrate how hard the Hard Problem really is. A famous one is Mary the Color Scientist, a colorful (as it were) instantiation of what's known as the *knowledge argument*. It was introduced by Australian philosopher Frank Jackson in the 1980s, with the goal of showing that there must be something in the world other than just physical facts. It's right up there with Searle's Chinese Room on the list of famous thought experiments in which philosophers lock people into strange rooms in order to illustrate some feature of consciousness.

Mary is a brilliant scientist who has been brought up under certain bizarre circumstances. She lives in a room that she has never left, and that room is completely devoid of color. Everything in the room is black, white, or some shade of gray. Her own skin is painted white, and all of her clothes are black. Curiously, given her environment, Mary grows up to become a specialist in the science of color. She has access to all of the equipment she would want, as well as to the entirety of the scientific literature on the subject of color. All of the color illustrations have been reduced to grayscale.

Eventually, Mary knows everything there is to know about color, from a physical point of view. She knows about the physics of light, and about the neuroscience of how the eye transmits signals to the brain. She's read up on art history, color theory, and the agricultural expertise involved in growing a perfect red tomato. She's just never *seen* the color red.

Jackson asks, what happens when Mary decides to leave her room and actually sees colors for the first time? In particular, does she *learn* anything new? He claims she does.

> What will happen when Mary is released from her black and white room or is given a color television monitor? Will she learn anything or not? It seems just obvious that she will learn something about the world and our visual experience of it. But then is it inescapable that her previous knowledge was incomplete.

But she had all the physical information. Ergo there is more to
have than that, and Physicalism is false.

Mary can know all of the physical facts about color, but there is still
something she doesn't know: "what it is like" to experience the color red.
Therefore, there are more kinds of things in the world than merely physical
things. The argument is not merely saying that we don't yet know how
to explain Mary's new experience in physical terms. The claim is that no
such explanation can possibly exist.

Like the Chinese Room, Mary's predicament relies on a thought-
experiment setup that sounds relatively innocent, but is wildly implausible
in practice. "All of the physical facts about color" is an awful lot of facts.
Here is a physical fact about color: when I cut my finger while chopping
onions last week, my blood was red. Does Mary know that I cut my finger
while chopping onions last week? Does she know the position and momentum and frequency of every photon of visible light in the whole universe?
What about the past and future of the universe? Like "an omniscient, omnipotent, and omnibenevolent being," the phrase "all the physical facts
about color" conjures a certain vague impression in our minds, but it's far
from clear that this expression corresponds to any well-defined concept.

Vagueness about physical facts isn't the biggest problem with citing Mary
as evidence for the existence of features of the universe that aren't purely
physical. The real issue is with slipperiness in the definitions of "knowledge" and "experience."

Let's consider Mary's predicament from a poetic-naturalism perspective.
There is some fundamental description of our world, in terms of an evolving quantum wave function or perhaps something deeper. The other concepts we appeal to, such as "rooms" and "red," are part of vocabularies that
provide useful approximate models for certain aspects of that underlying
reality in an appropriate domain of applicability. So we invent, for example,
the concept of a "person," which maps onto the underlying reality in a particular way—a way that might be difficult to precisely define in principle
but is easy to recognize in practice.

These "people" have different attributes, such as "age" and "height." One such attribute is "knowledge." A person has knowledge of something if they can (more or less) answer questions about it correctly, or carry out the actions associated with it effectively. If a reliable person tells us, "Linda knows how to change the tires on a car," we should have a high credence that the person labeled "Linda" is able to answer certain questions and perform certain actions, including helping us with our flat tire. The existence of knowledge in a person corresponds to the existence of certain networks of synaptic connections between the neurons in that person's brain.

So we are told there is a person named "Mary" who has some particular knowledge—all of the physical facts about color. Does she "gain new knowledge" when she steps out of the room and experiences color for the first time?

That depends on what you mean. If Mary knows all of the physical facts about color, that corresponds at the level of her brain to possessing the right synaptic connections to be able to correctly answer questions that we ask her concerning physical facts about color. Were she to actually see the color red, that would correspond to the firing of certain neurons in her visual cortex, which would in turn generate other synaptic connections, "memories of having seen red." By the assumptions of the thought experiment, this hasn't actually happened to Mary—the appropriate collections of neurons have never fired.

When she walks outside her room and those neurons do finally fire, does Mary "learn something new"? In one sense, surely yes—she now has memories that she hadn't previously possessed. Knowledge is related to our capacity to answer questions and do things, and Mary can now do something she couldn't before: recognize red things by sight.

Is this an argument that there is more to the universe than its physical aspects? Surely not. We have merely introduced an artificial distinction between two kinds of collections of synaptic connections: "ones induced by reading literature and doing scientific experiments in black and white," and "ones induced by stimulating the visual cortex by seeing red photons." This is a possible way to carve up our knowledge of the universe, but not a necessary one. It's a difference in the way the knowledge got to your brain, not in the kind of knowledge it is. This is not an argument that should induce us to start adding wholly new conceptual categories to our successful models of the natural world.

Mary could have experienced the color red. She could have rigged a probe that she could insert into her skull, which would send the appropriate electrochemical signal directly to her visual cortex, triggering precisely the experience we think of as "seeing the color red." (Mary was postulated to be a brilliant scientist, after all.) We can choose not to allow her to do such a thing, as part of her "learning all the physical facts about red"—but that's an arbitrary restriction on our part, not a deep insight into the structure of reality.

Mary's situation is related to the old chestnut "Is my color red the same as your color red?" Not the wavelengths, but is the experience of redness the same for you as it is for me? In some strict sense, no: my experience of the color red is a way of talking about certain electrochemical signals traveling through my brain, while yours is a way of talking about certain electrochemical signals traveling through your brain. So they can't be exactly the same, in a very boring reading: the same as "My pencil is not the same as your pencil, even though they look just alike, because this one belongs to me." But my experience of red is probably pretty similar to yours, simply because our brains are pretty similar. Interesting to think about, but not exactly a vortex of confusion that should lead us to reject the Core Theory as the underlying description for the whole business.

Frank Jackson himself has subsequently repudiated the original conclusion of the knowledge argument. Like most philosophers, he now accepts that consciousness arises from purely physical processes: "Although I once dissented from the majority, I have capitulated," he writes. Jackson believes that Mary the Color Scientist helps pinpoint our intuition about why conscious experience can't be purely physical, but that this isn't enough to qualify as a compelling argument for such a conclusion. The interesting task is to show how our intuition has led us astray—as, science keeps reminding us, it so often does.

41

Zombies and Stories

David Chalmers, who coined the phrase "Hard Problem of consciousness," is arguably the leading modern advocate for the possibility that physical reality needs to be augmented by some kind of additional ingredient in order to explain consciousness—in particular, to account for the kinds of inner mental experience pinpointed by the Hard Problem. One of his favorite tools has been yet another thought experiment: the *philosophical zombie*.

Unlike undead zombies, which seek out brains and generate movie franchises, philosophical zombies look and behave exactly like ordinary human beings. Indeed, they are perfectly physically identical to non-zombie people. The difference is that they are lacking in any inner mental experience. We can ask, and be puzzled about, what it is like to be a bat, or another person. But by definition, there is no "what it is like" to be a zombie. Zombies don't experience.

The possible existence of zombies hinges on the idea that one can be a naturalist but not a physicalist—we can accept that there is only the natural world, but believe that there is more to it than its physical properties. There are not, according this view, nonphysical kinds of things, such as immaterial souls. But the physical things with which we are familiar can have other kinds of properties—there can be a separate category of mental properties. This view is *property dualism*, as distinct from good old-fashioned Cartesian *substance dualism*, which holds that there are physical and nonphysical substances.

The idea is that you can have a collection of atoms, and tell me everything there is to say about the physical properties of those atoms, and yet you haven't told me everything. The system has various possible mental states. If the atoms make up a rock, those states might be primitive and unobservable, essentially irrelevant. But if they make up a person, a rich variety of mental states come to life. To understand consciousness, on this view, we need to take those mental properties seriously.

If these mental properties affected the behavior of particles in the same way that physical properties like mass and electric charge do, then they would simply be another kind of physical property. You are free to postulate new properties that affect the behavior of electrons and photons, but you're not simply adding new ideas to the Core Theory; you are saying that it is wrong. If mental properties affect the evolution of quantum fields, there will be ways to measure that effect experimentally, at least in principle—not to mention all of the theoretical difficulties with regard to conservation of energy and so on that such a modification would entail. It's reasonable to assign very low credence to such a complete overhaul of the very successful structure of known physics.

Alternatively, we could imagine that mental properties just go along for the ride, as far as physical systems are concerned. The Core Theory can be a complete description of the physical behavior of the quantum fields of which we are made, but not a complete description of us. Such a description would need to specify our mental properties as well.

Zombies would be collections of particles in exactly the same arrangement as would ordinarily make up a person, obeying the same laws of physics and therefore behaving in precisely the same way, but lacking the mental properties that account for inner experience. As far as you can tell by talking to them, all of your friends and loved ones are secretly zombies. And they can't be sure you're not a zombie. Perhaps they have suspicions.

The big question about zombies is a simple one: can they possibly exist? If they can, it's a knockout argument against the idea that consciousness can be explained in completely physical terms. If you can have two identical collections of atoms, both of which take the form of a human being, but one has consciousness and the other does not, then consciousness cannot be purely physical. There must be something else going on, not necessarily a disembodied spirit, but at least a mental aspect in addition to the physical configuration.

When we talk about whether zombies are possible, we don't necessarily mean *physically* possible. We don't need to imagine that we could find an honest-to-goodness zombie here in our real world, made out of the same particles that you and I are made from (if you're not a zombie, which I'm going to assume henceforth). We're just imagining a possible world, with a different fundamental ontology, even though it might have very similar-seeming particles and forces. What it would be lacking is mental properties.

As long as zombies are conceivable or logically possible, Chalmers argues, then we know that consciousness is not purely physical, regardless of whether zombies could exist in our world. Because then we would know that consciousness can't simply be attributed to what matter is doing: the same behavior of matter could happen with or without conscious experience.

Of course, Chalmers also then says that zombies are conceivable. He has no trouble conceiving of them, and maybe you feel the same way. Can we then conclude that there is more to the world than just the physical universe?

Deciding whether something is "conceivable" is harder than it might seem at first glance. We can conjure up an image in our mind of someone that looks and acts just like a human being, but who is dead inside, with no inner experiences. But can we really do so without imagining any differences in the physical behavior of them versus an ordinary person?

Imagine a zombie stubbed its toe. It would cry out in pain, because that's what a human would do, and zombies behave just like humans. (Otherwise we would be able to recognize zombies by observing their external behavior.) When you stub your toe, certain electrochemical signals bounce around your connectome, and the exact same signals bounce around the zombie connectome. If you asked it why it cried out, it could say, "Because I stubbed my toe and it hurts." When a human says something like that, we presume it's telling the truth. But the zombie must be lying, because zombies have no mental states such as "experiencing pain." Why do zombies lie all the time?

For that matter, are you *sure* you're not a zombie? You think you're not, because you have access to your own mental experiences. You can write about them in your journal or sing songs about them in a coffee shop. But a zombie version of you would do those things as well. Your zombie doppelgänger would swear in all sincerity that it had inner experiences, just as you would. You don't think you're a zombie, but that's just what a zombie would say.

The problem is that the notion of "inner mental states" isn't one that merely goes along for the ride as we interact with the world. It has an important role to play in accounting for how people behave. In informal speech, we certainly imagine that our mental states influence our physical actions. I am happy, and therefore I am smiling. The idea that mental properties are both separate from physical properties, and yet have no influence on them whatsoever, is harder to consistently conceive of than it might first appear.

According to poetic naturalism, philosophical zombies are simply inconceivable, because "consciousness" is a particular way of talking about the behavior of certain physical systems. The phrase "experiencing the redness of red" is part of a higher-level vocabulary we use to talk about the emergent behavior of the underlying physical system, not something separate from the physical system. That doesn't mean it's not real; my experience of redness is perfectly real, as is yours. It's real in exactly the same way as fluids and chairs and universities and legal codes are real—in the sense that they play an essential role in a successful description of a certain part of the natural world, within a certain domain of applicability.

It might seem strange that the logical possibility of a concept depends on whether this or that ontology turns out to be true, but we can't decide whether "humanlike beings without consciousness" is a sensible concept until we know what consciousness is.

In 1774, British clergyman Joseph Priestley isolated the element of oxygen. If you asked him whether he could imagine water without any oxygen, he presumably would have had no problem, since he didn't know that water is made of molecules with one oxygen atom and two hydrogens. (Water was first decomposed into hydrogen and oxygen in 1800.) But now we know better, and realize that "water without oxygen" is not conceivable. In some possible world with somewhat different laws of physics, there may be another substance that is not H_2O, yet has all the phenomenological properties of water—liquid at room temperature, transparent to visible light, and so on. But it wouldn't be the water that we know and love. Likewise, if you think that conscious experience is something truly distinct from the physical behavior of matter, you should have no trouble imagining zombies; but if consciousness is just a concept we use to describe certain physical behaviors, zombies become inconceivable.

The idea that our mental experiences or qualia are not actually separate *things*, but instead are useful parts of certain *stories we tell* about ordinary physical things, is one that many people find hard to swallow.

Even with the best of intentions on both sides, a dialogue between a property dualist who believes in the separate reality of mental properties (call him M) and a poetic naturalist who believes they are just ways of talking about physical states (call her P) can be frustrating. It might go something like this:

M: I grant you that, when I am feeling some particular sensation, it is inevitably accompanied by some particular thing happening in my brain—a "neural correlate of consciousness." What I deny is that one of my subjective experiences simply *is* such an occurrence in my brain. There's more to it than that. I also have a feeling of *what it is like* to have that experience.

P: What I'm suggesting is that the statement "I have a feeling..." is part of an emergent way of talking about those signals appearing in your brain. There is one way of talking that speaks in a vocabulary of neurons and synapses and so forth, and another way that speaks of people and their experiences. And there is a map between these ways: when the neurons do a certain thing, the person feels a certain way. And that's all there is.

M: Except that it's manifestly not all there is! Because if it were, I wouldn't have any conscious experiences at all. Atoms don't

have experiences. You can give a *functional* explanation of
what's going on, which will correctly account for how I actually behave, but such an explanation will always leave out the
subjective aspect.

P: Why? I'm not "leaving out" the subjective aspect, I'm suggest-
     ing that all of this talk of our inner experiences is a useful way
     of bundling up the collective behavior of a complex collection
     of atoms. Individual atoms don't have experiences, but macro-
     scopic agglomerations of them might very well, without invok-
     ing any additional ingredients.

M: No, they won't. No matter how many non-feeling atoms you
     pile together, they will never start having experiences.

P: Yes, they will.

M: No, they won't.

P: Yes, they will.

And you can imagine how it continues from there.

Nevertheless, let's make one more good-faith effort to explain to an
open-minded property dualist how a poetic naturalist thinks about qualia.
What do we mean when we say "I am experiencing the redness of red"? We
mean something like this:

> There is a part of the universe I choose to call "me," a collec-
> tion of atoms interacting and evolving in certain ways. I attri-
> bute to "myself" a number of properties, some straightforwardly
> physical, and others inward and mental. There are certain pro-
> cesses that can transpire within the neurons and synapses of my
> brain, such that when they occur I say, "I am experiencing red-
> ness." This is a useful thing to say, since it correlates in predict-
> able ways with other features of the universe. For example, a
> person who knows I am having that experience might reliably

infer the existence of red-wavelength photons entering my eyes, and perhaps some object emitting or reflecting them. They could also ask me further questions such as "What shade of red are you seeing?" and expect a certain spectrum of sensible answers. There may also be correlations with other inner mental states, such as "seeing red always makes me feel melancholy." Because of the coherence and reliability of these correlations, I judge the concept of "seeing red" to be one that plays a useful role in my way of talking about the universe as described on human scales. Therefore the "experience of redness" is a real thing.

It's a mouthful, and nobody would ever mistake it for a Shakespearean sonnet. But there's a kind of poetry there, if you look closely enough.

There are two points of view relevant to consciousness that are close cousins of poetic naturalism, but different in important ways.

One view is to argue that all of these so-called qualia or inner experiences simply *don't exist*—they are illusions. Maybe you thought you had inner experiences, but that is an antiquated part of our intuitive view of the world, a relic of a prescientific age. Now we know better, and should use a more updated and appropriate set of concepts.

The other perspective is a strong form of reductionism that insists that subjective experiences simply *are* physical processes happening in the brain. They exist, but they can be identified with specific neural correlates. A famous example along these lines comes from philosopher Hilary Putnam, who contemplated—to refute the idea, not to defend it—the position that "pain" is to be literally identified with "the firing of C-fibers." (C-fibers are a part of the nervous system that carries pain signals.)

A poetic naturalist has no trouble saying that conscious experiences exist. They are not part of the fundamental architecture of reality, but they serve as essential pieces of an emergent effective theory. The best way we have of talking about people and their behaviors makes important reference to their inner mental states; therefore, by the standards of poetic naturalism, those states are real, existing things.

There is a relationship between the different ways we have of talking about the world, including the human-level vocabulary that includes our subjective experiences, and the cell-biological level that includes firing nerve fibers, and the particle-physics level that includes fermions and bosons. The relationship is that certain states in the more comprehensive theories (particles, cells) correspond to unique states in the coarse-grained theories (people, experiences). The reverse relationship is typically not unique; there may be a large number of arrangements of atoms that correspond to "me being in pain."

A subtle but important distinction lurks between "there is a map between the concepts of different theories" and "the concepts of the coarse-grained theories are to be identified with certain states in the more comprehensive theories," such as "pain is to be identified with the firing of C-fibers." The difference is important because granting the latter, stronger formulation gets us in trouble. Putnam, for example, would then want to ask, "Do you mean to say there can be no such thing as pain without the existence of C-fibers? That artificial beings, or aliens, or even very different animals here on Earth, are by definition incapable of feeling pain?"

We don't want to say that, and we don't have to. There are certain configurations of atoms that correspond to "a human being feeling pain," but there could be other configurations of atoms that correspond to "a Wookiee feeling pain," or any related instantiation of the concept. (There is nothing in principle that prevents a computer from feeling pain.) Poetic naturalism is "poetic" because there are different stories we can tell about the world, many of them capturing some aspects of reality, and all useful in their appropriate context.

There's no reason for us to pretend that subjective experiences don't exist, or on the other hand that they "are" something happening in the brain. They are essential concepts within a way of talking about things happening in our brains, and that makes all the difference.

42

Are Photons Conscious?

I f consciousness were something over and above the physical properties of matter, there would be a puzzle: what was it doing for all those billions of years before life came along?

Poetic naturalists have no problem with this question. The appearance of consciousness is a phase transition, like water boiling. The fact that sufficiently hot water is in the form of a gas doesn't mean that there was always something gaslike about the water, even when it was in the form of liquid; the system simply acquired new properties as its situation changed.

But if you believe that mental properties are an additional ingredient, over and above the underlying physical substrate, then the question of what they were doing for most of the history of the universe is a pointed one. The most straightforward answer is that those mental properties were always there, even before there were brains or even organisms. Even the individual atoms and particles that were bumping into one another in the early universe, or are currently doing so at the center of the sun or in the desolate cold of intergalactic space, are equipped with mental properties of their own. They would be, in this sense, a little bit conscious.

The suggestion that consciousness pervades the universe, and is a part of every piece of matter, goes by the name of *panpsychism*. It's an old idea, going back arguably as far as Thales and Plato in ancient Greece, as well as in certain Buddhist traditions. In its modern guise it has been contemplated seriously by philosophers like David Chalmers and neuroscientists such as

Giulio Tononi and Christof Koch. Here is Chalmers, admirably biting the bullet and accepting the consequences of what such a view would imply:

> Even a photon has some degree of consciousness. The idea is not that photons are intelligent or thinking. It's not that a photon is wracked with angst because it's thinking, "Aww, I'm always buzzing around near the speed of light. I never get to slow down and smell the roses." No, not like that. But the thought is maybe photons might have some element of raw, subjective feeling, some primitive precursor to consciousness.

Consciousness, or at least protoconsciousness, could be analogous to "spin" or "electric charge"—one of the basic properties characterizing each bit of matter in the universe.

It's worth taking the implications of this idea seriously, and seeing how well it fits in with what we know about the physics of photons.

Unlike brains, which are complicated and hard to explain, elementary particles such as photons are extraordinarily simple, and therefore relatively easy to study and understand. Physicists talk about different kinds of particles having different "degrees of freedom"—essentially, the number of different kinds of such particles that there are. An electron, for example, has two degrees of freedom. It has both electric charge and spin, but the electric charge can take on only one value (-1), while the spin comes in two possibilities: clockwise or counterclockwise. One times two is two, for two total degrees of freedom. An up quark, by contrast, has six degrees of freedom; like an electron, it has a fixed charge and two possible ways of spinning, but it also has three possible "colors," and one times two times three is six. Photons have an electric charge fixed at zero, but they do have two possible spin states, so they have two degrees of freedom just like electrons do.

We could interpret the supposed existence of mental properties in the most direct way possible, as introducing new degrees of freedom for each elementary particle. In addition to spinning clockwise or counterclockwise, a photon could be in one of (let's say) two mental states. Call them "happy" and "sad," although the labels are more poetic than authentic.

This overly literal version of panpsychism cannot possibly be true. One of the most basic things we know about the Core Theory is exactly how many degrees of freedom each particle has. Recall the Feynman diagrams from chapter 23, describing particles scattering off of one another by exchanging other particles. Each diagram corresponds to a number that we can compute, the total contribution of that particular process to the end result, such as two electrons scattering off of each other by exchanging photons. Those numbers have been experimentally tested to exquisite precision, and the Core Theory has passed with flying colors.

A crucial ingredient in calculating these processes is the number of degrees of freedom associated with each particle. If photons had some hidden degrees of freedom that we didn't know about, they would alter all of the predictions we make for any scattering experiment that involves such photons, and all of our predictions would be contradicted by the data. That doesn't happen. So we can state unambiguously that photons do not come in "happy" and "sad" varieties, or any other manner of mental properties that act like physical degrees of freedom.

Advocates of panpsychism would probably not go as far as to imagine that mental properties play roles similar to true physical degrees of freedom, so that the preceding argument wouldn't dissuade them. Otherwise these new properties would just be ordinary physical properties.

That leaves us in a position very similar to the zombie discussion: we posit new mental properties, and then insist that they have no observable physical effects. What would the world be like if we replaced "protoconscious photons" with "zombie photons" lacking such mental properties? As far as the behavior of physical matter is concerned, including what you say when you talk or write or communicate nonverbally with your romantic partner, the zombie-photon world would be exactly the same as the world where photons have mental properties.

A good Bayesian can therefore conclude that the zombie-photon world is the one we actually live in. We simply don't gain anything by attributing the features of consciousness to individual particles. Doing so is not a useful way of talking about the world; it buys us no new insight or predictive power. All it does is add a layer of metaphysical complication onto a description that is already perfectly successful.

Consciousness seems to be an intrinsically collective phenomenon, a way of talking about the behavior of complex systems with the capacity for representing themselves and the world within their inner states. Just because it is here full-blown in our contemporary universe doesn't mean that there was always some trace of it from the very start. Some things just come into being as the universe evolves and entropy and complexity grow: galaxies, planets, organisms, consciousness.

Regardless of whether individual particles possess a form of protoconscious awareness, there is a long history of attempts to link the mystery of consciousness to another famous mystery, that of quantum mechanics. In part these efforts can be attributed to what Chalmers has jokingly called the "Law of Minimization of Mystery": consciousness is confusing, and quantum mechanics is confusing, so maybe they're somehow related.

There is no doubt that there are real mysteries associated with quantum mechanics, especially what precisely happens when an observer measures a quantum system. In Everett's Many-Worlds Interpretation, the answer is simple: nothing special. Everything continues to smoothly evolve according to a deterministic set of equations, but the interaction of the macroscopic observer with a vast environment around them causes the way we talk about the system to evolve from "one universe in a quantum superposition" to "two separate universes." The fact that observers happen to be conscious plays precisely zero role; measurements can be easily carried out by nematodes, video cameras, or rocks.

Sadly, not everyone accepts the advantages of this approach. In the textbook version of quantum mechanics, there is a moment during the observation process at which wave functions "collapse." Before collapse, a particle might have been in a superposition of two different states, like spinning clockwise and spinning counterclockwise; after collapse, only one alternative remains. So what precisely leads to the collapse event? It is not completely crazy to speculate that it might have something to do with the presence of a conscious observer, and a number of respectable physicists have done so over the years.

The possibility that consciousness plays a role in understanding quantum mechanics has lost almost all of whatever support it may have once enjoyed. These days we understand quantum mechanics a lot better than the pioneers did; we have very specific and quantitative theories that can plausibly explain exactly what happens during the process of measurement, without any need to invoke consciousness. We don't know which if any of these theories is right, so mysteries remain—but even without having the final answer, the very existence of respectable alternatives tends to make the way-out ones seem less attractive.

Some people have an inordinate fondness for way-out possibilities, and will grab on to their associated buzzwords and use them for their own ends. Such is the situation with most of what goes by the label of "quantum consciousness" in popular conversation. Quantum mechanics says that superpositions evolve into definite outcomes during the process of measurement, at least for any one observer; it's not hard to twist that into the claim that conscious observation literally brings reality into existence.

It's the ultimate anti-Copernican move, a way of restoring the central importance of humanity to our picture of the universe. Sure, you might feel insignificant in the vastness of the cosmos, and perhaps you become alienated by thinking that your atoms obey impersonal laws of physics, but hey, don't worry: you are personally creating the world at every moment, just by looking at it. Advocates of this approach will sometimes throw in something about "entanglement"—which isn't even a mystery, just an interesting feature of quantum mechanics—to make you feel like you are connected to everything else in the universe. As a final flourish, they might suggest that quantum mechanics has discarded the physical world entirely, leaving us with idealism, where everything is a projection of the mind.

There is nothing in anything we know about physics that suggests any of that is true. Quantum mechanics may be mysterious, but it is still—in all of its suggested formulations—an ordinary physical theory, governed by impersonal laws expressed in the form of equations. In particular, even in interpretations where wave functions really do collapse when systems are observed, the person doing the observing has no influence whatsoever on what the measurement outcome turns out to be. That just follows a rule, the Born rule for quantum probabilities, which says the probability of each outcome is given by the value of the wave function squared. Nothing spooky, nothing personal, nothing intrinsically human. Just physics.

✲

"Quantum consciousness" in this disreputable formulation is distinct from an idea that is speculative, but at least physically sensible: that quantum processes play an important role in the actual workings of the brain. At some level this is trivially true. The brain is made of particles, which are vibrations of quantum fields, which obey the rules of quantum mechanics. But most neuroscience starts with the assumption that important processes in the brain are well described by the approximation of classical physics. We don't need wave functions or entanglement to get a rocket to the moon, and it seems reasonable to imagine that we don't need them to understand the brain either.

The brain is a warm, wet environment, not a cold, precise laboratory setup. Every particle in your head is constantly being jostled by other particles, leading to an ongoing process of "collapse" (or branching of the wave function, for fearless Everettians like me). There's not much time for particles to linger in a superposition, become entangled with other particles, and so on. Maintaining quantum coherence inside the brain would seem to be analogous to building a house of cards outside during a hurricane.

Nevertheless, recent discoveries in biology have indicated that living organisms do seem to take advantage of certain quantum effects that go beyond what classical physics could do. Photosynthesis, in particular, involves transfers of energy by particles in quantum superposition. (Darwinian evolution stumbled across quantum mechanics long before human beings discovered it.) So we can't discard the possibility that quantum effects are important in the brain simply on the basis of pure thought—we have to do the usual empiricist Bayesian procedure of inventing hypotheses and testing them against the data.

Physicist Matthew Fisher has identified one very specific set of quantum objects in the brain that could become entangled with one another, and remain so for a relatively long time: the nuclei of certain phosphorous atoms that are found in subgroups of ATP molecules and elsewhere. In Fisher's model, the rate at which chemical reactions involving these atoms will occur depends on whether their nuclei share quantum entanglement with other nearby phosphorous nuclei. As a result, quantum mechanics could play a very real role in brain processes, perhaps even allowing the brain to act as a "quantum computer." Or not—these are all new and speculative ideas. They do remind us not to jump to conclusions when we're talking about a system as subtle and complicated as a brain.

When most people think of quantum effects in the brain, however, they're not imagining something as prosaic as accounting for how the brain performs computations. They want to invoke new physics to help us explain consciousness.

The most famous proponent of this approach is Roger Penrose, the British physicist and mathematician renowned for his contributions to our modern understanding of Einstein's general relativity. Penrose is one of those scientists who rattles off brilliant ideas like most of us brush bread crumbs from our shirts. And he is convinced that human brains do things that computers can't do. But computers can simulate anything that could happen according to the known laws of physics. So we need some genuinely new physical phenomena at work in the brain—in particular, something special about the collapse of the wave function.

Penrose's argument is elaborate and ingenious, but ultimately unconvincing to the vast majority of researchers studying physics, neuroscience, or consciousness. He starts with Gödel's Incompleteness Theorem, a celebrated result by Austrian logician Kurt Gödel. At the risk of dramatic oversimplification, the gist of the Incompleteness Theorem is that within any consistent mathematical *formal system*—a set of axioms, and rules for deriving consequences from them—there will be statements that are true but cannot be proven within that system. (Gödel's basic trick was to invent a way of expressing "This statement cannot be proven" within any sufficiently powerful formal system. Either you can prove it and it is therefore false, showing that your system is inconsistent, or you can't prove it and it's true.) A computer working with the appropriate set of formal rules wouldn't be able to prove such a statement.

But, Penrose says, human mathematicians have no trouble perceiving the truth of statements like that. Therefore, what's going on inside the brain of a human mathematician must be something over and above a formal mathematical system. The known laws of physics don't grant us such powers.

As we discussed in chapter 24, if there is going to be a loophole in the audacious claim that the laws of physics underlying everyday life are completely known, the leading candidate would be some alteration in how we think about quantum measurement. Penrose has some specific ideas about what those alterations might be—quantum gravity is involved, and filamentary structures in the brain called microtubules—but the upshot is that the wave functions of structures in our brains collapse in just the right way to grant human beings powers of insight and cognition that computers will never achieve.

There are a number of objections one could raise, and people have had fun raising them against Penrose for years now. The best ones center on the leap from "Human cognition doesn't work like a formal mathematical system" to "The human brain doesn't obey the known laws of physics." What we call "thinking" is a way of talking about a very high-level emergent phenomenon. It may emerge out of underlying processes that are absolutely rigid and logical, and yet itself not show those characteristics very much at all. Indeed, rigid logic (or even the ability to multiply big numbers accurately) is something that human beings are notoriously bad at. Our thoughts leap around, we make mistakes, we have hunches. The fact that we can reach conclusions that wouldn't be reached by a specific formal system doesn't seem particularly surprising.

Gödel's Incompleteness Theorem doesn't *quite* say there are true statements that can't be proven. Rather, it says that such statements exist for any *consistent* formal system. How do we know that some particular set of axioms defines a consistent system? Or—putting it another way—how can we be sure that we are accurately "perceiving" the truth of Gödel's self-referential sentences?

As Scott Aaronson has pointed out, it's more accurate to say that we *believe* certain systems are consistent, though Gödel has shown that we can never prove it. If we allow a computer to assume that the system is consistent, it would have no trouble at all proving statements like "This statement cannot be proven." (Proof: if it could be proven, the system would be inconsistent!) He quotes Alan Turing: "If we want a machine to be intelligent, it can't also be infallible. There are theorems that say almost exactly that." Humans certainly satisfy the criterion of not being infallible.

Putting on our Bayesian hats, the fact that the convoluted minds of human beings naïvely seem to be able to perceive truths that can't be directly proven by completely rigorous computer programs doesn't seem nearly strong enough to warrant modifying our best understanding of quantum mechanics. Especially because the use to which such modifications are being put has nothing directly to do with the mysteries of quantum mechanics themselves—it's just a way to grant powers of insight and cognitive wizardry to the human brain. And at the end of the day, there's nothing about the brain's ability to see the truth of unprovable statements that helps us understand the Hard Problem, the issue of inner mental experiences. If you think the Hard Problem is hard, quantum mechanics is unlikely to help you; if you think it's not so bad, you probably don't feel the need to change the laws of physics to help us understand the brain.

43

What Acts on What?

The idea that we are part of the natural world can lead to a sense of profound loss if the reasons and causes for our actions aren't what we thought they were. We're not human beings, equipped with intentions and goals, so the worry goes; we're bags of particles mindlessly bumping into one another as time chugs forward. It's not love that will keep us together, it's just the laws of physics. A version of this concern was articulated by philosopher Jerry Fodor:

> If it isn't literally true that my wanting is causally responsible for my reaching, and my itching is causally responsible for my scratching, and my believing is causally responsible for my saying . . . if none of that is literally true, then practically everything I believe about anything is false and it's the end of the world.

Don't worry! It's not the end of the world.

We live in a reality that can be fruitfully talked about in many different ways. We have an extravagant assortment of theories, models, vocabularies, stories, whatever you prefer to call them. When we speak about a human being, we can describe them as a person with desires and tendencies and inner mental states; or we can describe them as a collection of biological cells interacting via electrochemical signals; or we can describe them as an agglomeration of elementary particles following the rules of the Core Theory. The question is, how do we fit these different stories together? In particular, what acts on what? Does the existence of the particle-physics description, in which "causality" is nowhere to be found, imply that it is illegitimate to talk about scratching being caused by itching?

The poetic-naturalist answer is that any of the stories we have stands or falls on its own terms as a description of reality. To evaluate a model of the world, the questions we need to ask include "Is it internally consistent?," "Is it well-defined?," and "Does it fit the data?" When we have multiple distinct theories that overlap in some regime, they had better be compatible with one another; otherwise they couldn't both fit the data at the same time. The theories may involve utterly different kinds of concepts; one may have particles and forces obeying differential equations, and another may have human agents making choices. That's fine, as long as the predictions of the theories line up in their overlapping domains of applicability. The success of one theory doesn't mean that another one is wrong; that only happens when a theory turns out to be internally incoherent, or when it does a bad job at describing the observed phenomena.

Developing a theory of human thought and behavior in terms of neural signals or interacting particles doesn't in any way imply that your wanting is not responsible for your reaching. There is no obstacle to that kind of vocabulary of desire and intentionality being "true," as long as its predictions are compatible with those of other successful vocabularies.

It's possible that what Fodor means by "literally true" is something like "an essential element of every possible description of nature," or perhaps "of our best and most comprehensive description of nature." In other words, there can't exist any successful vocabulary that doesn't include "wanting" and "believing" as fundamental concepts. In that case, it is not literally true—the physical and biological descriptions of human beings are perfectly adequate on their own terms, and don't invoke concepts like wants and beliefs.

But that's an unnecessarily constraining notion of "literally true." Thermodynamics and the fluid description of air didn't stop being true once we discovered atoms and molecules. Both ways of talking are true. Likewise, human thoughts and intentions haven't disappeared just because we obey the laws of physics.

This issue seems more complicated than it is because of an understandable tendency, in a world described by multiple distinct but mutually compatible stories, to jumble up the concepts of one story with those of another—to cross the lines separating distinct ways of talking.

Rather than acknowledging that there is one way of talking about the world in terms of the quantum fields and interactions of the Core Theory, and another way in terms of electrochemical signals traveling between cells, and yet another way in terms of human agents with desires and mental states, we fall into the trap of using multiple vocabularies at the same time. When told that every mental state corresponds to various physical states of one's brain, one wants to complain, "Do you really think the reason why I'm scratching is only because of some synaptic signaling, and not because I feel an itch?" The complaint is misplaced. You can describe what's happening in terms of electrochemical signals in your central nervous system, *or* in terms of your mental states and the actions they cause you to perform; just don't trip up by starting a sentence in one language and attempting to finish it in another one.

One of the most common arguments against Cartesian dualism (or mental properties that influence physical ones) is *causal closure of the physical*. The laws of physics as we know them—the Core Theory, in the domain we're interested in—are complete and self-consistent. You give me a quantum state of a system, and there are unambiguous equations that will tell me what it will do next. (We've written down one such equation in the Appendix.) There is no ambiguity, no secret fudge factors, no opportunity for differing interpretations of what is happening. If you give me the precise and complete quantum state corresponding to "a person feeling an itch," and I have the calculational abilities of Laplace's Demon, I could predict with extraordinary accuracy that the quantum state will evolve into a different state corresponding to "a person scratching themselves." No further information is needed, or allowed.

✳

In chapter 13 we discussed the idea of "strong emergence," according to which the behavior of a system with many parts is not reducible to the aggregate behavior of all those parts. A related idea is *downward causation*: behavior of the parts is actually caused by the state of the whole, in a way not interpretable as due to the parts themselves.

Poetic naturalists tend to view downward causation as a deeply misguided idea. Then again, they view upward causation as equally misguided. "Causation," which after all is itself a derived notion rather than a fundamental one, is best thought of as acting within individual theories that rely on the concept. Thinking of behavior in one theory as *causing* behavior in a completely different theory is the first step toward a morass of confusion from which it is difficult to extract ourselves.

It's certainly possible that behavior in coarse-grained macroscopic theories might be *entailed by* features of more comprehensive theories, and we certainly want them to be *consistent with* such theories when the descriptions overlap. We might even, as long as we're careful, say that features of an underlying theory can help *explain* features of an emergent one. But we get in trouble if we try to say that phenomena in one theory are *caused by* phenomena in a different one. I know that I cannot use my mental powers to reach across space and bend spoons, since the fields and interactions of the Core Theory don't accommodate that kind of capacity. But I can describe that feature purely in the macroscopic language: human beings don't possess the power of telekinesis. The microscopic explanation might aid my understanding, but it's not a necessary part of how I talk about human-scale behavior.

And the converse, downward causation of human-scale properties influencing the microscopic behavior of particles, is misguided. A standard example is the formation of snowflakes. Snowflakes are made of water molecules, interacting with other molecules to form a crystalline structure. There are many possible structures, determined by the initial configuration of the seed from which the snowflake grows. Therefore, it is claimed, the macroscopic shape of the snowflake is acting "downward" to determine the precise location of individual water molecules.

It's bad form to mix vocabularies in such a vulgar way. Water molecules interact with other water molecules, and other molecules in the air, in precise ways that are specified by the rules of atomic physics. Those rules are unambiguous: you tell me what other molecules any individual water molecule is interacting with, and the rules will say precisely what will happen next. The relevant molecules may be part of a larger crystalline structure, but that knowledge is of zero import when studying the behavior of the water molecule under consideration. The environment in which the molecule is embedded is relevant, but there is no obstacle to describing that environment in terms of its own molecular structure. The individual molecule has no idea it's part of a snowflake, and could not care less.

Something like downward causation is possible in principle, even if there's no evidence for it in the real universe. We could imagine a possible world in which electrons and atoms obeyed the rules of the Core Theory in situations of very low numbers of particles, but started obeying different rules when the numbers became large (such as in a human being). Even then, the right way to think about the situation would not be "the larger structure is influencing the smaller particles"; it's "the rules we thought were obeyed by particles were wrong." In other words, we could discover that the domain of applicability of the Core Theory was smaller than we thought it was. There is no evidence that anything along those lines is true, and it would violate everything we know about effective quantum field theories—but many things are possible.

The way we talk about human beings and their interactions is going to end up being less crisp and precise than our theories of elementary particles. It might be harmless, and even useful, to borrow terms from one story because they are useful in another one—"diseases are caused by microscopic germs" being an obvious example. Drawing relations between different vocabularies, such as when Boltzmann suggested that the entropy of a gas was related to the number of indistinguishable arrangements of the molecules of which it was composed, can be extremely valuable and add important insights. But if a theory is any good, it has to be able to speak sensibly about the phenomena it purports to describe all by itself, without leaning on causes being exerted to or from theories at different levels of focus.

Mental states are ways of talking about particular physical states. To say that a mental state causes a physical effect is precisely as legitimate as saying that any macroscopic physical situation is the cause of some macroscopic physical event. There is nothing incorrect about attributing your scratching to the existence of your itching; there's simply more than one story we can legitimately tell about what's going on.

44

Freedom to Choose

Once we see how mental states can exert physical effects, it's irresistible to ask, "Who is in charge of those mental states?" Am I, my emergent self, actually making choices? Or am I simply a puppet, pulled and pushed as my atoms jostle amongst themselves according to the laws of physics? Do I, at the end of the day, have free will?

There's a sense in which you do have free will. There's also a sense in which you don't. Which sense is the "right" one is an issue you're welcome to decide for yourself (if you think you have the ability to make decisions).

The usual argument against free will is straightforward: We are made of atoms, and those atoms follow the patterns we refer to as the laws of physics. These laws serve to completely describe the evolution of a system, without any influences from outside the atomic description. If information is conserved through time, the entire future of the universe is already written, even if we don't know it yet. Quantum mechanics predicts our future in terms of probabilities rather than certainties, but those probabilities themselves are absolutely fixed by the state of the universe right now. A quantum version of Laplace's Demon could say with confidence what the probability of every future history will be, and no amount of human volition would be able to change it. There is no room for human choice, so there is no such thing as free will. We are just material objects who obey the laws of nature.

It's not hard to see where that argument violates our rules. Of course there is no such notion as free will when we are choosing to describe human beings as collections of atoms or as a quantum wave function. But that says nothing about whether the concept nevertheless plays a useful role when we choose to describe human beings as people. Indeed, it pretty clearly does play a useful role. Even the most diehard anti–free will partisans are constantly speaking about choices that they and other people make in their daily activities, even if they afterward try to make light of it by adding, "Except of course the concept of choice doesn't really exist."

The concept of choice does exist, and it would be difficult indeed to describe human beings without it. Imagine you're a high school student who wants to go to college, and you've been accepted into several universities. You look at their web pages, visit campuses, talk to students and faculty at each place. Then you say yes to one of them, no to the others. What is the best way to describe what just happened, the most useful vocabulary for talking about our human-scale world? It will inevitably involve some statements along the lines of "you made a choice," and the reasons for that choice. If you had been a simplistic robot or a random-number generator, there might have been a better way of talking. But it is artificial and counterproductive to deny ourselves the vocabulary of choice when we talk about human beings, regardless of how well we understand the laws of physics. This stance is known in the philosophical literature as *compatibilism*, and refers to the compatibility between an underlying deterministic (or at least impersonal) scientific description and a macroscopic vocabulary of choice and volition. Compatibilism, which traces its roots back as far as John Locke in the seventeenth century, is the most popular way of thinking about free will among professional philosophers.

From this perspective, the mistake made by free-will skeptics is to carelessly switch between incompatible vocabularies. You step out of the shower in the morning, walk to your closet, and wonder whether you should put on the black shirt or the blue shirt. That's a decision that you have to make; you can't just say, "I'll do whatever the atoms in my body were going to deterministically do anyway." The atoms are going to do whatever they were going to do; but you don't know what that is, and it's irrelevant to the question of which decision you should make. Once you frame the question in terms of you and your choice, you can't *also* start talking about your atoms and the laws of physics. Either vocabulary is perfectly legitimate, but mixing them leads to nonsense.

✳

You may be willing to accept that oceans and temperature are real, even though they are nowhere to be found among the fundamental ingredients of the Core Theory, but feel unwilling to apply the same logic to free will. After all, the ability to make choices isn't just a macroscopic collection of many microscopic pieces; it's an entirely different kind of thing. If it's not there in our best comprehensive description of nature, why is it helpful to act like it's there in our human-scale vocabulary?

The answer comes down to the arrow of time. In chapter 8 we talked about how we have epistemic access to the past—memories—that we don't have when it comes to the future. That's because there is a special boundary condition, the Past Hypothesis, according to which entropy was very low near the Big Bang. That's a powerful bit of information about the past, which enables us to pin it down in a way that we can't pin down the future. This temporal asymmetry arises only because of the distribution of matter in the universe on macroscopic scales; there is no analogue of it in the Core Theory itself.

There is a crucial role played by the leverage that features of our current state exert over our knowledge of events in the past or future. When a feature of our current state implies (given the Past Hypothesis, and all else being equal) something about the past, that's a memory; when a feature of our current state implies something about the future, that's a cause of some future effect. The small differences in a person's brain state that correlate with different bodily actions typically have negligible correlations with the past state of the universe, but they can be correlated with substantially different future evolutions. That's why our best human-sized conception of the world treats the past and future so differently. We remember the past, and our choices affect the future.

Laplace's Demon discerns no such imbalance; he sees the whole history of the world with perfect clarity. But none of us is Laplace's Demon. None of us knows the exact state of the universe, or has the calculational power to predict the future even if we did. The unavoidable reality of our incomplete knowledge is responsible for why we find it useful to talk about the future using a language of choice and causation.

One popular definition of free will is "the ability to have acted differently." In a world governed by impersonal laws, one can argue that there is no such ability. Given the quantum state of the elementary particles that make up me and my environment, the future is governed by the laws of physics. But in the real world, we are not given that quantum state. We have incomplete information; we know about the rough configuration of our bodies and we have some idea of our mental states. Given only that incomplete information—the information we actually have—it's completely conceivable that we could have acted differently.

This is the point at which free-will doubters will object that the stance we've defended here isn't really free will at all. All we've done is redefine the notion to mean something completely different, presumably because we are too cowardly to face up to the desolate reality of an impersonal cosmos.

I have no problem with the desolate reality of an impersonal cosmos. But it's important to explore the most accurate and useful ways of talking about the world, on all relevant levels.

Admittedly, some formulations of "free will" go well beyond anything that a poetic naturalist would be willing to countenance. There is what is called libertarian freedom. This has nothing to do with the political free-market idea of libertarianism. Rather, it's the position that human agency introduces an element of indeterminacy into the universe; people are not governed by the impersonal laws of physics; they have a distinct ability to shape their own futures. It's a denial that there could be anything like Laplace's Demon, who could know the future before it happened.

There's no reason to accept libertarian freedom as part of the real world. There is no direct evidence for it, and it violates everything we know about the laws of nature. In order for libertarian freedom to exist, it would have to be possible for human beings to overcome the laws of physics just by thinking.

A poetic naturalist says that we can have two very different-sounding ways of describing the world, a physics-level story and a human-level story, which invoke separate sets of concepts and yet end up being compatible in their predictions concerning what happens in the world. A libertarian thinks that the right way to talk about human beings ends up making predictions that are *incompatible* with the known laws of physics. We don't need to do such dramatic violence to our understanding of reality just to make peace with the fact that we make choices as we go through the day.

In a famous experiment in the 1980s, physiologist Benjamin Libet measured brain activity in subjects as they decided to move their hands. The volunteers were also observing a clock, and could report precisely when they made their decisions. Libet's results seemed to indicate that there was a telltale pulse of brain activity before the subjects became consciously aware of their decision. To put it dramatically: part of their brain had seemingly made the decision before the people themselves became aware of it.

Libet's experiment, and various follow-ups, have become controversial. Some claim that they are evidence against the existence of free will, since obviously our consciousness is a bit behind the curve when it comes to decision making. Others have raised technical concerns about whether the signal Libet measured is truly a sign of a decision having been made, and whether the subjects were reliable in reporting when their decisions occurred.

If you already accept that the world is fundamentally physical, nothing in the Libet experiments or their successors should have much of an influence on your attitude toward free will. You weren't going to believe in libertarian free will anyway, and these experiments have no bearing on one's stance toward compatibilism. Our brains are messy places, with many small subsystems churning along beneath the surface, only occasionally poking their way up into our conscious attention. There is no question that we sometimes make decisions unconsciously, whether it's steering our car on the way to work or turning onto our side while we sleep. There's also no question that other decisions, like whether to write a book and whether to include a discussion of downward causation in that book, are essentially conscious ones. There are fascinating detailed questions that are worth addressing about the specific ways in which our brain goes about its business, but none of that alters the basic truth that we are collections of elementary particles interacting through the rules of the Core Theory. And it's okay to talk about us as human beings making decisions.

If you accept the universal applicability of the laws of nature, and therefore deny libertarian freedom, the argument between compatibilists and incompatibilists can seem a bit tiresome. We basically agree on what's happening—particles obeying the laws of physics, and a macroscopic description of people making choices—and whether we decide to label it "free will" might not seem like the most important question.

Where the issue becomes more than merely academic is when we confront the notions of blame and responsibility. Much of our legal system, and much of the way we navigate the waters of our social environment, hinges on the idea that individuals are largely responsible for their actions. At extreme levels of free-will denial, the idea of "responsibility" is as problematic as that of human choice. How can we assign credit or blame if people don't choose their own actions? And if we can't do that, what is the role of punishment or reward?

Poetic naturalists and other compatibilists don't need to face up to these questions, since they accept the reality of human volition, and therefore have no difficulty in attributing responsibility or blame. There are cases that are not so clear, however.

We attribute reality to our ability to make choices because thinking that way provides the best description we know of for the human-scale world. In some circumstances, though, that ability seems to be absent, or at least downgraded. One well-known example involved an anonymous patient in Texas who developed a brain tumor after being operated on to help alleviate his epilepsy. Once the tumor occurred, the patient started exhibiting symptoms of Klüver-Bucy syndrome, a disease that appears in rhesus monkeys but is very rare in humans. Among the symptoms are hyperphagia (excessive appetite and eating) and hypersexuality, including compulsive masturbation.

Eventually, the patient started downloading child pornography, which led to his arrest. At his trial, neurosurgeon Orrin Devinsky testified that the patient was not actually in control of his actions—he lacked free will. His compulsion to download pornography, in Devinsky's view, could be completely attributed to the effects of his previous surgery, leaving him without any volition in the matter. The court disagreed, and found him guilty, although he received a relatively light sentence. One of the arguments against him was that he was able to avoid pornography when he was at work, so he evidently was able to exert some degree of control over his own actions.

What matters here is not the extent to which this particular patient actually lost control over his choices, but the fact that such loss is possible. What that does to our notions of personal responsibility is a pressing real-world question, not an academic abstraction.

If our belief in free will is predicated on the idea that "agents making choices" is part of the best theory we have of human behavior, then the existence of a better and more predictive understanding could undermine that belief. To the extent that neuroscience becomes better and better at predicting what we will do without reference to our personal volition, it will be less and less appropriate to treat people as freely acting agents. Predestination will become part of our real world.

It doesn't seem likely, however. Most people do maintain a certain degree of volition and autonomy, not to mention a complexity of cognitive functioning that makes predicting their future actions infeasible in practice. There are gray areas—drug addiction is an obvious case where volition can be undermined, even before we go all the way to considering tumors and explicit brain damage. This is a subject in which the basics are far from settled, and much of the important science has yet to be established. What seems clear is that we should base our ideas about personal responsibility on the best possible understanding of how the brain works that we can possibly achieve, and be willing to update those ideas whenever the data call for it.

PART SIX

CARING

45

Three Billion Heartbeats

Carl Sagan, who introduced so many people to the wonders of the cosmos, died in 1996. At an event in 2003, his wife, Ann Druyan, was asked about him. Her response is worth quoting at length:

> When my husband died, because he was so famous and known for not being a believer, many people would come up to me—it still sometimes happens—and ask me if Carl changed at the end and converted to a belief in an afterlife. They also frequently ask me if I think I will see him again.
>
> Carl faced his death with unflagging courage and never sought refuge in illusions. The tragedy was that we knew we would never see each other again. I don't ever expect to be reunited with Carl. But, the great thing is that when we were together, for nearly twenty years, we lived with a vivid appreciation of how brief and precious life is. We never trivialized the meaning of death by pretending it was anything other than a final parting.
>
> Every single moment that we were alive and we were together was miraculous—not miraculous in the sense of inexplicable or supernatural. We knew we were beneficiaries of chance. . . . That pure chance could be so generous and so kind. . . . That we could find each other, as Carl wrote so beautifully in *Cosmos*, you know, in the vastness of space and the immensity of time. . . .

That we could be together for twenty years. That is something
which sustains me and it's much more meaningful. . . .

The way he treated me and the way I treated him, the way we
took care of each other and our family, while he lived. That is so
much more important than the idea I will see him someday. I
don't think I'll ever see Carl again. But I saw him. We saw each
other. We found each other in the cosmos, and that was wonderful.

There are few issues of greater importance than the question of whether
our existence continues on after we die. I believe in naturalism, not because
I would prefer it to be true, but because I think it provides the best account
of the world we see. The implications of naturalism are in many ways uplift-
ing and liberating, but the absence of an afterlife is not one of those ways.
It would be nice to keep on living in some fashion, assuming my personal
continuation would be relatively pleasant, rather than being tortured by
ornery demons. Perhaps not for eternity, but I can easily imagine keeping
things interesting for a few hundred thousand years. Regrettably, that's not
the way the evidence points.

The longing for life to continue beyond our natural span of years is part
of a deeper human impulse: the hope, and expectation, that our lives mean
something, that there is some point to it all. The notion of "reasons why" is
often useful in our human-scale world, but might not apply when we start
talking about the origin of the universe or the nature of the laws of physics.
Does it apply to our lives? Are there reasons why we are here, why things
happen the way they do?

It takes courage to face up to the finitude of our lives, and even more
courage to admit the limits of purpose in our existence. The most telling
part of Druyan's reflection is not the acknowledgment that she won't see
Carl again, but where she affirms that it was pure chance that they ever
found each other in the first place.

Our finite life-span reminds us that human beings are part of nature,
not apart from it. Physicist Geoffrey West has studied a remarkable series
of *scaling laws* in a wide range of complex systems. These scaling laws are
patterns that describe how one feature of a system responds as some other
feature is changed. For example, in mammals, the expected lifetime scales

as the average mass of an individual to the 1/4 power. That means that a mammalian species that is sixteen times heavier will live twice as long as a smaller species. But at the same time, the interval between heartbeats in mammalian species also scales as their mass to the 1/4 power. As a result, the two effects cancel out, and the number of heartbeats per typical lifetime is roughly the same for all mammals—about 1.5 billion heartbeats.

A typical human heart beats between sixty and a hundred times a minute. In the modern world, where we are the beneficiaries of advanced medicine and nutrition, humans live on average for about twice as long as West's scaling laws would predict. Call it 3 billion heartbeats.

Three billion isn't such a big number. What are you going to do with your heartbeats?

Ideas like "meaning" and "morality" and "purpose" are nowhere to be found in the Core Theory of quantum fields, the physics underlying our everyday lives. The same could be said about "bathtubs" and "novels" and "the rules of basketball." That doesn't prevent these ideas from being real—they each play an essential role in a successful higher-level emergent theory of the world. The same goes for meaning, morality, and purpose. They aren't built into the architecture of the universe; they emerge as ways of talking about our human-scale environment.

But there is a difference; the search for meaning is not another kind of science. In science we want to describe the world as efficiently and accurately as possible. The quest for a good life isn't like that: it's about evaluating the world, passing judgment on the way things are and could be. We want to be able to point to different possible events and say, "That's a worthy goal to strive for," or "That's the way we ought to behave." Science couldn't care less about such judgments.

The source of these values isn't the outside world; it's inside us. We're part of the world, but we've seen that the best way to talk about ourselves is as thinking, purposeful agents who can make choices. One of those choices, unavoidably, is what kind of life we want to live.

We're not used to thinking that way. Our folk ontology treats meaning as something wholly different from the physical stuff of the world. It might be given by God, or inherent in life's spiritual dimension, or part of a teleological inclination built into the universe itself, or part of an ineffable, transcendent aspect of reality. Poetic naturalism rejects all of those possibilities, and asks us to take the dramatic step of viewing meaning in the same way we view other concepts that human beings invent to talk about the universe.

Rick Warren's bestselling book *The Purpose-Driven Life* opens with a simple admonition: "It's not about you." It might come as a surprise that a book so many people have turned to for comfort and advice begins on such a down note. But Warren's strategy is to appeal precisely to people's sense of being overwhelmed at life's challenges. He offers them a direct way out: it's not about you; it's about God.

You don't have to accept Warren's theology to sympathize with the impulse. There are many ways it could be about something other than us: we could be spiritually inclined without belonging to a traditional organized religion, or we could feel devoted to a culture or nation or family, or we could believe in objective forms of meaning based on scientific grounds. Any such strategy can be both challenging, in the sense that it can be hard to live up to the standards that are imposed on you, but also comforting, because at least there are standards, darn it.

Poetic naturalism offers no such escape from the demands of meeting life in a creative and individual way. It *is* about you: it's up to you, me, and every other person to create meaning and purpose for ourselves. This can be a scary prospect, not to mention exhausting. We can decide that what we want is to devote ourselves to something larger—but that decision comes from us.

The ascendance of naturalism has removed the starting point for much of how we used to conceive of our place in the universe. We're Wile E. Coyote, and we've just looked down. We need some new ground to stand on— or we need to learn how to fly.

There are two legitimate worries about the idea that we construct meaning for our lives.

The first worry is that it's cheating. Maybe we are fooling ourselves if we think we can find fulfillment once we accept that we are part of the physical world, patterns of elementary particles beholden to the laws of physics. Sure, you can *say* you are leading a rich and rewarding life based on your love for your family and friends, your dedication to your craft, and your work to make the world a better place. But are you *really*? If the value we place in such things isn't objectively determined, and if you won't be around to witness any of it in a hundred years or so, how can you say your life truly matters?

This is just grumpiness talking. Say you love somebody, genuinely and fiercely. And let's say you also believe in a higher spiritual power, and think of your love as a manifestation of that greater spiritual force. But you're also an honest Bayesian, willing to update your credences in light of the evidence. Somehow, over the course of time, you accumulate a decisive amount of new information that shifts your planet of belief from spiritual to naturalist. You've lost what you thought was the source of your love—do you lose the love itself? Are you now obligated to think that the love you felt is now somehow illegitimate?

No. Your love is still there, as pure and true as ever. How you would explain your feelings in terms of an underlying ontological vocabulary has changed, but you're still in love. Water doesn't stop being wet when you learn it's a compound of hydrogen and oxygen.

The same goes for purpose, meaning, and our sense of right and wrong. If you are moved to help those less fortunate than you, it doesn't matter whether you are motivated by a belief that it's God's will, or by a personal conviction that it's the right thing to do. Your values are no less real either way.

The second worry about creating meaning within ourselves is that there isn't any place to start. If neither God nor the universe is going to help us attach significance to our actions, the whole project seems suspiciously arbitrary.

But we do have a starting place: who we are. As living, thinking organisms, we are creatures of motion and motivation. At a basic, biological level, we are defined not by the atoms that make us up but by the dynamic patterns we trace out as we move through the world. The most important thing about life is that it occurs out of equilibrium, driven by the second law. To stay alive, we have to continually move, process information, and interact with our environment.

In human terms, the dynamic nature of life manifests itself as *desire*. There is always something we want, even if what we want is to break free of the bonds of desire. That's not a sustainable goal; to stay alive, we have to eat, drink, breathe, metabolize, and generally continue to ride the wave of increasing entropy.

Desire has a bad reputation in certain circles, but that's a bum rap. Curiosity is a form of desire; so are helpfulness and artistic drive. Desire is an aspect of *caring*: about ourselves, about other people, about what happens to the world.

People are not inanimate rocks, accepting what goes on around them with serene indifference. Different people might exhibit different levels of care, and they might care in different ways, but caring itself is ubiquitous. They might care in an admirable way, watching out for the well-being of others, or their caring might be purely selfish, guarding their own interests. But people are inescapably characterized by what they care about: their enthusiasms, inclinations, passions, hopes.

When our lives are in good shape, and we are enjoying health and leisure, what do we do? We play. Once the basic requirements of food and shelter have been met, we immediately invent games and puzzles and competitions. That's a lighthearted and fun manifestation of a deeper impulse: we enjoy challenging ourselves, accomplishing things, having something to show for our lives.

That makes sense, in light of evolution. An organism that didn't give a crap about anything that happened to it would be at a severe disadvantage in the struggle for survival when compared to one that looked out for itself, its family, and its compatriots. We are built from the start to care about the world, to make it matter.

Our evolutionary heritage isn't the whole story. The emergence of consciousness means that what we care about, and how we behave in response to those impulses, can change over time as a result of our learning, our interaction with others, and our own self-reflection. Our instincts and unreflective desires aren't all we have; they're just a starting point for building something significant.

Human beings are not blank slates at birth, and our slates become increasingly rich and multidimensional as we grow and learn. We are bubbling cauldrons of preferences, wants, sentiments, aspirations, likes, feelings, attitudes, predilections, values, and devotions. We aren't slaves to our desires; we have the capacity to reflect on them and strive to change them. But they make us who we are. It is from these inclinations within ourselves that we are able to construct purpose and meaning for our lives.

The world, and what happens in the world, matters. Why? Because it matters to me. And to you.

The personal desires and cares that we start with may be simple and self-regarding. But we can build on them to create values that look beyond ourselves, to the wider world. It's our choice, and the choice we make can be to expand our horizons, to find meaning in something larger than ourselves.

The movie *It's a Wonderful Life* has unmistakable religious underpinnings—it's Christmas Eve, and George Bailey is saved from killing himself by the intervention of a guardian angel. But as author Chris Johnson has pointed out, what changes George's mind isn't words of angelic wisdom; it's the demonstration that his life had a tangible, positive effect on the lives of other people in the town of Bedford Falls. Real stuff, here on Earth, the lives we actually lead. In the end, that's the only place meaning can possibly reside.

The construction of meaning is a fundamentally individual, subjective, creative enterprise, and an intimidating responsibility. As Carl Sagan put it, "We are star stuff, which has taken its destiny into its own hands."

The finitude of life lends poignancy to our situations. Each of us will have a last word we say, a last book we read, a last time we fall in love. At each moment, who we are and how we behave is a choice that we individually make. The challenges are real; the opportunities are incredible.

46

What Is and What Ought to Be

David Hume, the eighteenth-century Scottish thinker whom we've encountered before as a forefather of poetic naturalism, is widely regarded as a central figure of the Enlightenment. When he was only twenty-three years old, he began work on a book that would turn out to be extraordinarily influential, *A Treatise of Human Nature*. At least, it would be judged so by history; at the time, Hume's ambition to write a bestseller fell somewhat short, as he lamented that the book "fell dead-born from the press."

We should give Hume credit for trying to be a lively writer, even if the reading public didn't necessarily agree. In one famous passage, he sardonically remarks on what he sees as a curious tendency among his fellow philosophers: a predilection for suddenly declaiming what *ought* to be true when they had previously been describing only what *is* true.

> In every system of morality, which I have hitherto met with, I have always remark'd, that the author proceeds for some time in the ordinary way of reasoning, and establishes the being of a God, or makes observations concerning human affairs; when of a sudden I am surpriz'd to find, that instead of the usual copulations of propositions, *is*, and *is not*, I meet with no proposition that is not connected with an *ought*, or an *ought not*. This change is imperceptible; but is, however, of the last consequence. For as this *ought*, or *ought not*, expresses some new relation or

David Hume. (Painting by Allan Ramsay)

affirmation, 'tis necessary that it shou'd be observ'd and explain'd; and at the same time that a reason should be given, for what seems altogether inconceivable, how this new relation can be a deduction from others, which are entirely different from it.

While it's amusing to think of propositions copulating with each other, Hume's sentences admittedly do go on a bit. But his main point is clear: talking about "oughts" is an entirely different kind of thing from simply talking about what "is." The former is passing a judgment, saying what should be the case; the latter is merely descriptive, saying what actually happens. If you're going to perform such a magic trick and call it philosophy, you should at least have the consideration to tell us how the trick is done. Modern thought has distilled the point down to a maxim: "You can't derive ought from is."

There is an apparent problem here for naturalism: if you can't derive ought from is, then you're in trouble, because "is" is all there is. There isn't anything outside the natural world to which we can turn for guidance about how to behave. The temptation to somehow extract such guidance from the natural world itself is incredibly strong.

But it doesn't work. The natural world doesn't pass judgment; it doesn't provide guidance; it doesn't know or care about what ought to happen. We are allowed to pass judgment ourselves, and we're part of the natural world, but different people are going to end up with different judgments. So be it.

To see why it's impossible to derive ought from is, it's useful to think about how we can ever derive anything from anything else. There are many such ways, but let's focus in on one of the simplest: the *logical syllogism*, paradigm of deductive reasoning. Syllogisms look like this:

1. Socrates is a living creature.
2. All living creatures obey the laws of physics.
3. Therefore, Socrates obeys the laws of physics.

This is just one example of the general form, which can be expressed as:

1. X is true.
2. If X is true, then Y is true.
3. Therefore, Y is true.

Syllogisms are not the only kind of logical argument—they're just a particularly simple form that will suffice to make our point.

The first two statements in a syllogism are the premises of the argument, while the third statement is the conclusion. An argument is said to be *valid* if the conclusion follows logically from the premises. In contrast, an argument is said to be *sound* if the conclusion follows from the premises and the premises themselves are true—a much higher standard to achieve.

Consider: "Pineapples are reptiles. All reptiles eat cheese. Therefore, pineapples eat cheese." Any logician will explain to you that this is a completely valid argument. But it's not very sound. An argument can be valid, and even interesting, without telling us much that is true about the real world.

If we were to try to put a derivation of ought from is into the form of a syllogism, it might look something like this:

1. I would like to eat the last slice of pizza.
2. If I don't move quickly, someone else will eat the last slice of pizza.
3. Therefore, I ought to move quickly.

At a casual glance this seems like a good argument, but it's not a logically valid syllogism. The two premises are both "is" statements—my desire to eat the last slice, and the likelihood that I will miss that chance if I don't move quickly, are both factual claims about the world, whether or not they are actually true. And the conclusion is undeniably an "ought" statement. But if you look past the everyday meaning of the sentences to their underlying logical content, something is missing. Premises 1 and 2 don't actually imply the conclusion 3; what they imply is "Therefore, if I don't move quickly, I will not get what I like."

To make the conclusion follow validly, we would need to add another premise, along the lines of:

2a. I ought to act in such a way as to bring about what I would like.

With this addition, the argument becomes valid. It's also no longer a candidate for deriving ought from is—an "ought" statement appears right there in the new premise. All we've done is to derive an ought from an ought plus a few is's, which isn't nearly as impressive.

That's the problem with attempting to derive ought from is: it's logically impossible. If someone tells you they have derived ought from is, it's like someone telling you that they've added together two even numbers and obtained an odd number. You don't have to check their math to know that they've made a mistake.

And yet, it happens all the time. Over and over again, before and after the appearance of Hume's famous passage, many people have triumphantly declared they have finally cracked the code and shown how to derive ought from is. Smart, knowledgeable people, with interesting things to say. But somehow they have all gone wrong.

Physicist Richard Feynman liked to tell the story of meeting a painter and asking him about his craft. The painter boasted that he could mix red and white paint together and get yellow. Feynman knew enough about how color works to be skeptical, so the painter fetched some paint and commenced with mixing. After a bit of effort and nothing but pink paint to show for it, the painter mumbled that he should probably add a touch of yellow to the mix, to "sharpen it up a bit." At that point Feynman understood the trick—to get yellow out, you put a bit of yellow in.

The painter's gambit is the same basic move that has been used to perform the logically impossible, deriving ought from is, many times over the centuries. One presents a set of incontrovertible "is" statements, then sneaks in an implied "ought" statement that seems so extremely reasonable that nobody could possibly deny it. Sadly, all statements about what ought to happen can (and will) be denied by somebody, and even if not, that doesn't prevent them from being ought statements.

A classic example was offered up by John Searle, of Chinese Room fame. Here is Searle's version of the kind of deductive argument we examined above:

1. Jones uttered the words "I hereby promise to pay you, Smith, five dollars."
2. Jones promised to pay Smith five dollars.
3. Jones placed himself under (undertook) an obligation to pay Smith five dollars.
4. Jones is under an obligation to pay Smith five dollars.
5. Jones ought to pay Smith five dollars.

You see the magical appearance of "ought" in the last line, even though all of the other lines were about "is." Where did the sleight-of-hand occur?

It's not that hard to find. Just as we had to imagine a new premise 2a above, Searle is relying on a hidden premise between 4 and 5:

4a. All else being equal, one ought to do what one is under an obligation to do.

Searle actually admits the need for a premise like this, right in the text of his paper. But he thinks it doesn't count as a premise, since it's a "tautology"—something that is automatically true by the definitions of the terms involved. Searle is claiming that what it means to say, "Jones made a promise to do something," is simply "Jones ought to do something" (all else being equal).

That's not true. Hopefully the equivocation is clear. Up in premises 1–3, the idea of "placing himself under an obligation" referred to a certain fact about the world, a sentence that Jones uttered. But now in 4–5, Searle wants us to treat an "obligation" as a moral command, a statement about what ought to happen. He's using the same word in two different senses, to trick us into thinking that factual statements about what happens can somehow lead to evaluative conclusions about right and wrong.

This example is worth belaboring because it stands in for an impressive number of attempts to derive ought from is over the years. Inevitably, the argument introduces just a tiny bit of prescription into their list of descriptions: the painter sharpens things up with a touch of yellow.

This inherent flaw in deriving ought from is has been pointed out many times. The list of thinkers who claim to have successfully pulled off the trick is long and distinguished; they aren't simply making elementary mistakes. Lurking in the back of their minds is usually some kind of justification along the lines of "Okay, there is some hidden premise that introduces an ought into my list of is's, but surely we agree that this particular hidden premise isn't so bad, right?"

It wouldn't be so bad if it weren't for the fact that, when brought out into the clear light of day, the hidden evaluative premises don't seem to be universally true. Quite the contrary; they tend to be conspicuously contentious. The reason why deriving ought from is should be thought of as a philosophical felony, rather than a simple misdemeanor, is because these hidden premises deserve our closest scrutiny. They are, more often than not, where most of the action is.

You might be tempted to think that Searle's hidden premise 4a seems pretty unobjectionable, but let's examine it more closely. Surely there are some kinds of obligations that one ought not to carry out—when they were made under duress, or when they would grossly violate some other moral precept. Searle would say that such examples don't count, because of the "all else being equal" clause. So what exactly does that clause mean? He tells us:

> The force of the expression "other things being equal" in the present instance is roughly this. Unless we have some reason for supposing the obligation is void (step 4) or the agent ought not to keep the promise (step 5), then the obligation holds and he ought to keep the promise.

So you ought to do what you are under an obligation to do—unless there is some reason you ought not to do it. This doesn't seem like a useful foundation for moral reasoning.

We shouldn't hide or downplay the assumptions we make in order to get moral reasoning off the ground. Our attempts to be better people are best served if those assumptions are brought out into the open, interrogated, and evaluated as carefully as we can manage.

A modern twist on the ought-from-is campaign is to claim that morality can be reduced to, or absorbed by, the practice of science. The idea is something like this:

1. Condition X would make the world a better place.
2. Science can tell us how to achieve condition X.
3. Therefore, we ought to do what science tells us to do.

In this case, the hidden assumption would appear to be:

2a. We ought to make the world a better place.

This might seem like a tautology, depending on your definition of the word "better." But whether we put the hidden assumption into a statement such as this one, or bury it in the definition of "better," we are still making some positive claim that something ought to be done. Such claims cannot be grounded on factual statements alone. Who decides what is "better"?

Proponents of this technique will sometimes argue that all we're doing is making some reasonable assumptions, and science makes reasonable assumptions all the time, so what we're doing really isn't any different. That's missing an important aspect of what science is. Consider the following statements:

- The universe is expanding.
- Humans and chimpanzees share a common ancestor.
- We should work to allow people to lead happier and longer lives.

All of these statements are, by some lights, true. But only the first two are "scientific." The reason is that each of them *could have been false*. They are not true by definition or assumption. We can imagine possible worlds in which the universe was contracting, or in which there were species like humans and chimpanzees that had not evolved from a common ancestor. We decide whether such statements are true or not by empiricism, abduction, and Bayesian reasoning—we go out and observe the world, and update our credences appropriately.

We don't imagine carrying out experiments to decide whether we should work to allow people to lead happier and longer lives. We assume that it's so, or we try to derive it from a related set of assumptions. That crucial extra ingredient separates how science works from how we think about right and wrong. Science does require assumptions; there are certain epistemological precepts, like our trust in our basic sensory inputs, that play an important role in constructing stable planets of belief for working scientists. But the assumptions that suffice to get science off the ground don't do the same trick for morality.

☀

None of this is to say that we can't address "ought" issues using the tools of reason and rationality. There is an entire form of logical thought called *instrumental rationality*, devoted to answering questions of the form "Given that we want to attain a certain goal, how do we go about doing it?" The trick is deciding what we want our goal to be.

One attractive suggestion was put forward by Bill Preston and Ted Logan, as played by Alex Winter and Keanu Reeves in the movie *Bill & Ted's Excellent Adventure*. They proposed the timeless moral axiom, "Be excellent to each other."

As foundational precepts for moral theorizing go, you could do worse. It's tempting to brush aside concerns about the foundation of morality on the grounds that we know moral goodness when we see it, and what's really important is how we go about achieving it.

But there are important reasons why we have to do a little bit better than Bill-and-Ted-level philosophizing. The truth is that we don't ultimately all agree on what constitutes happiness, or pleasure, or justice, or other forms of being excellent to each other. Morality and meaning are areas where foundational disagreement doesn't arise just by someone making a mistake; it's real and inevitable, and we need to figure out how to deal with it.

It's tempting to say, "Everyone agrees that killing puppies is wrong." Except that there are people who do kill puppies. So maybe we mean "Every *reasonable* person agrees . . ." Then we need to define "reasonable," and realize we haven't really made much progress at all.

The lack of an ultimate objective scientific grounding for morality can be worrisome. It implies that people with whom we have moral disagreements—whether it's Hitler, the Taliban, or schoolyard bullies who beat up smaller children—aren't *wrong* in the same sense that it's wrong to deny Darwinian evolution or the expansion of the universe. We can't do an experiment, or point to data, or construct a syllogism, or write a stinging blog post, that would persuade them of why their actions are bad. And if that's true, why should they ever stop?

But that's how the world is. We should recognize that our desire for an objective grounding for morality creates a cognitive bias, and should compensate by being especially skeptical of any claims in that direction.

47

Rules and Consequences

Abraham heard God commanding him to take Isaac, his only son, to the region of Moriah and sacrifice him there as a burnt offering. The next morning Abraham and Isaac, along with two servants and a donkey, began the arduous three-day journey. Arriving at the site, Abraham built an altar and arranged the wood atop it. He bound his son and drew a heavy knife. At the last moment he faltered; he could not bring himself to sacrifice his boy. Isaac, however, had seen the despair in his father's eyes. By the time they returned to his mother, Sarah, Isaac had completely lost his faith.

This isn't the usual telling of the Abraham and Isaac story, familiar from Genesis. It's one of four alternative imaginings offered by Søren Kierkegaard in his book *Fear and Trembling*. In the original, God intercedes at the last minute and offers Abraham a ram to sacrifice in place of his son. Kierkegaard suggests a number of different twists, each harrowing in its own way: Abraham tricks Isaac into thinking Abraham is a monster, so Isaac wouldn't lose faith in God; Abraham sees a ram and decides to sacrifice it rather than his son, in contravention of his orders; Abraham begs God to forgive him that he would have even contemplated sacrificing his son; and Abraham falters at the last moment, causing Isaac to lose faith.

There are many readings of the tale of Abraham and Isaac. A traditional explanation casts it as a lesson about the strength of faith: God wanted to test Abraham's loyalty by making the strongest possible demand. Martin Luther held that Abraham's willingness to kill Isaac was correct, given one's fundamental need to defer to God's will. Immanuel Kant held that Abraham should have realized that there are no conditions under which it would have been justified to sacrifice his son—and therefore the command could not actually have come from God. Kierkegaard, concerned that a proliferation of interpretations was diluting the impact of this clash of apparent absolutes, wanted to emphasize the impossibility of finding a simple answer to Abraham's dilemma, and highlight the demands placed by true faith.

From a broader perspective, the story highlights the issue of competing moral commitments: what do we do when something that seems utterly wrong at a visceral level (killing your own son) runs into a foundational rule to which you are devoted (obeying God's word)? When it is not clear what is right and wrong, what are the most basic principles that should ultimately decide?

In modern manifestations of moral argument, hearing commands from God doesn't have the same force it once did. But the fundamental dichotomy lives on. The descendant of Abraham's dilemma in our secularized, technological world is something called the *trolley problem*.

Introduced by philosopher Philippa Foot in the 1960s, the trolley-problem thought experiment aims to sharpen the conflict between competing moral sentiments. A group of five people is tied to some trolley tracks. Unfortunately, a speeding trolley has lost its brakes, and is barreling toward them. If no action is taken, they will surely die. But you have the option of taking an action: you are standing by a switch that will divert the trolley onto another track. This alternate track, by unfortunate coincidence, has a single person tied down on it, who will surely be killed if you pull the switch. (Trolley-track security is remarkably lax in this hypothetical world.) What do you do?

It's not quite sacrificing-your-only-son-because-of-God's-command level stuff, but the dilemma is real. On the one hand, there is a choice between five people dying and one person dying. All else being equal, it would seem to be better, or at least less bad, if only one person died. On the other hand, you have to actively do something in order to divert the train. Instinctively, if the trolley barrels forward and kills the five people, it's not really our fault, whereas if we choose of our own volition to pull the switch, we bear the responsibility for the death of the one person on the other track.

This is where we see Bill and Ted's "Be excellent to each other" falling short when it comes to providing the basis for a fully articulated ethical system. Moral quandaries are real, even if they are usually not as stark as the trolley problem. How much of our income should we spend on our own pleasure, versus putting it toward helping the less fortunate? What are the best rules governing marriage, abortion, and gender identity? How do we balance the goal of freedom against that of security?

As Abraham learned, having an absolute moral standard such as God can be extraordinarily challenging. But without God, there is no such standard, and that is challenging in its own way. The dilemmas are still there, and we have to figure out a way to face them. Nature alone is no help, as we can't extract ought from is; the universe doesn't pass moral judgments.

And yet we must live and act. We are collections of vibrating quantum fields, held together in persistent patterns by feeding off of ambient free energy according to impersonal and uncaring laws of nature, and we are also human beings who make choices and care about what happens to ourselves and to others. What's the best way to think about how we should live?

Philosophers find it useful to distinguish between *ethics* and *meta-ethics*. Ethics is about what is right and what is wrong, what moral guidelines we should adopt for our own behavior and that of others. A statement like "killing puppies is wrong" belongs to ethics. Meta-ethics takes a step back, and asks what it *means* to say that something is right or wrong, and why we should adopt one set of guidelines rather than some other set. "Our system of ethics should be based on improving the well-being of conscious creatures" is a meta-ethical claim, from which "killing puppies is wrong" might be derived.

Poetic naturalism has little to say about ethics, other than perhaps for a few inspirational remarks. But it does have something to say about meta-ethics, namely: our ethical systems are things that are constructed by us human beings, not discovered out there in the world, and should be

evaluated accordingly. To help with that kind of evaluation, we can contemplate some of the choices we have when it comes to ethics.

Two ideas serve as a useful starting point: *consequentialism* and *deontology*. At the risk of vastly oversimplifying thousands of years of argument
and contemplation, consequentialists believe that the moral implications of
an action are determined by what consequences that action causes, while
deontologists feel that actions are morally right or wrong in and of themselves, not because of what effects they may lead to. "The greatest good for
the greatest number," the famous maxim of utilitarianism, is a classic consequentialist way of thinking. "Do unto others as you would have them do
unto you," the Golden Rule, is an example of deontology in action. Deontology is all about rules. (The word "deontology" comes from the Greek
deon, for "duty," while "ontology" comes from the Greek *on*, for "being."
Despite the similarity of the words, the two ideas are unrelated.)

Bill and Ted were deontologists. Had they been consequentialists, their
motto would have been something like "Make the world a more excellent
place."

The problem is that both consequentialism and deontology seem perfectly reasonable at first glance. "The greatest good for the greatest number"
sounds like a splendid idea, as does "Do unto others as you would have
them do unto you." The point of the trolley problem is that these approaches
can come into conflict. The idea that it would be reasonable to sacrifice one
person in order to save five people is consequentialist at its core, while our
reluctance to actually pull the switch stems from deep deontological
impulses—diverting the trolley and killing an innocent person just seems
wrong, even if it does save lives. The standard moral sentiments of most
people include both consequentialist and deontological impulses.

The operation of these competing ethical inclinations can be traced to
different parts of our babbling brains. Our minds have a System 1 that is
built on heuristics, instincts, and visceral reactions, as well as a System 2
that is responsible for cognition and higher-level thoughts. Roughly speaking, System 1 tends to be responsible for our deontological impulses, and
System 2 kicks in when we start thinking as consequentialists. In the words
of psychologist Joshua Greene, we not only have "thinking fast and slow";
we also have "morality fast and slow." System 2 thinks we should pull the
switch, while System 1 is appalled by the idea.

✳

Philosophers have thought up many modifications of the original trolley problem. A famous one is the "footbridge problem," proposed by Judith Jarvis Thomson. Let's say you are a committed consequentialist, and are sure you would pull the switch in the original problem. But this time there is no switch: rather, the only way to stop the trolley from killing the five unfortunate people on the track is to push a large man off of a footbridge and into the path of the trolley. (All such thought experiments imagine we are able to predict the future with uncanny accuracy; this one also assumes that you yourself are too tiny to stop the trolley on its course, so self-sacrifice is not an option.)

As before, either one person will die or five will die. To a consequentialist, there is no difference between the footbridge scenario and the original trolley problem. But to a deontologist there might be. In the first problem we are not actively trying to kill the one person on the side track; that's just an unfortunate repercussion of our attempts to save the five people. But up on that footbridge, we are purposely forcing someone to their death. Our emotions recoil at the prospect; it's one thing to pull a switch, quite another to push someone off of a bridge.

Greene has studied volunteers hooked up to an MRI machine while being asked to contemplate various moral dilemmas. As expected, contemplation of "personal" situations (like pushing someone off of a bridge) led to increased activity in areas of the brain that are associated with emotions and social reasoning. "Impersonal" situations (like pulling a switch) engaged the parts of the brain associated with cognition and higher reasoning. Different modules within ourselves spring to life when we're forced to deal with slightly different circumstances. When it comes to morality, the unruly parliament that constitutes our brain includes both deontological and consequentialist factions.

Sticking someone inside an imposing medical scanner and asking them to consider philosophical thought experiments might not tell us much about how that person would actually react in the situation described. The real world is messy—are you *sure* you could stop the trolley by pushing that guy off the footbridge?—and people's predictions about how they would act in stressful situations aren't always reliable. That's okay; our goal here isn't to understand how people behave, it's to get a better idea for how they think about how they *should* behave.

Consequentialism and deontology aren't the only kinds of ethical systems we can consider. Another popular approach is *virtue ethics*, which traces its roots back to Plato and Aristotle. If deontology is about what you do, and consequentialism is about what happens, virtue ethics is about who you are. To a virtue ethicist, what matters isn't so much how many people you save by diverting a trolley, or the intrinsic good of your actions; what matters is whether you made your decision on the basis of virtues such as courage, responsibility, and wisdom. The virtue-ethical versions of Bill and Ted would have simply said, "Be excellent."

Virtue sounds like a good thing to strive toward. Like consequentialism and deontology, it's an ostensibly attractive moral stance. Sadly, all of these attractive approaches end up offering different advice in important cases. How should we decide what ethical system to abide by?

That's a trick question. Knowing how we "should" decide something requires that we already have some normative stance, a way of judging different approaches. Let's instead contemplate how we possibly could go about choosing an ethical system at all.

There are many distinct ways of talking that can each capture some important truth about reality. Not all vocabularies capture truth; some are simply incorrect. Our goal is to describe the world in useful ways, where "useful" is always relative to some stated purpose. In the case of scientific theories, "useful" means things like "able to make accurate predictions on the basis of minimal input," and "providing insight into the behavior of a system."

Morality adds an evaluative component to how we talk about the world. This or that person or behavior is bad or good, right or wrong, admirable or reprehensible. The criteria for usefulness that help us choose between alternative scientific theories are insufficient when it comes to constructing moral principles. The point of moral reasoning is not to help us make predictions or provide insight into a person's behavior.

Happily, there are other senses of usefulness besides "helping us fit the data." Each of us comes into the meta-ethical game with a preexisting set of commitments. We have desires, we have feelings, we have things that we care about. There are things that naturally attract us, and those that repel us. Long before we have ever started thinking reflectively about what our ethical stance should be, we already have some kind of nascent moral sensibility.

Primatologist Frans de Waal has done studies to probe the origins of empathy, fairness, and cooperation in primates. In one famous experiment, he and collaborator Sarah Brosnan placed two capuchin monkeys in separate cages, each able to see the other one. When the monkeys performed a simple task, they were rewarded with a slice of cucumber. The capuchins were quite content with this setup, doing the task over and over, enjoying their cucumber. The experimenters then began rewarding one of the monkeys with grapes—a sweeter food than cucumbers, preferable in every way. The monkey who didn't get the grapes, who was previously perfectly content with cucumbers, saw what was going on and refused to do the assigned task, outraged at the inequity of the new regime. Recent work by Brosnan's group with chimpanzees shows cases where even the chimp who gets the grapes is unhappy—their sense of fairness is insulted. Some of our most advanced moral commitments have very old evolutionary roots.

One approach to moral philosophy is to think of it as simply a method for making sense of those commitments: making sure that we are true to our own self-proclaimed morals, that our justifications for our actions are internally consistent, and that we take into account the values of other people where appropriate. Rather than fitting data in a scientific sense, we can choose our ethical theories by how well they conform to our own existing sentiments. A moral framework is "useful" to a poetic naturalist to the extent that it reflects and systematizes our moral commitments in a logically coherent way.

A nice feature of this perspective is that it is resolutely practical: it is what people actually do when they try to think carefully about morality. We have a feeling for what distinguishes right from wrong, and we try to make it systematic. We talk to other people to learn how they feel, and take that into account when developing the rules for functioning in society.

It can also be terrifying. You're telling me that judging right from wrong is just a matter of our personal feelings and preferences, grounded in nothing more substantial than our own views, with nothing external to back it up? That there are no objectively true moral *facts* out there in the world?

Yes. But admitting that morality is constructed, rather than found lying on the street, doesn't mean that there is no such thing as morality. All hell has not broken loose.

The idea that moral guidelines are things invented by human beings based on their subjective judgments and beliefs, rather than being grounded in anything external, is known as *moral constructivism*. (When I say "human beings" in this context, feel free to substitute "conscious creatures." I'm not trying to discriminate against animals, aliens, or hypothetical artificial intelligences.) Constructivism is a bit different from "relativism." A moral relativist thinks that morality is grounded in the practices of particular cultures or individuals, and therefore cannot be judged from outside. Relativism is sometimes derided as an overly quietist stance—it doesn't permit legitimate critique of one system by another.

A moral constructivist, by contrast, acknowledges that morality originates in individuals and societies, but accepts that those individuals and societies will treat the resulting set of beliefs as "right," and will judge others accordingly. Moral constructivists have no qualms about telling other people that they're doing the wrong thing. Furthermore, the fact that morals are constructed doesn't mean that they are arbitrary. Ethical systems are invented by human beings, but we can all have productive conversations about how they could be improved, just as we do with all sorts of things that human beings put together.

Philosopher Sharon Street distinguishes between Kantian constructivism, after Immanuel Kant, and Humean constructivism, after David Hume. These are two enormously influential thinkers who tended to come at problems from very different perspectives, perhaps in part due to their differing personalities. Kant, whose strict personal schedule was such that residents of Königsberg were known to set their timepieces by his daily walks, was part of a long tradition within philosophy of trying make everything precise, rigorous, and certain. He would brook no fuzziness in his ethical philosophy. Kant was the deontologist par excellence, and he founded his views on morality on the *categorical imperative*: act in such a way that your actions could become a universal law. At one point Kant seemed to suggest that it would be wrong to lie to a murderer who was at your door in order to protect their potential victim, because lying shouldn't be a universal law. Scholars debate whether Kant really thought that it was always wrong to lie, but one certainly gets the impression of strict deontological rectitude in his thought.

Hume, meanwhile, was much more at home in a world of skepticism, empiricism, and uncertainty. He rejected absolute moral principles, and instead of an objective imperative he proudly proclaimed that "Reason is, and ought only to be, the slave of the passions." Reason, that is, can help us get what we want; but what we actually do want is defined by our passions. Hume was dubious of the natural philosophical tendency to make things look just a bit tidier and more exact than they really are.

A Kantian constructivist accepts that morality is constructed by human beings, but believes that every rational person would construct the same moral framework, if only they thought about it clearly enough. A Humean constructivist takes one more step: morality is constructed, and different people might very well construct different moral frameworks for themselves.

Hume was right. We have no objective guidance on how to distinguish right from wrong: not from God, not from nature, not from the pure force of reason itself. Alive in the world, individual and contingent, we are burdened and blessed with all of the talents and inclinations and instincts that evolution and our upbringings have bequeathed to us. Those are the raw materials from which morals are constructed. Judging what is good and what is not is a quintessentially human act, and we need to face up to that reality. Morality exists only insofar as we make it so, and other people might not pass judgments in the same way that we do.

48

Constructing Goodness

S o then, fellow humans. What kind of morality shall we construct?
There is no unique answer to this question that applies equally
well to all persons. But that shouldn't stop each of us from doing the
best we can to expand and articulate our own moral impulses into systematic positions.

Perhaps the most well-known approach to ethics is the consequentialist
theory of *utilitarianism*. It imagines that there is some quantifiable aspect
of human existence, which we can label "utility," such that increasing it is
good, decreasing it is bad, and maximizing it would be best of all. The issue
then becomes how we should define utility. A simple answer is "happiness"
or "pleasure," but that can seem a bit superficial and self-centered. Other
options include "well-being" and "preference satisfaction." What matters is
that there is something we can, in principle, quantify into a number (the
total amount of utility in the world), and then we can work to make that
number as big as possible.

This kind of utilitarianism runs into a number of well-known problems.
The attractive idea of "quantifying utility" becomes slippery when we try to
put it into practice. What does it really mean to say that one person has
0.64 times the well-being of another person? How do we combine well-
beings—is one person with a utility of 23 better or worse than two people
with utilities of 18 each? As Derek Parfit has pointed out, if you believe that
there is some positive utility in the very existence of a somewhat-satisfied
human being, it follows that having a huge number of somewhat-satisfied

people has more utility than a relatively smaller number of exquisitely happy people. It seems counter to our moral intuitions to think that utility can be increased just by making more people, even if they are less happy ones.

Another challenge for utilitarianism was offered by philosopher Robert Nozick: the "utility monster," a hypothetical being with incredibly refined sensibilities and an enormous capacity for pleasure. At face value, standard utilitarianism might lead us to think that the most moral actions are those that keep the utility monster happy, no matter how sad that might make the rest of us, because the monster is so incredibly good at being happy. Relatedly, we could imagine technology progressing to the point where we could place people in machines that would render them immobile, but generate in their brains maximal feelings of happiness or preference satisfaction or a feeling of flourishing or whatever other utility measure we dreamed up. Should we work toward a world where everyone is hooked up to such machines?

Finally, the utilitarian calculus tends to not discriminate between utility of ourselves and those we know and love, versus the utility associated with anyone else in the world, or at any other time in history. For the majority of people in the developed world, utilitarianism would seem to insist that we give away a large fraction of our wealth to the cause of ridding the world of disease and poverty. That may be a laudable goal, but it reminds us that utilitarianism can be an exceedingly demanding taskmaster.

Utilitarianism doesn't always do a good job of embodying our moral sentiments. There are some things we tend to think are just wrong, even if they increase the net happiness of the world, like going around and secretly murdering people who are lonely and unhappy. There are other things we think are laudable, even if happiness is slightly decreased thereby. Utilitarians know about such examples, and are able to adjust the rules to make them seem less problematic. The basic issue remains: the notion of attaching a single value of "utility" to every action, and working to increase it, is a hard one to pull off in practice.

Deontological approaches run into their own problems. Psychologists have suggested that moral reasoning in general, and deontological reasoning in particular, functions primarily to rationalize opinions that we reach intuitively, rather than leading us to novel moral conclusions. Thalia

Wheatley and Jonathan Haidt did a study in which they hypnotized subjects to feel a strong sense of revulsion at certain innocuous words such as "often" and "take." They were then told simple stories about people who did nothing particularly wrong from any reasonable ethical perspective. When those stories contained the words they had been primed to react to, not only did they feel disgust, but they also judged the actions of the people in the stories to be somehow morally wrong. Without being able to articulate exactly why, the subjects were convinced that the people being described were somehow up to no good.

Clashes between universal ethical guidelines and our personal moral sentiments would be okay, if we thought that our sentiments were merely crude approximations to the more transcendent truths captured by those guidelines. In that case, so much the worse for our sentiments. But if we envision the project of moral philosophy as systematizing and rationalizing our sentiments, rather than replacing them with an objective truth, then such approaches have a bigger problem. Talking about morality might not be so cut-and-dried.

Deontology and consequentialism, and for that matter virtue ethics and various other approaches, all capture something real about our moral impulses. We want to act in good ways; we want to make the world a better place; we want to be good people. But we also want to make sense and be internally consistent. That's hard to do while accepting all of these competing impulses at once. In practice, moral philosophies tend to pick one approach and apply it universally. And as a result of that, we often end up with conclusions that don't sit easily with the premises we started with.

It may be that the kind of moral code that fits most people the best isn't based on a strict construal of any one approach, but takes bits and pieces from all of them. Consider a kind of "soft consequentialism," where the value of actions depends on their consequences, but also to some degree on the actions themselves. Or imagine that we allow ourselves to place greater value on helping people we know and care about than on helping those farther away. These need not be seen as "mistakes"; they could be part of a complex and multifaceted, but internally consistent, way of realizing our basic moral inclinations.

Or—someone could be a perfectly moral person who based their behavior on a small set of absolute rules, whether it was a particular flavor of utilitarianism or adherence to the categorical imperative, because that's what they felt was the best fit to their inner convictions. And that's okay. The moral systems we construct serve our own purposes.

Abraham was commanded by God to do something horrible. It was a great challenge to his humanity, but given his view of the world, the correct course of behavior was clear: if you are certain that God is telling you to do something, that's what you do. Poetic naturalism refuses to offer us the consolation of objective moral certainty. There is no "right" answer to the trolley problem. How you should act depends on who you are.

Ay, there's the rub. We *want* there to be objective solutions to our dilemmas, as surely as there are theorems in mathematics or experimental discoveries in science. As good Bayesians, aware of our bias toward claims that we would like to be true, this desire should make us especially skeptical of attempts to found objective morality on a natural basis. But as human beings, it often makes us all too readily accepting.

The worry is that, if morality is constructed, everyone will construct whatever they like, and what they like won't actually be very good. It's an ancient concern, usually directed at believers in the wrong religion or no religion at all. Tertullian, an early Christian thinker from Africa who is recognized as a Father of the Church, explained that an atomist like the Greek philosopher Epicurus couldn't be a good person. The problem is that for Epicurus, life ends at death, so suffering is ephemeral, while Christians believe in hell, so for them it's forever. Why should anyone strive to be good if there were no promise of an eternal reward, nor threat of eternal punishment?

> Think of these things, too, in the light of the brevity of any punishment you can inflict—never to last longer than till death. On this ground Epicurus makes light of all suffering and pain, maintaining that if it is small, it is contemptible; and if it is great, it is not long-continued. No doubt about it, we, who receive our awards under the judgment of an all-seeing God, and

who look forward to eternal punishment from Him for sin,—
we alone make real effort to attain a blameless life.

The modern version of this worry is that, if we were to accept that morality is constructed, individuals will run around giving in to their worst
instincts, and we would have no basis on which to condemn obviously bad
things like the Holocaust. After all, somebody thought it was a good idea,
and without objective guidance how can we say they were wrong?

The constructivist answers that just because moral rules are invented by
human beings, that doesn't make them any less real. The rules of basketball
are also invented by human beings, but once invented they really exist.
People even argue over what the "right" rules should be. When James Nai-
smith invented the game, the ball was thrown into peach baskets, and had
to be retrieved by hand each time a shot was made. Only later did they realize that the game would be improved by replacing the basket with a hoop.
That made the game "better," in the sense that it did a better job at fulfilling
its purpose as a game. The rules of basketball aren't objectively defined,
waiting out there in the universe to be discovered; but they aren't arbitrary
either. Morality is like that: we invent the rules, but we invent them for
sensible purposes.

The problem arises when we imagine people whose purposes—whose
foundational moral sentiments and commitments—are radically at odds
with ours. What are we to do with someone who just wants to play hockey
rather than basketball? In sports we might seek out different people to play
with, but when it comes to morality we all have to live together here on this
Earth.

We might hope, in the spirit of Kant, that simple logical requirements
of internal consistency would lead every rational person to construct the
same moral rules, even starting from slightly different initial feelings. But
that hope seems slim indeed. Sharon Street imagines an "internally coherent Caligula," who takes pleasure in the suffering of others. Such a monster
need not be illogical or inconsistent; they just have fundamental attitudes
with which we cannot agree. We're not going to reason them out of their
stance. If they act on their impulses in ways that bring harm to others, we
should respond as we actually do in the real world: by preventing them
from doing so. When criminals refuse to be deterred, we put them in jail.

As a practical matter, the worries associated with constructivism are somewhat overblown. Most people, in most circumstances, want to think of themselves as doing good rather than evil. It's not clear what operational benefit would be gained by establishing morality as an objective set of facts. Presumably we envision a person or group who was relatively rational, but disagreed with us about morality, whom we could sit down with over coffee and convince of the mistake they were making. In practice the recommended strategy for a constructivist would be essentially the same: sitting down and talking with the person, appealing to our common moral beliefs, attempting to work out a mutually reasonable solution. Moral progress is possible because most people share many moral sentiments; if they don't, reasoning with them wouldn't help much no matter what.

If instead the worry is that we can't justify stepping in to prevent immoral actions, that simply isn't an issue for constructivists. If, upon rational reflection, we decide that something is deeply wrong, there is no reason why we cannot work to prevent it from happening, regardless of whether our decision is based on external criteria or our own inner convictions. Again, this is no more or less than what really happens in the world.

Deciding how to be good isn't like solving a math puzzle, or discovering a new fossil. It's like going out to dinner with a group of friends. We think about what we want for our individual selves, talk to others about their desires and how we can work together, and reason about how to make it happen. The group may include both vegetarians and omnivores, but with a good-faith effort there's no reason everyone can't be satisfied.

I once found myself on a panel at a large interdisciplinary meeting, attended by people from the worlds of business, science, politics, and the arts. The purpose of the panel was to discuss morality in the modern world. I had been invited not because of any particular expertise in moral matters, but it was a conference where most of the participants tended to be religious, and I was known not to be; my job was to be the token atheist. And when the time came for me to speak, the single question I was asked was: "What do you think would be the best argument *against* your atheism?" The other panelists, by contrast, were offered a chance to say something positive and constructive about their moral positions. There is a lurking suspicion in many corners that naturalists are objects of curiosity but not to be taken seriously when it comes to talking about values.

Here in the early years of the twenty-first century, a majority of philosophers and scientists are naturalists. But in the public sphere, at least in the United States, on questions of morality and meaning, religion and spirituality are given a preeminent place. Our values have not yet caught up to our best ontology.

They had better start catching up. When it comes to deciding how to live, we're like that first fish flapping up onto land: faced with a new world of challenges and opportunities, and not yet really adapted to it. Technology has given us enormous power to shape our world for better or for worse, and by any reasonable estimate we are only at the very beginning of the associated changes. We're going to be faced with the kinds of moral questions that our ancestors could not possibly have contemplated, from human-machine interfacing to the exploration of new planets. Engineers working on self-driving cars have already begun to realize that the software is going to have to be programmed to solve certain kinds of trolley problems.

Poetic naturalism doesn't tell us how to behave, but it warns us away from the false complacency associated with the conviction that our morals are objectively the best. Our lives are changing in unpredictable ways; we need to be able to make judgments with clear eyes and an accurate picture of how the world operates. We don't need an immovable place to stand; we need to make our peace with a universe that doesn't care what we do, and take pride in the fact that we care anyway.

49

Listening to the World

The idea of "Ten Commandments" is a deeply compelling one. It combines two impulses that are ingrained in our nature as human beings: making lists of ten things, and telling other people how to behave.

The most famous such list is found in the Hebrew Bible. It's a compilation of instructions for the Israelite people, handed from God to Moses atop Mount Sinai. The commandments are found twice, once in Exodus and once in Deuteronomy. In neither case is the list numbered, and the wording between the two appearances is slightly different. As a result, there is no agreement on what "The Ten Commandments" actually are. Jews, Orthodox Christians, Catholics, and different Protestant denominations quote slightly different lists. Lutherans, for example, don't include the traditional prohibition against graven images, and split the coveting of thy neighbor's house into a commandment all its own, rather than grouping it with the coveting of thy neighbor's wife and thy neighbor's servants. What matters is that there are ten of them.

Inevitably, schools of thought outside the traditional religious mainstream have borrowed the Ten Commandments idea, and proposed their own lists. There are atheist commandments, secular commandments, and so on. The Socialist Sunday Schools, an organization that began in the United Kingdom as an alternative to Christian Sunday schools, proposed a list of socialist commandments. ("Remember that all good things of the earth are produced by labour. Whoever enjoys them without working for them is stealing the bread of the workers.")

A good poetic naturalist will resist the temptation to hand out commandments. "Give someone a fish," the saying goes, "and you feed them for a day. Teach them to fish, and you feed them for a lifetime." When it comes to how to lead our lives, poetic naturalism has no fish to give us. It doesn't even really teach us how to fish. It's more like poetic naturalism helps us figure out that there are things called "fish," and perhaps investigate the various possible ways to go about catching them, if that were something we were inclined to do. It's up to us what strategy we want to take, and what to do with our fish once we've caught them.

It makes sense, then, to put aside the concept of "commandments" and instead propose *Ten Considerations*: a list of things we think are true, that might be useful to keep in mind as we shape and experience our own ways of valuing and caring about our lives. We can draw inspiration from the universe by listening to it carefully.

1. Life Isn't Forever.

Julian Barnes, in his novel *A History of the World in 10 1/2 Chapters*, imagines a version of what heaven would be like. A man, who had been a working-class Englishman, wakes up after his death in a new environment, where everything is wonderful. He can have anything he asks for, with one implicit catch: he has to have the imagination to ask for it. Being who he is, he has sex with countless attractive women, eats meal after amazing meal, meets up with famous celebrities and politicians, and becomes so good at playing golf that he scores a hole in one more often than not.

Inevitably, he begins to grow fidgety and bored. After inquiring a bit from one of heaven's staff members, he discovers there is an option to simply end it all and die. And do people in heaven actually choose to die, he asks?

"Everyone takes the option," the staffer answers, "sooner or later."

Humanity has always imagined ways that life might continue on after our bodily deaths. None of them holds up very well under close examination. What the stories fail to account for is that change, including death, isn't an optional condition to be avoided; it's an integral part of life itself. You don't really want to live forever. Eternity is longer than you think.

Life ends, and that's part of what makes it special. What exists is here, in front of us, what we can see and touch and affect. Our lives are not dress rehearsals in which we plan and are tested in anticipation of the real show to come. This is it, the only performance we're going to get to give, and it is what we make of it.

2. Desire Is Built into Life.

Imagine trying to achieve perfect stillness. Close your eyes, slow your bodily rhythms, let your mind go quiet. While some are better at it than others, no person can ever be truly motionless. You will always be breathing; your heart will be pumping; billions of ATP molecules are being synthesized inside you, then used to power invisible processes inside your body. There is no perfect stillness this side of the grave. (And not even then, though we may be permitted a bit of poetic license.)

Compare this with a computer. Build a machine with immense processing power, turn it on, and watch what it does all by itself: nothing at all. It will just sit there. We can program it, give it some task and ask it to do something. But if we don't, the machine won't have volition just because it has the capacity to crunch numbers. You can ignore it and it won't get impatient; cause it damage and it won't defend itself; belittle it and it won't be annoyed.

Life is characterized by motion and change, and these characteristics manifest themselves in human beings as forms of desire. From our evolutionary origins we have things that we want, from enjoying a good meal to helping other people to creating an affecting work of art. It's those desires that shape us, and cause us to care about ourselves and others. But they don't enslave us; we are reflective and self-aware, with the ability to shape what it is we care about. We can, if we choose, focus our caring on making the world a better place.

3. What Matters Is What Matters to People.

The universe is an intimidating place. Compared to its smallest pieces, we are quite large; there are about 10^{28} atoms in a typical human body. But compared to its overall size, we are absurdly small; it would take more than 10^{26} people holding hands to stretch across the span of the observable cosmos. Long after the human race has vanished from existence, the universe will still be here, trundling along in placid accord with the underlying laws of nature.

The universe doesn't care about us, but we care about the universe. That's what makes us special, not any immaterial souls or special purpose in the grand cosmic plan. Billions of years of evolution have created creatures capable of thinking about the world, forming a picture of it in our minds and holding it up to scrutiny.

We are interested in the world, in its physical manifestations and in our fellow humans and other creatures. That caring, contained inside us, is the only source of "mattering" in any cosmic sense.

Whenever we ask ourselves whether something matters, the answer has to be found in whether it matters to some person or persons. We take the world and attach value to it, an achievement of which we can be justly proud.

4. We Can Always Do Better.

Understanding develops through the process of making mistakes. We make guesses about the world, test them against what we observe, learn more often than not that we were wrong, and try to improve our hypotheses. To err is human, and that's about it.

We can make our fallibility into a virtue by recognizing it and cherishing it, by always working to do better at whatever it is we are attempting. Mathematical proofs can be perfect in their logic, but scientific discoveries are typically the conclusion of a long series of trials and errors. When it comes to valuing, caring, loving, and being good, perfection is even more of a chimera, since there isn't even an objective standard against which to judge our successes.

We nevertheless make progress, both at understanding the world and at living within it. It may seem strange to claim the existence of *moral* progress when there isn't even an objective standard of morality, but that's exactly what we find in human history. Progress comes, not from new discoveries in an imaginary science of morality, but from being more honest and rigorous with ourselves—from uncovering our rationalizations and justifications for behavior that, if we admit it, was pretty reprehensible from the start. Becoming better people is hard work, but by sifting through our biases and being open to new ideas, our ability to be good advances.

5. It Pays to Listen.

If we admit that we can always be mistaken, it makes sense to open our minds to our fellow human beings to hear what they have to say. We all have our biases, so getting a bit of outside perspective isn't a bad thing. If purpose and morality aren't out there to be discovered, we might be able to learn something from our compatriots in the ongoing creation of meaning.

That includes ancient wisdom. Over thousands of years, people have struggled intensely with the question of how to be a good person. For the large majority of history, that work has been carried out within religious or spiritual traditions. There's no reason to throw out everything associated with the great thinkers of the past just because we have a more updated and accurate ontology. Nor is there any reason to stick with ethical commandments that have become unmoored from their original justification. We can take inspiration from ancient teachings, not to mention from great literature and art, without being bound by them.

Consciousness gives us an inner model of ourselves. It also allows us to model other people, opening the door for empathy and ultimately to love. To not only listen to others but also to imagine ourselves as them, to consider what they care about, is a powerful driver of moral progress. Once we see that mattering comes from inside people, understanding others becomes more important than ever.

6. There Is No Natural Way to Be.

Evolution is extraordinarily ingenious, inventing mechanisms that human designers would be hard-pressed to match. But there was no designer, which has its drawbacks. There is no simplistic, undivided self, no tiny homunculus in the brain steering us around on the basis of unbendable rules. We are the final product of a cacophony of competing impulses, and so are other people.

If we are part of nature, it can be tempting to valorize "being natural." That's backward: we can't help but be natural, since we are unavoidably part of nature. But nature doesn't guide us or lay down rules, or even offer exemplars of good behavior. Nature is kind of a mess. We can be inspired by it, and occasionally horrified by it, but nature simply is.

Searching for clues to the nature of human caring and morality in the behavior of our animal cousins reveals a mixed bag. Chimpanzee social groupings are dominated by males, while bonobos are dominated by females. Elephants mourn for their dead comrades, and species as diverse as rats and ants have been known to rescue friends who are in trouble. Biologists Robert Sapolsky and Lisa Share studied a group of Kenyan baboons who fed off the garbage from a nearby tourist lodge. The clan was dominated by high-status males, and females and lesser males would often go hungry. Then at one point, the clan ate infected meat from the garbage dump, which led to the deaths of most of the dominant males. Afterward, the "personality" of the troop completely changed: individuals were less aggressive, more likely to groom one another, and more egalitarian. This behavior persisted as long as the study continued, for over a decade.

The lesson is not that we should learn from the baboons (although if they can improve their lifestyles, maybe there is hope for us). It's that we are not simple, unified, fixed creatures. We have inclinations and desires, partly born of our innate dispositions, but we also have the opportunity to change, as individuals and as a society.

7. It Takes All Kinds.

If our lives are to have meaning and purpose, we are going to have to create them. And people are different, so they're going to create different things. That's a feature to be celebrated, not an annoyance to be eradicated.

Much of what has been written about the quest to lead a meaningful life has been produced by people who (1) enjoy thinking deeply and carefully about such things, and (2) enjoy writing down what they have thought about. Consequently, we see certain kinds of virtues celebrated: imagination, variety, passion, artistic expression. And these are all worth celebrating. But a fulfilled life might alternatively be characterized by reliability, obedience, honor, contentment. Some might find fulfillment in devoting their efforts to helping others; others will concentrate on their own daily practice of being. The right way to live for one person might not suit someone else.

Poetic naturalism doesn't provide much comfort for those who take joy in telling other people the proper way to live their lives. It allows for pluralism in purpose and meaning, a rich ecosystem of virtues and lives well lived.

We are faced with both an opportunity and a challenge. There is no single right way to live, an objectively best life out there to be discovered by reason or revelation. We have the opportunity to shape our lives in many ways, and count them as true and good.

8. The Universe Is in Our Hands.

We are collections of atoms and particles, bumping into one another and interacting through the forces of nature. We are also collections of biological cells, passing electricity and chemicals back and forth as we metabolize free energy from our environments. And we are also thinking, feeling, caring beings, capable of contemplating our actions and making decisions about how to behave.

It's the last bit that sets us apart. We are made of the same stuff as the rest of the universe, but our stuff is assembled in just the right way that a new way of talking about ourselves becomes appropriate. We have the capacity to contemplate alternatives and make choices. It's not a mystical or supernatural ability, giving us the right to flout the laws of physics; it's a way of talking about who we are that captures some of the power of the complex systems we call "human beings." And with great power comes great responsibility.

Our ability to think has given us enormous leverage over the world around us. We won't be able to stave off the heat death of the universe, but we can alter bodies, transform our planet, and someday spread life through the galaxy. It's up to us to make wise choices and shape the world to be a better place.

9. We Can Do Better Than Happiness.

We live at a time when the search for happiness has taken center stage as never before. Books, TV shows, and websites are constantly offering pointers about how to finally achieve and sustain this elusive and sought-after state of being. If only we were happy, everything would be okay.

Imagine a drug that would make you perfectly happy, but remove any interest you might have in doing anything more than simple survival. You would lead a thoroughly boring treadmill of a life, from the outside—but inside you would be blissfully happy, romping through imaginary adventures and always-successful romantic escapades. Would you take the drug?

Think of Socrates, Jesus, Gandhi, Nelson Mandela. Or Michelangelo, Beethoven, Virginia Woolf. Is "happy" the first word that comes to mind when you set out to describe them? They may have been—and surely were, from time to time—but it's not their defining characteristic.

The mistake we make in putting emphasis on happiness is to forget that life is a process, defined by activity and motion, and to search instead for the one perfect state of being. There can be no such state, since change is the essence of life. Scholars who study meaning in life distinguish between *synchronic meaning* and *diachronic meaning*. Synchronic meaning depends on your state of being at any one moment in time: you are happy because you are out in the sunshine. Diachronic meaning depends on the journey you are on: you are happy because you are making progress toward a college degree. If we permit ourselves to take inspiration from what we have learned about ontology, it might suggest that we focus more on diachronic meaning at the expense of synchronic. The essence of life is change, and we can aim to make change part of how we find meaning in it.

At the end of the day, or the end of your life, it doesn't matter so much that you were happy much of the time. Wouldn't you rather have a good story to tell?

10. Reality Guides Us.

In 1988, psychologists Shelley Taylor and Jonathon Brown coined the term "positive illusions" to describe beliefs people have that aren't true but that make them happy. The average person thinks they are above average; we tend to be much more optimistic about future events than past experience would actually warrant. It's part of our standard complement of cognitive biases.

The effect is real: there is little doubt that certain illusions make us happier. We can even come up with evolutionary-psychological explanations for why a bit of overenthusiastic self-regard might be helpful for our survival. One might imagine a program designed to make people feel better through targeted falsehoods. But is that what we want?

While having such illusions might make us happier, very few people knowingly seek out false beliefs. When we think we're better than average, it's not because we're saying to ourselves, "I'm going to consider myself better than I am because it will make me feel better." It's because we really do think that.

The upshot is that getting things right—being honest with ourselves and others, facing up to the world and looking it right in the eyeball—doesn't just happen. It requires a bit of effort. When we want something to be true, when a belief makes us happy—that's precisely when we should be questioning. Illusions can be pleasant, but the rewards of truth are enormously greater.

We have aspirations that reach higher than happiness. We've learned so much about the scope and workings of the universe, and about how to live together and find meaning and purpose in our lives, precisely because we are ultimately unwilling to take comforting illusions as final answers.

50

Existential Therapy

My family and I were regular churchgoers while I was growing up. It was probably my grandmother's influence that enforced the weekly discipline. Her parents had been born in England, and she was devoted to the Episcopal Church. We attended services at Trinity Cathedral in Trenton, New Jersey; while not anyone's idea of a leading example of sacred architecture, it did boast high Gothic stained-glass windows, which loomed impressively from the perspective of a young boy.

I liked going to church. Probably my favorite part was that we got to go for pancakes afterward, at a local place that offered strawberry syrup—the pinnacle of culinary excellence, if you had asked me at the time. But I enjoyed the hymns, the imposing wood pews, even the ritual of getting dressed up in the morning. More than anything else, I loved the mysteries and the doctrine. Going to Sunday school, reading the Bible, trying to figure out what it was all about. The most interesting part of the Bible was the Book of Revelation, prophesying what was to come. I became confused when I read somewhere that modern readers tended to find Revelation off-putting and even embarrassing. As a kid it was the coolest stuff in the book. There were angels, beasts, seals, trumpets; what's not to like?

We stopped going to church after my grandmother died when I was ten. I remained the kind of casual believer you find in many American households. My transformation to naturalism wasn't dramatic or life-shaking; it just kind of crept up on me. It was a smooth phase transition, not a sudden one.

Two incidents in particular stand out, however. The first happened when I was quite young. We were at church and a couple of the volunteers were chatting about recent alterations in the sequence of the service. They were pleased with the new arrangement, because the previous version of the liturgy required too much standing and kneeling, without enough breaks to sit down. I found this to be scandalously heretical. How is it possible that we can just mess around with what happens in the service? Isn't all that decided by *God*? You mean to tell me that *people* can just change things around at a whim? I was still a believer, but doubts had been sown.

Eventually, I found myself as an undergraduate astronomy major at a Catholic university, Villanova, just outside Philadelphia. By that point I had thought enough about how the universe works that I had become a naturalist by anyone's definition, though I still wasn't "out," to myself or to anyone else. Villanova had an enormous set of required courses, including three semesters each of philosophy and theology. I was enthralled by the former, and had a good time in the latter—my professors were incredibly smart—and loved talking through the ideas, regardless of whether I personally believed in them.

The second incident was when I heard a song, "The Only Way," from the Emerson, Lake & Palmer album *Tarkus*. (The Villanova astronomy department at the time was a hotbed of progressive-rock fandom.) In addition to some nifty pipe-organ work from Keith Emerson, the song featured something I hadn't ever heard: an unmistakable, in-your-face atheist message. "Don't need the word/Now that you've heard/Don't be afraid/Man is manmade." As poetry, it's not that great. As a reasoned philosophical argument, it falls well short. But this silly song made me think, for the first time, that it was okay to be a nonbeliever—that it wasn't something I should be ashamed of, something I should keep hidden. For a shy kid at a Catholic university, this was a big deal.

A number of atheists are driven to unbelief by a repressive religious upbringing. Not me; my experience could not have been less repressive, at least once they fixed the services so that there wasn't so much kneeling. Our brand of Episcopalianism was as mellow as churchgoing ever gets, and Villanova made no religious demands on its students outside of the theology classes.

I was always curious about the world, and fascinated by science. We talk about "awe and wonder," but those are two different words. I am in awe of the universe: its scope, its complexity, its depth, its meticulous precision. But my primary feeling is wonder. Awe has connotations of reverence: "this fills me with awe and I am not worthy." Wonder has connotations of curiosity: "this fills me with wonder and I am going to figure it out." I will take wonder over awe every day.

Many things about our world are mysterious to us, and there is something seductive and exciting about mysteries. It's a mistake to start embracing mystery for its own sake, and to take refuge in a conviction that the universe is fundamentally inscrutable. It would be like buying a big stack of detective novels and reading only the first halves of each of them. The real attraction of mysteries isn't that they represent something truly unknowable but that they promise an exciting journey to go figure them out.

Like Princess Elisabeth, I always thought it was crucial that different aspects of the world fit together and make sense. Everything we've experienced about the universe suggests that it is *intelligible*: if we try hard enough we can come to understand it. There is so much we still don't know about how reality works, but at the same time there's a great deal that we have figured out. Mysteries abound, but there's no reason to worry (or hope) that any of them are unsolvable.

Thinking like this eventually led me to abandon my belief in God and become a cheerful naturalist. But I hope I never make the mistake of treating people who disagree with me about the fundamental nature of reality as my enemies. The important distinction is not between theists and naturalists; it's between people who care enough about the universe to make a good-faith effort to understand it, and those who fit it into a predetermined box or simply take it for granted. The universe is much bigger than you or me, and the quest to figure it out unites people with a spectrum of substantive beliefs. It's us against the mysteries of the universe; if we care about understanding, we're on the same side.

Here's a story one could imagine telling about the nature of the world. The universe is a miracle. It was created by God as a unique act of love. The splendor of the cosmos, spanning billions of years and countless stars, culminated in the appearance of human beings here on Earth—conscious, aware creatures, unions of soul and body, capable of appreciating and returning God's love. Our mortal lives are part of a larger span of existence, in which we will continue to participate after our deaths.

It's an attractive story. You can see why someone would believe it, and work to reconcile it with what science has taught us about the nature of reality.

Here's a different story. The universe is not a miracle. It simply is, unguided and unsustained, manifesting the patterns of nature with scrupulous regularity. Over billions of years it has evolved naturally, from a state of low entropy toward increasing complexity, and it will eventually wind down to a featureless equilibrium. We are the miracle, we human beings. Not a break-the-laws-of-physics kind of miracle; a miracle in that it is wondrous and amazing how such complex, aware, creative, caring creatures could have arisen in perfect accordance with those laws. Our lives are finite, unpredictable, and immeasurably precious. Our emergence has brought meaning and mattering into the world.

That's a pretty darn good story too. Demanding in its own way, it may not give us everything we want, but it fits comfortably with what science has taught us about nature. It bequeaths to us the responsibility and opportunity to make life into what we would have it be.

Poetic naturalism offers a rich and rewarding way to apprehend the world, but it's a philosophy that calls for a bit of fortitude, a willingness to discard what isn't working. In the enthusiasm of my first public acknowledgment of my atheism, I tended to embrace the idea that science would eventually solve all of our problems, including answering questions about why we are here and how we should behave. The more I thought about it, the less sanguine I became about such a possibility; science describes the world, but what we're going to do with that knowledge is a different matter.

Facing up to reality can make us feel the need for some existential therapy. We are floating in a purposeless cosmos, confronting the inevitability of death, wondering what any of it means. But we're only adrift if we choose to be. Humanity is graduating into adulthood, leaving behind the comfortable protocols of its childhood upbringing and being forced to fend for itself. It's intimidating and wearying, but the victories are all the more sweet.

Albert Camus, the French existentialist novelist and philosopher, outlined some of his approach to life in his essay "The Myth of Sisyphus." The title refers to the Greek legend that describes a man who was cursed by Zeus to spend eternity pushing a rock up a mountain, only to have it fall back down, where he would have to start pushing it up again. The metaphor for life in a universe without purpose should be clear. But Camus turns the obvious lesson of the myth on its head, making Sisyphus into a hero who creates his own purpose.

> I leave Sisyphus at the foot of the mountain! One always finds one's burden again. But Sisyphus teaches the higher fidelity that negates the gods and raises rocks. He too concludes that all is well. This universe henceforth without a master seems to him neither sterile nor futile. Each atom of that stone, each mineral flake of that night-filled mountain, in itself forms a world. The struggle itself toward the heights is enough to fill a man's heart. One must imagine Sisyphus happy.

I'm not sure whether Sisyphus was actually happy, but I suspect he found meaning in his task, and perhaps took pride in pushing rocks like nobody else. We work with what life gives us.

Earlier in his essay, Camus described the universe as "unintelligible." It's actually the opposite of that—the fact that the universe is so gloriously knowable is perhaps its most remarkable feature. It's one of the aspects of reality that helps make our Sisyphean struggles so ultimately rewarding.

While writing this final chapter of the book, thinking about my late grandmother and going to church and having pancakes, I became hungry. I needed to refill my body's supply of free energy. There were no pancakes available, and certainly no strawberry syrup, so I got up and made one of my grandmother's favorite breakfast recipes, a "bird's nest." A simpler dish could not be imagined: use a shot glass (there was always one nearby in my grandparents' house) to carve out a circular hole from the center of a piece of bread, drop it in a frying pan, and follow with an egg, the yolk nestled snugly into the hole. Salt, pepper, butter, that's it.

Delicious. I love fine dining, and this was not that, but it hit the spot. A fond memory, simple tastes and smells fulfilling a basic need, the uncomplicated pleasure of cooking for yourself. This is *life*—a tiny sliver of the tangible, real experience of the world.

I miss my grandmother, but I don't need to imagine that she's still alive somewhere. She lives on in memories, but eventually even that will pass. Change and passage are part of life—not just a part we reluctantly accept, but its very essence, enabling our hopeful anticipation of what is to come. I care about my remembrances of the past, hopes for the future, the state of the wider world, and the life I have now, with a wife I love more than all of the galaxies in the sky and an abiding joy in puzzling out the nature of reality.

All lives are different, and some face hardships that others will never know. But we all share the same universe, the same laws of nature, and the same fundamental task of creating meaning and of mattering for ourselves and those around us in the brief amount of time we have in the world.

Three billion heartbeats. The clock is ticking.

Appendix: The Equation Underlying You and Me

The world of our everyday experience is based on the Core Theory: a quantum field theory describing the dynamics and interactions of a certain set of matter particles (fermions) and force particles (bosons), including both the standard model of particle physics and Einstein's general theory of relativity (in the weak-gravity regime). Though we don't need it for the rest of the book, in this appendix we're going to very briefly dig into some of the specifics of those fields and interactions in the Core Theory. The discussion will be telegraphically concise, full of buzzwords and jargon and tricky ideas. You can think of this either as extra credit that you are welcome to skip, or a welcome reward for making it this far.

The capstone of our discussion will be a single formula, the Feynman path integral for the Core Theory. It encapsulates all there is to know about the quantum dynamics of this model: starting from one configuration of fields, how probable is it that the fields end up in some other configuration at a later time? If you know that, you can calculate anything you want to about the behavior of the Core Theory. It's worth putting on a T-shirt.

There are two kinds of quantum fields: fermions and bosons. Fermions are the particles of matter; they take up space, which helps explain the solidity of the ground beneath your feet or the chair you are sitting on. Bosons are the force-carrying particles; they can pile on top of one another, giving rise to macroscopic force fields like those of gravity and electromagnetism. Here is the complete list, as far as the Core Theory is concerned:

Fermions

1. Electron, muon, tau (electric charge -1).
2. Electron neutrino, muon neutrino, tau neutrino (neutral).
3. Up quark, charm quark, top quark (charge $+2/3$).
4. Down quark, strange quark, bottom quark (charge $-1/3$).

Bosons

5. Graviton (gravity; spacetime curvature).
6. Photon (electromagnetism).
7. Eight gluons (strong nuclear force).
8. W and Z bosons (weak nuclear force).
9. Higgs boson.

In quantum field theory, it doesn't take that much information to specify the properties of a particular field or, equivalently, the particle with which it is associated. Each particle has a mass, and it also has a "spin." We can think of the particles almost like little spinning tops, except elementary particles (which are really vibrations of quantum fields) don't actually have any size; their spin is an intrinsic property, not the revolution of their bodies around an axis. Every particle associated with a particular field has exactly the same spin; all electrons are "spin $-1/2$," while all gravitons are "spin -2," for example.

How particles interact with one another is governed by their *charges*. When used without modification, the word "charge" is short for "electric charge," but the other forces—gravity and the nuclear forces—also have charges associated with them. The charge of a particle tells us how it interacts with the field that carries the associated force. So electrons, which have electric charge -1, interact directly with photons, which carry the electromagnetic force; neutrinos, which have electric charge 0, don't interact directly with photons at all. (They can interact indirectly, since neutrinos interact with electrons, which then interact with photons.) Photons are neutral themselves, so they don't interact directly with one another.

The gravitational "charge" is just the energy of the particle, which is equal to the mass times the speed of light squared when the particle is at rest. Every single particle has a gravitational charge; Einstein taught us that gravity is universal. All of the fermions we know about have a weak nuclear charge, so they interact with W and Z bosons. Half of the fermions we know about interact with the gluons that carry the strong force, and we call those fermions *quarks*; the other half do not, and we call them *leptons*. There are up-type quarks, with (electric) charge $+2/3$, and down-type quarks, with charge $-1/3$. The strong force is so strong that quarks and gluons are confined inside particles like protons and neutrons, so we never see them directly. The charged leptons are the electron and its heavier cousins, the muon and the tau, and there are three neutrinos associated with them, imaginatively labeled the electron neutrino, the muon neutrino, and the tau neutrino.

Then there is the Higgs field and its associated particle, the Higgs boson. Proposed in the 1960s, the Higgs boson was finally discovered at the Large Hadron Collider in Geneva in 2012. Although it's a boson, we don't usually talk about a "force" associated with the Higgs field—we could, but the Higgs is so massive that the corresponding force is extremely weak and short-range. What makes the Higgs special is that its field has a nonzero value even in empty space. All of the particles of which you are made are constantly swimming in a Higgs bath, and that affects their properties. Most important, it gives mass to the quarks and charged leptons, as well as to the W and Z bosons. Discovering it put the final touches on the Core Theory.

I know what you're thinking. "Sure, all of these fields are colorful and enchanting. But what we really want is an *equation*."

Here you go.

$$\overbrace{}^{\text{quantum mechanics}} \quad \overbrace{}^{\text{spacetime}} \overbrace{}^{\text{gravity}}$$

$$W = \int_{k<\Lambda} [Dg][DA][D\psi][D\Phi]\, \exp\left\{ i \int d^4x\, \sqrt{-g}\left[\frac{m_p^2}{2}R \right.\right.$$

$$\left.\left. -\frac{1}{4}F_{\mu\nu}^{a}F^{a\mu\nu} + i\psi^i\gamma^\mu D_\mu\psi^i + \left(\bar{\psi}_L^i V_{ij}\Phi\psi_R^j + \text{h.c.} \right) - |D_\mu\Phi|^2 - V(\Phi) \right]\right\}$$

$$\underbrace{}_{\text{other forces}} \qquad \underbrace{}_{\text{matter}} \qquad\qquad \underbrace{}_{\text{Higgs}}$$

The essence of the Core Theory—the laws of physics underlying everyday life—expressed in a single equation. This equation is the quantum amplitude for undergoing a transition from one specified field configuration to another, expressed as a sum over all the paths that could possibly connect them.

To be compatible with our earlier discussion of how quantum mechanics works, what I really should give you is the Schrödinger equation for the Core Theory. That's what tells you how the wave function of a given quantum system evolves from one moment of time to the next. But there are many ways of encapsulating that information, and the one shown here is an especially compact and elegant one. (Though it might not appear that way to the naked eye.)

This is what's called the *path-integral formulation of quantum mechanics*, pioneered by Richard Feynman. The wave function describes a superposition of every possible configuration of the system you are working with. For the Core Theory, a configuration is a particular value for every field, at every point in space. Feynman's version of quantum evolution (which is equivalent to Schrödinger's, just written differently) tells you how likely it is that the system will end up in a particular configuration within the wave function, given that it started at some previous time in a different configuration within an earlier wave function. Or you can start with a later wave function and work backward; Feynman's equation, like Schrödinger's, is perfectly reversible in the Laplacian sense. It's only when we start observing things that quantum mechanics violates reversibility.

That's what the quantity W is; it's what we call the "amplitude" to go from one field configuration to another. It's given by a Feynman path integral over all of the ways the fields could evolve in between. An integral, as you may remember if you ever took calculus, is a way of summing up an infinite number of infinitely small things, such as when we add up infinitesimal regions to calculate the area under a curve. Here, we're summing up contributions from each possible thing the fields can do in between the starting and ending points, which we simply call a "path" the field configuration can take.

So what exactly is it that we are integrating, or summing up? For every possible path a system can take, there's a number we calculate called the *action*, traditionally written as S. If the system is jumping willy-nilly all over the place, its action will be very large; if it moves more smoothly, the action will be relatively small. The concept of the action along a path plays an important role even in classical mechanics; among all of the possible paths we can imagine the system taking, the one it actually does take (that is, the one that obeys the classical equations of motion) will be the one that has the *least* action. Every classical theory can be defined by saying what the action for the system is, and then looking for motions that minimize it.

In quantum mechanics the action appears again, but with a twist. Feynman put forward an approach in which we can think of a quantum system as taking *every* path, not just the classically allowed one. To each path we associate a certain *phase factor*, $\exp\{iS\}$. This notation tells us to take a constant called Euler's number, $e = 2.7181\ldots$ and raise it to the power of i, the imaginary number given by the square root of -1, times the action S for the path.

The phase factor $\exp\{iS\}$ is a complex number, with both a real part and an imaginary part. Each will sometimes be positive and sometimes be negative. Summing up all the contributions for all the paths will generally involve a bunch of positive numbers and a bunch of negative numbers, and everything will cancel out, or nearly so, leaving us with a small answer. The exception is when a group of nearby paths have very similar values for the action; then their phase factors will be similar, and adding them up will accumulate rather than canceling out. This happens exactly when the action is near a minimum value, which corresponds to the classically allowed path. So the largest quantum probability gets associated with evolution that looks almost classical. That's why our everyday world is well modeled by classical mechanics; it's classical behavior that gives the largest contributions to the probability of quantum transitions.

We can take our equation apart, piece by piece.

Look at the part of the equation labeled "quantum mechanics." That's where the amplitude is being written as an integral (the $\int$ symbol) over a collection of fields, followed by "$\exp i \ldots$" The fields that we're including are indicated in the notation $[Dg][DA][D\psi][D\Phi]$. The letter D just means "here are the infinitesimal quantities we're going to add up in our integral," and the other symbols stand for the fields themselves. The gravitational field is g; the other bosonic force fields (electromagnetism, strong and weak nuclear forces) are grouped under A; all of the fermions are collectively labeled ψ (Greek letter psi); and the Higgs boson is Φ (Greek letter phi). The notation "exp" means "e to the power $\ldots$"; i is the square root of -1; and everything following i is the action S for the Core Theory. So quantum mechanics enters our expression by saying, "Integrate, over all of the paths that all of the fields can take, a quantity given by raising e to the power of i times the action."

The action itself is where all of the fun is happening. Many professional particle physicists spend a good fraction of their lives writing down different possible actions for different collections of fields. But everyone starts with this one, for the Core Theory.

The action is an integral over all of space, and over the time period in between the initial configuration and the final configuration. That's what the notation $\int d^4x$ is doing; x stands for the coordinates on all the dimensions of spacetime, and the 4 is reminding us that spacetime is four-dimensional. There's an extra factor lurking under the "spacetime" label, which is the square root of something called $-g$. As you might guess from the letter g, this has something to do with gravity, and in particular the fact that spacetime is curved; this piece accounts for the fact that the volume of spacetime (over which we are integrating) is affected by how spacetime is curved.

All of the terms inside the square brackets [] are the different contributions to the action from all of our various fields: both their intrinsic properties and how they are interacting with one another. They fall into the categories of "gravity," "other forces," "matter," and "Higgs."

The "gravity" term is pretty simple; that reflects the pristine elegance of Einstein's theory of general relativity. The quantity R is called the *curvature scalar*, and characterizes how much of a certain kind of spacetime curvature is present at any one point. It's multiplied by a constant, $m_p^2/2$, where m_p is the Planck mass. That's just a funny way of expressing Newton's gravitational constant G, which characterizes the strength of gravity: $m_p^2 = 1/(8\pi G)$. I'm using "natural units," in which both the speed of light and Planck's constant of quantum mechanics are set equal to unity. The curvature scalar R can be calculated from the gravitational field, and the action for general relativity is simply proportional to the integral of R over a region of spacetime. Minimizing that integral gives you Einstein's field equation for gravity.

Next up, we have the term labeled "other forces," which includes two appearances of a quantity F and a few superscripts and subscripts. F is called the *field strength tensor*, and in our notation it includes contributions from electromagnetism and the strong and weak nuclear forces. Essentially the field strength tensor tells us how much the field is twisting and vibrating through spacetime, much as the curvature scalar tells us how much the geometry of spacetime itself is twisting and vibrating. For electromagnetism, the field strength tensor incorporates both the electric and the magnetic field.

Here, and elsewhere in the equation, those superscripts and subscripts label different subquantities, such as which field we're talking about (photon, gluon, W or Z boson), but also which part of the field, such as "the part of the electric field pointing along the x-axis." When you see two quantities (like the two F's in this term) with the same indices on them, that's code for "sum over all of the possibilities." This is a very compact notation, allowing us to hide great complexity in just a few symbols; that's why this one term encompasses the contributions from all the different force fields.

Things get a bit more complicated when we look at the part of our equation labeled "matter." The matter fields are fermions, represented collectively by the letter ψ. As with the bosons, this one symbol includes all of the fermions at once. The first term has two appearances of ψ, one of the Greek letter γ (gamma), and another D. That γ stands for the Dirac matrices, introduced by British physicist Paul Dirac; they play an essential role in how fermions behave, including the fact that fermions generally have antiparticles as well as particles. The D in this case stands for a derivative, or rate of change, of the field. So this term is doing the same thing for the fermions that the previous terms did for the force bosons: it tells us how much the fields are changing through space and time. But there is something hidden in that derivative (the magic of compact notation, again): a *coupling*, or interaction, between the fermions and the force bosons, which depends on how the fermions are charged. The way that an electron interacts with a photon, for example, is characterized by this term in the action.

The term next to it involves a different kind of coupling, between the fermions and the Higgs field Φ. Unlike the rest of the action of the Core Theory, the interaction between the Higgs and fermions is somewhat baroque and unappealing. But there it is: two ψ's and one Φ, telling us that this term encapsulates how fermions and the Higgs field interact with each other. Two things make it complicated. One is that symbol V_{ij}, known as a mixing matrix. It keeps track of the fact that fermions can "mix" with one another—when a top quark decays, for example, it actually decays into a particular mixture of down, strange, and bottom quarks.

The other complication is that you see that one fermion field has a subscript L and the other a subscript R. These stand for "left-handed" and "right-handed" fields. Think of lining up the thumb of your left hand along the direction of motion of a spinning particle. Your other fingers define a possible direction that the spin could be in; if that's what the particle is doing, it's left-handed, while if it's spinning the other way, it's right-handed. The appearances of these subscripts in this term in the Core Theory is an indication that the theory treats left-handedness differently from right-handedness, at least at the subatomic level. That's a remarkable feature, but a necessary one, since nature treats left-handed and right-handed particles differently. That phenomenon, *parity violation*, came as a shock to particle physicists when it was first discovered, but is now seen as simply the sort of thing that can happen when you get these kinds of fields interacting with one another.

The last bit of this term, "h.c.," stands for *hermitian conjugate*. It's a fancy way of saying that the first term is a complex number, but the action needs to be a real number, so we're going to subtract out the imaginary part and be left with a purely real quantity.

Finally we have the part of the action that is devoted to the Higgs field Φ. It's pretty simple; the first part is the "kinetic" term, representing how much the field is changing, and the second is the "potential" term, representing how much energy is locked up in the field even when it's not changing at all. It's that second term that makes the Higgs field special. Like any other field, it wants to be sitting peacefully with the lowest energy it can have; unlike the other known fields, in its minimum-energy state the Higgs field itself does not vanish but has a nonzero value. Its potential energy is higher when the field is zero than it is when the field is not zero. That's what gives the Higgs field a presence even in "empty space," and allows it to affect all the other particles that move through it.

✵

So there we have it: the Core Theory in a nutshell. One equation that tells us the quantum amplitude for the complete set of fields to go from some starting configuration (part of a superposition inside a wave function) to some final configuration.

We know that the Core Theory, and therefore this equation, can't be the final story. There is dark matter in the universe, which doesn't fit comfortably into any of the known fields. Neutrinos have mass, which can be accommodated by the equation we wrote down, but we haven't experimentally verified that the terms we wrote are actually the ones that are responsible for neutrino masses. Moreover, almost every physicist believes there are more particles and fields to be found, at higher masses and energies—but they must be ones that either interact with us very weakly (like dark matter) or decay away very quickly.

The Core Theory doesn't even provide a complete theory of the fields that we know are there. That's the problem, for example, with quantum gravity. The equation we wrote is okay if the gravitational field is very weak, but it doesn't work when gravity becomes strong, such as near the Big Bang or inside a black hole.

That's okay. Indeed, the theory's limitations are built into the formalism. There is one piece of notation in our equation that we haven't yet mentioned: in the very first integral sign, indicating that we'll be summing over all the different field configurations through time, there is a subscript reading $k < \Lambda$. Here k is the *wave number* of a particular mode of a field, and Λ is called the *ultraviolet cutoff*. Remember the viewpoint that Ken Wilson advocated, as we discussed in chapter 24: we can think of every field as a combination of modes, each constituting a vibration with a specific wavelength. The wave number is a way of labeling these modes; larger k corresponds to smaller wavelength, and therefore higher energies. So this notation is limiting the field configurations we include in the path integral to those that "don't vibrate too energetically." That means low-energy, weak-field situations—but still enough to include all of the buzzing and bouncing of the particles and fields that describe the world you see around you every day.

The Core Theory, in other words, is an effective field theory. It has a very specific, well-defined regime of applicability—particles interacting with energies well below the ultraviolet cutoff Λ—and we don't pretend that it's accurate past that. It can describe the gravitational pull of the sun on the Earth, but not what was happening at the Big Bang.

There's a lot going on here, material usually relegated to graduate-level physics courses. This condensed presentation can't reasonably be expected to convey much understanding to anyone who isn't already pretty familiar with the concepts.

But it's useful to see how the Core Theory underlying our everyday lives is extremely precise, rigid, and well defined. There is no ambiguity in it, no room to introduce important new aspects that we simply haven't noticed yet.

As science continues to learn more about the universe, we will keep adding to it, and perhaps we will even find a more comprehensive theory underlying it that doesn't refer to quantum field theory at all. But none of that will change the fact that the Core Theory is an accurate description of nature in its claimed domain. The fact that we have successfully put together such a theory is one of the greatest triumphs of human intellectual history.

References

In this section, references are provided for quotations and the sources of various specific topics in the text. When I thought it might not be obvious, the topic is defined by a word or phrase immediately preceding the reference. The list is organized in chapter sequence but not all chapters have references.

3. The World Moves by Itself

History of momentum: Freely, J. (2010). *Aladdin's Lamp: How Greek Science Came to Europe through the Islamic World*. Vintage Books.

5. Reasons Why

Just World Fallacy: Lerner, M. J., and C. H. Simmons. (1966). "Observer's Reaction to the 'Innocent Victim': Compassion or Rejection?" *Journal of Personality and Social Psychology* 4 (2): 203.

8. Memories and Causes

Russell quote: Russell, B. (1913). "On the Notion of Cause." *Proceedings of the Aristotelian Society* 13: 1–26.

14. Planets of Belief

Dorothy Martin: Tavris, C., and E. Aronson. (2006). *Mistakes Were Made (But Not by Me): Why We Justify Foolish Beliefs, Bad Decisions, and Hurtful Acts*. Houghton Mifflin Harcourt.

15. Accepting Uncertainty

Catechism of the Catholic Church: "Catechism of the Catholic Church—The Transmission of Divine Revelation." Accessed December 10, 2015. http://www.vatican.va/archive/ccc_css/archive/catechism/p1s1c2a2.htm.

16. What Can We Know about the World without Actually Looking at It?

National Academy of Sciences on methodological naturalism: National Academy of Sciences. (1998). *Teaching about Evolution and the Nature of Science.* National Academy Press.

Huxley, A. (1957). *The Doors of Perception.* Chatto & Windus.

Carhart-Harris and Nutt: Halberstadt, A., and M. Geyer. (2012). "Do Psychedelics Expand the Mind by Reducing Brain Activity?" *Scientific American.* Accessed December 10, 2015. http://www.scientificamerican.com/article/do-psychedelics-expand-mind -reducing-brain-activity/.

17. Who Am I?

National Catholic Bioethics Center: "Resources." FAQ. Accessed December 10, 2015. http://www.ncbcenter.org/page.aspx?pid=1287.

18. Abducting God

Nietzsche, F. (1882). *The Gay Science.* Walter Kaufmann, trans. with commentary. (Vintage Books, March 1974).

19. How Much We Know

Newcomb, S. (1888). *Sidereal Messenger* 7, 65.

Michelson, A. A. (1894). Speech delivered at the dedication of the Ryerson Physics Lab, University of Chicago. Quoted in *Annual Register* 1896, 159.

Born, M. (1928). Remarks to visitors to Göttigen University. Quoted by S. W. Hawking. (1988). *A Brief History of Time.* Bantam Books.

Hawking, S. W. (1980). "Is the End in Sight for Theoretical Physics? An Inaugural Lecture." Cambridge University Press.

Hume, D. (1748). *An Enquiry Concerning Human Understanding.* Reprinted by Oxford University Press, 1999.

21. Interpreting Quantum Mechanics

Petersen, A. (1963). "The Philosophy of Niels Bohr." *Bulletin of the Atomic Scientists* 19, no. 7 (September 1963).

22. The Core Theory

Wilczek, F. (2015). *A Beautiful Question: Finding Nature's Deep Design.* Penguin Press.

23. The Stuff of Which We Are Made

Limits on new forces: Long, J. C., et al. (2003). "Upper Limits to Submillimeter-Range Forces from Extra Space-Time Dimensions." *Nature* 421: 922.

25. Why Does the Universe Exist?

Leibniz, G. (1697). "On the Ultimate Origination of Things." Reprinted in *Philosophical Essays* (1989). R. Ariew, trans. D. Garber, ed. Hackett Classics.

Parfit, D. (1998). "Why Anything? Why This?" *London Review of Books* 20, 24.

26. Body and Soul

Princess Elisabeth's correspondence with Descartes: Nye, A. (1999). *The Princess and the Philosopher*. Rowman & Littlefield.

27. Death Is the End

Studies testing out-of-body experiences: Lichfield, G. "The Science of Near-Death Experiences." *The Atlantic*. March 10, 2015. Accessed December 16, 2015. http://www.theatlantic.com/magazine/archive/2015/04/the-science-of-near-death-experiences/386231/.

28. The Universe in a Cup of Coffee

Aaronson, S., et al. (2014). "Quantifying the Rise and Fall of Complexity in Closed Systems: The Coffee Automaton."

29. Light and Life

NASA definition of life: Joyce, G. F. (1995). *The RNA World: Life Before DNA and Protein*. Cambridge University Press.

Schrödinger, E. (1944). *What Is Life?* Cambridge University Press.

30. Funneling Energy

Hoffman, P. (2012). *Life's Ratchet: How Molecular Machines Extract Order from Chaos*. Basic Books.

31. Spontaneous Organization

Schelling, T. C. (1969). "Models of Segregation." *American Economic Review* 59 (2): 488.

Friston, K. (2013). "Life As We Know It." *Journal of the Royal Society Interface* 10: 20130475.

32. The Origin and Purpose of Life

Watson, J. D., and H.F.C. Crick. (1953). "A Structure for Deoxyribose Nucleic Acid." *Nature* 171: 737.

Bartel, D. P., and J. W. Szostak. (1993). "Isolation of New Ribozymes from a Large Pool of Random Sequences." *Science* 261 (5127): 1411.

Lincoln, T. A., and G. F. Joyce. "Self-Sustained Replication of an RNA Enzyme." *Science* 323 (5918): 1229.

Hoyle, F. (1981). "Hoyle on Evolution." *Nature* 294 (5837): 105.

33. Evolution's Bootstraps

Lenski's experiment: Barrick, J. E., et al. (2009). "Genome Evolution and Adaptation in a Long-Term Experiment with Escherichia Coli." *Nature* 461 (7268): 1243.

34. Searching through the Landscape

Evolution as a search strategy: Chastain, E., et al. (2014). "Algorithms, Games, and Evolution." *Proceedings of the National Academy of Sciences* 111 (29): 10620.

Robby the Robot: Mitchell, M. (2009). *Complexity: A Guided Tour.* Oxford University Press.

Reducibly complex mousetraps: McDonald, J. A. (n.d.). "A Reducibly Complex Mousetrap." Accessed December 10, 2015. http://udel.edu/~mcdonald/mousetrap.html.

Fidelibus, A. "Mousetrap Evolution through Natural Selection." Accessed December 10, 2015. http://www.fidelibus.com/mousetrap/.

Dagg, J. L. (2011). "Exploring Mouse Trap History." *Evolution: Education and Outreach* 4: 397.

35. Emergent Purpose

NABT statement and Smith/Plantinga letter: "Science and Religion, Methodology and Humanism | NCSE." Accessed December 10, 2015. http://ncse.com/religion/science-religion-methodology-humanism.

Plantinga, A. (2011). *Where the Conflict Really Lies: Science, Religion, and Naturalism.* Oxford University Press.

36. Are We the Point?

Adams, F. C. (2008). "Stars in Other Universes: Stellar Structure with Different Fundamental Constants." *Journal of Cosmology and Astroparticle Physics* 8: 010.

37. Crawling into Consciousness

MacIver, M. A. (2009). "Neuropathology: From Morphological Computation to Planning." In *The Cambridge Handbook of Situated Cognition.* P. Robbins and M. Aydede, eds. Cambridge University Press.

Becker, E. (1975). *The Denial of Death.* The Free Press.

Kahneman, D. (2011). *Thinking, Fast and Slow.* Farrar, Straus and Giroux.

Eagleman, D. (2011). *Incognito: The Secret Lives of the Brain.* Pantheon.

C. elegans: *Wikipedia.* Accessed December 10, 2015. https://commons.wikimedia.org/wiki/File:Adult_Caenorhabditis_elegans.jpg.

Bridgeman quote: "On the Evolution of Consciousness and Language: Target Article on Consciousness." *Psycoloquy* 3(15). Accessed December 10, 2015. http://www.cogsci.ecs.soton.ac.uk/cgi/psyc/newpsy?3.15.

Imagining and remembering: Schacter, D. L., D. R. Addis, and R. L. Buckner. (2007). "Remembering the Past to Imagine the Future: The Prospective Brain." *Nature Reviews Neuroscience* 8: 657.

Tulving, E. (2005). "Episodic Memory and Autonoesis: Uniquely Human?" In *The Missing Link in Cognition: Origins of Self-Reflective Consciousness.* H. S. Terrace and J. Metcalfe, eds. Oxford University Press.

38. The Babbling Brain

Mouse memories: de Lavilléon, G., et al. (2015). "Explicit Memory Creation during Sleep Demonstrates a Causal Role of Place Cells in Navigation." *Nature Neuroscience* 18: 493.

Anesthetized patients: Casali, A. G., et al. (2013). "A Theoretically Based Index of Consciousness Independent of Sensory Processing and Behavior." *Science Translational Medicine* 198RA105.

Quote from Dante Chialvo: Ouellette, J. (2014). "A Fundamental Theory to Model the Mind." *Quanta Magazine*. Accessed December 10, 2015. https://www.quantamag azine.org/20140403-a-fundamental-theory-to-model-the-mind/.

fMRI image reconstruction: Nishimoto, S., et al. (2011). "Reconstructing Visual Experiences from Brain Activity Evoked by Natural Movies." *Current Biology* 21: 1641.

Capgras delusion: Passer, K. M., and J. K. Warnock. (1991). "Pimozide in the Treatment of Capgras' Syndrome. A Case Report." *Psychosomatics* 32 (4): 446–48.

39. What Thinks?

Heinlein, R. A. (1966). *The Moon Is a Harsh Mistress*. G. P. Putnam's Sons.

Turing, A. (1950). "Computing Machinery and Intelligence." *Mind* LIX (236): 433–60.

Searle, J. (1980). "Minds, Brains, and Programs." *Behavioral and Brain Sciences* 3 (3): 417–57.

Cole. D. (2004). "The Chinese Room Argument." *Stanford Encyclopedia of Philosophy*. Stanford University. Accessed December 10, 2015. http://plato.stanford.edu/entries/chinese-room/.

Removing one neuron: Chalmers, D. (n.d.). "A Computational Foundation for the Study of Cognition." Accessed December 10, 2015. http://consc.net/papers/computation .html.

Dennett, D. C. (1987). *The Intentional Stance*. MIT Press.

Rats: "Rats Dream Path to a Brighter Future." *ScienceDaily*. Accessed December 10, 2015. http://www.sciencedaily.com/releases/2015/06/150626083433.htm.

Leonardo in Breazeal's lab: "Leonardo—Social Cognition." Personal Robots Group. Accessed December 10, 2015. http://robotic.media.mit.edu/portfolio/leonardo -social-cognition/.

40. The Hard Problem

Nagel, T. (2012). *Mind and Cosmos: Why the Materialist Neo-Darwinian Conception of Nature Is Almost Certainly False*. Oxford University Press.

Churchland, P. Quoted in Ouellette, J. (2014). *Me, Myself, and Why: Searching for the Science of Self*. Penguin Books, 256.

Hankins, P. (2015). *The Shadow of Consciousness*.

Jackson, F. (1982). "Epiphenomenal Qualia." *Philosophical Quarterly* 32: 127–36.

Jackson, F. (2003). "Mind and Illusion." In *Minds and Persons*, Anthony O'Hear, ed. Cambridge University Press, 251–71.

41. Zombies and Stories

Chalmers, D. (1996). *The Conscious Mind*. Oxford University Press.

Putnam, H. (1975). *Mind, Language, and Reality: Philosophical Papers* (Vol. 2). Chapter 42. "Are Photons Conscious?" Cambridge University Press.

Chalmers, D. "How Do You Explain Consciousness?" Filmed March 2014. TED Talk 18:37. Posted July 2014. https://www.ted.com/talks/david_chalmers_how_do_you_explain_consciousness.

Fisher, M.P.A. (2015). "Quantum Cognition: The Possibility of Processing with Nuclear Spins in the Brain." *Annals of Physics* 362: 593–602.

Penrose, R. (1989). *The Emperor's New Mind: Concerning Computers, Minds, and the Laws of Physics*. Oxford University Press.

Aaronson, S. (2013). *Quantum Computing Since Democritus*. Cambridge University Press.

43. What Acts on What?

Fodor, J. (1990). "Making Mind Matter More." In *A Theory of Content and Other Essays*. Bradford Book/MIT Press.

44. Freedom to Choose

Libet, B. (1985). "Unconscious Cerebral Initiative and the Role of Conscious Will in Voluntary Action." *The Behavioral and Brain Sciences* 8: 529.

Tumor patient: "Brain Damage, Pedophilia, and the Law—Neuroskeptic." *Neuroskeptic*. November 23, 2009. Accessed December 10, 2015. http://blogs.discovermagazine.com/neuroskeptic/2009/11/23/brain-damage-pedophilia-and-the-law/.

45. Three Billion Heartbeats

Druyan, A. (2003). *Skeptical Inquirer* 27: 6.

West, G. B., W. H. Woodruff, and J. H. Brown. (2002). "Allometric Scaling of Metabolic Rate from Molecules and Mitochondria to Cells and Mammals." *Proceedings of the National Academy of Sciences* 99 (suppl 1): 2473.

46. What Is and What Ought to Be

Hume, D. (2012). *A Treatise of Human Nature*. Courier Corporation.

Feynman, R. P. (1985). *Surely You're Joking, Mr. Feynman! Adventures of a Curious Character*. W. W. Norton & Company.

Searle, J. (1964). "How to Derive 'Ought' from 'Is.'" *The Philosophical Review* 73: 43.

47. Rules and Consequences

Kierkegaard, S. (2013). *Kierkegaard's Writings, VI: Fear and Trembling/Repetition* (Vol. 6). Princeton University Press.

Greene, J. D., et al. (2001). "An fMRI Investigation of Emotional Engagement in Moral Judgment." *Science* 293 (5537): 2105.

Brosnan, S. F., and F.B.M. de Waal. (2003). "Monkeys Reject Unequal Pay." *Nature* 425: 297.

Brosnan, S. F., et al. (2010). "Mechanisms Underlying Responses to Inequitable Outcomes in Chimpanzees, *Pan troglodytes.*" *Animal Behavior* 79: 1229.

Street, S. (2010). "What Is Constructivism in Ethics and Metaethics?" *Philosophy Compass* 5 (5): 363.

48. Constructing Goodness

Wheatley, T., and J. Haidt. (2005). "Hypnotically Induced Disgust Makes Moral Judgments More Severe." *Psychological Science* 16: 780.

Tertullian: "Ante-Nicene Fathers/Volume III/Apologetic/Apology/Chapter XLV."—Wikisource, the Free Online Library. Accessed December 10, 2015. http://en.wikisource.org/wiki/Ante-Nicene_Fathers/Volume_III/Apologetic/Apology/Chapter_XLV.

49. Listening to the World

Barnes, J. (2012). *A History of the World in 10 1/2 Chapters.* Vintage Canada.

Sapolsky, R. M., and L. J. Share. (2004). "A Pacific Culture among Wild Baboons: Its Emergence and Transmission." *PLOS Biology* 2: 0534.

Taylor, S. E., and J. D. Brown. (1988). "Illusion and Well-Being: A Social Psychological Perspective on Mental Health." *Psychological Bulletin* 103 (2): 193.

50. Existential Therapy

Camus, A. (1955). *The Myth of Sisyphus, and Other Essays.* Vintage.

Further Reading

Part One: Cosmos

Adams, F., and G. Laughlin. (1999). *The Five Ages of the Universe: Inside the Physics of Eternity*. Free Press.

Albert, D. Z. (2003). *Time and Chance*. Harvard University Press.

Carroll, S. (2010). *From Eternity to Here: The Quest for the Ultimate Theory of Time*. Dutton.

Feynman, R. P. (1967). *The Character of Physical Law*. MIT Press.

Greene, B. (2004). *The Fabric of the Cosmos: Space, Time, and the Texture of Reality*. A. A. Knopf.

Guth, A. (1997). *The Inflationary Universe: The Quest for a New Theory of Cosmic Origins*. Addison-Wesley Pub.

Hawking, S. W., and L. Mlodinow. (2010). *The Grand Design*. Bantam.

Pearl, J. (2009). *Causality: Models, Reasoning, and Inference*. Cambridge University Press.

Penrose, R. (2005). *The Road to Reality: A Complete Guide to the Laws of the Universe*. A. A. Knopf.

Weinberg, S. (2015). *To Explain the World: The Discovery of Modern Science*. HarperCollins.

Part Two: Understanding

Ariely, D. (2008). *Predictably Irrational: The Hidden Forces That Shape Our Decisions*. HarperCollins.

Dennett, D. C. (2014) *Intuition Pumps and Other Tools for Thinking*. W. W. Norton.

Gillett, C., and B. Lower, eds. (2001). *Physicalism and Its Discontents*. Cambridge University Press.

Kaplan, E. (2014). *Does Santa Exist? A Philosophical Investigation*. Dutton.

Rosenberg, A. (2011). *The Atheist's Guide to Reality: Enjoying Life without Illusions*. W. W. Norton.

Sagan, C. (1995). *The Demon-Haunted World: Science as a Candle in the Dark*. Random House.

Silver, N. (2012). *The Signal and the Noise: Why So Many Predictions Fail—But Some Don't*. Penguin Press.

Tavris, C., and E. Aronson. (2006). *Mistakes Were Made (But Not by Me): Why We Justify Foolish Beliefs, Bad Decisions, and Hurtful Acts*. Houghton Mifflin Harcourt.

Part Three: Essence

Aaronson, S. (2013). *Quantum Computing Since Democritus*. Cambridge University Press.

Carroll, S. (2012). *The Particle at the End of the Universe: How the Hunt for the Higgs Boson Leads Us to the Edge of a New World*. Dutton.

Deutsch, D. (1997). *The Fabric of Reality: The Science of Parallel Universes and Its Implications*. Viking.

Gefter, A. (2014). *Trespassing on Einstein's Lawn: A Father, a Daughter, the Meaning of Nothing, and the Beginning of Everything*. Bantam.

Holt, J. (2012). *Why Does the World Exist? An Existential Detective Story*. Liveright.

Musser, G. (2015). *Spooky Action at a Distance: The Phenomenon That Reimagines Space and Time—and What It Means for Black Holes, the Big Bang, and Theories of Everything*. Scientific American / Farrar, Straus and Giroux.

Randall, L. (2011). *Knocking on Heaven's Door: How Physics and Scientific Thinking Illuminate the Universe and the Modern World*. Ecco.

Wallace, D. (2014). *The Emergent Multiverse: Quantum Theory according to the Everett Interpretation*. Oxford University Press.

Wilczek, F. (2015). *A Beautiful Question: Finding Nature's Deep Design*. Penguin Press.

Part Four: Complexity

Bak, P. (1996). *How Nature Works: The Science of Self-Organized Criticality*. Copernicus.

Cohen, E. (2012). *Cells to Civilizations: The Principles of Change That Shape Life*. Princeton University Press.

Coyne, J. (2009). *Why Evolution Is True*. Viking.

Dawkins, R. (1986). *The Blind Watchmaker: Why the Evidence of Evolution Reveals a Universe without Design*. W. W. Norton.

Dennett, D. C. (1995). *Darwin's Dangerous Idea: Evolution and the Meanings of Life*. Simon & Schuster.

Hidalgo, C. (2015). *Why Information Grows: The Evolution of Order, from Atoms to Economies*. Basic Books.

Hoffman, P. (2012). *Life's Ratchet: How Molecular Machines Extract Order from Chaos*. Basic Books.

Krugman, P. (1996). *The Self-Organizing Economy*. Wiley-Blackwell.

Lane, N. (2015). *The Vital Question: Energy, Evolution, and the Origins of Complex Life*. W. W. Norton.

Mitchell, M. (2009). *Complexity: A Guided Tour*. Oxford University Press.

Pross, A. (2012). *What Is Life? How Chemistry Becomes Biology*. Oxford University Press.

Rutherford, A. (2013). *Creation: How Science Is Reinventing Life Itself.* Current.

Shubin, N. (2008). *Your Inner Fish: A Journey into the 3.5-Billion-Year History of the Human Body.* Pantheon.

Part Five: Thinking

Alter, T., and R. J. Howell. (2009). *A Dialogue on Consciousness.* Oxford University Press.

Chalmers, D. J. (1996). *The Conscious Mind: In Search of a Fundamental Theory.* Oxford University Press.

Churchland, P. S. (2013). *Touching a Nerve: The Self as Brain.* W. W. Norton.

Damasio, A. (2010). *Self Comes to Mind: Constructing the Conscious Brain.* Pantheon.

Dennett, D. C. (1991). *Consciousness Explained.* Little, Brown & Co.

Eagleman, D. (2011). *Incognito: The Secret Lives of the Brain.* Pantheon.

Flanagan, O. (2003). *The Problem of the Soul: Two Visions of Mind and How to Reconcile Them.* Basic Books.

Gazzaniga, M. S. (2011). *Who's In Charge? Free Will and the Science of the Brain.* Ecco.

Hankins, P. (2015). *The Shadow of Consciousness.*

Kahneman, D. (2011). *Thinking, Fast and Slow.* Farrar, Straus and Giroux.

Tononi, G. (2012). *Phi: A Voyage from the Brain to the Soul.* Pantheon.

Part Six: Caring

de Waal, F. (2013). *The Bonobo and the Atheist: In Search of Humanism among the Primates.* W. W. Norton.

Epstein, G. M. (2009). *Good without God: What a Billion Nonreligious People Do Believe.* William Morrow.

Flanagan, O. (2007). *The Really Hard Problem: Meaning in a Material World.* MIT Press.

Gottschall, J. (2012). *The Storytelling Animal: How Stories Make Us Human.* Houghton Mifflin Harcourt.

Greene, J. (2013). *Moral Tribes: Emotion, Reason, and the Gap between Us and Them.* Penguin Press.

Johnson, C. (2014). *A Better Life: 100 Atheists Speak Out on Joy & Meaning in a World without God.* Cosmic Teapot.

Kitcher, P. (2011). *The Ethical Project.* Harvard University Press.

Lehman, J., and Y. Shemmer. (2012). *Constructivism in Practical Philosophy.* Oxford University Press.

May, T. (2015). *A Significant Life: Human Meaning in a Silent Universe.* University of Chicago Press.

Ruti, M. (2014). *The Call of Character: Living a Life Worth Living.* Columbia University Press.

Wilson, E. O. (2014). *The Meaning of Human Existence.* Liveright.

Acknowledgments

It stands to reason that the bigger the picture one addresses, the more input and advice one should seek out. I've been fortunate to be the beneficiary of the wisdom and good judgment of a group of people who have been extraordinarily generous with their time, whether in talking with me directly, answering questions over email, or reading sections of the manuscript and offering gentle suggestions. I'm extremely grateful to Scott Aaronson, David Albert, Dean Buonomano, David Chalmers, Clifford Cheung, Patricia Churchland, Tom Clark, Simon DeDeo, John de Lancie, Daniel Dennett, Owen Flanagan, Rebecca Goldstein, Joshua Greene, Veronique Greenwood, Kevin Hand, Eric Kaplan, Philip Kitcher, Eric Johnson, Richard Lenski, Barry Loewer, Malcolm MacIver, Tim and Vishnya Maudlin, Christina Ochoa, Taryn O'Neill, Laurie Paul, Steven Pinker, David Poeppel, Alex Rosenberg, Michael Russell, Mari Ruti, Chip Sebens, Walter Sinnott-Armstrong, John Skrentny, Sharon Street, Maia Szalavitz, Jack Szostak, Carol Tavris, John Timmer, Zach Weinersmith, Ed Yong, and Carl Zimmer for all their help. I owe them all a substantial debt.

Thanks also to my editor, Stephen Morrow, whom I've been very proud to work with for three projects now. He and my agent, Katinka Matson, played a big role in shaping the book into its current form. Special thanks to Nick and Susan Pritzker for their encouragement during the writing process. I owe a debt to my students and colleagues for their forbearance and understanding while I spent time writing this book rather than helping with our joint research projects. And thanks to the John Simon Guggenheim Foundation for a generous fellowship, and to the Gordon and Betty Moore Foundation, and the Walter Burke Institute for Theoretical Physics for support at Caltech.

Words are not up to the task of expressing my debt to Jennifer Ouellette, the best partner, writing tutor, and support system one could ask for. I'm glad to be sharing my heartbeats with her.

Index

Aaronson, Scott, 231–232, 370
abduction, 40–41, 72
abiogenesis (origin of life)
    amino acids for, 251
    compartmentalization in cells, 252–259
    Darwin on, 250–251
    metabolism-first camp, 260–264
    as phase transition, 274
    probability of, 271–272
    replication-first camp, 264–270
    in RNA world, 267–270
    spontaneous generation, 250
    supernatural means not needed,
        270–272
Abraham and Isaac tale, 403–404,
        405, 415
adenosine diphosphate (ADP), 245, 246
adenosine triphosphate (ATP), 244–247,
        248–249
aether, 111
aging, 54, 59, 60
agreement theorem, 121–122
Ajivika, 32
Albert, David, 58, 92
Alexander VII, Pope, 87
algorithmic complexity, 232
alkaline vents, 262–263
amino acids, appearance on Earth, 251
Anderson, Philip, 106
anthropic principle, 306

antiparticles, 182
antirealism, 167
Aquinas, Thomas, 28
argument from consciousness, the, 338
Aristotle, 24–25, 27–28, 116, 196
Aronson, Elliot, 120, 121
arrow of time. See time's arrow
atheism. See also God
    author's adoption of, 429–430
    Bayesian reasoning about, 145–149
    Church accusations of, 86–87
    story of the universe according to, 431
    values and, 417–418
atoms, 160–162, 174–175
Aumann, Robert, 121
Avicenna (Ibn Sina), 26, 27
awe, 430
axons, 328
Ayala, Francisco, 226

backfire effect, 120–121
Barnes, Julian, 420
Barres, Ben, 140
Bartel, David, 269
Bayes, Thomas, 69
Bayesian networks, 257
Bayesian reasoning, 71–83
    as abduction, 40–41, 72
    about God, 80–82, 145–149
    admitting all evidence in, 82–83

approaching consensus using, 80–81
Aumann's agreement theorem, 121–122
emphasis on credences in, 72–73,
79–80
experimental results and, 73–74
love and, 391
probabilities in, 70
skepticism and, 91
uncertainty in, 126–127
understanding of consciousness
and, 320
updating credences in, 71–72, 75–78,
81–82, 121–122
Bayes's Theorem, 71–72, 126–127
Becker, Ernest, 319
Behe, Michael, 288
being natural, 423–424
beliefs
definition in this book, 70
degrees of belief, 70–71
inconsistent or conflicting, 118
planets of, 116–121
testing, 117–119
biases, 120, 149
Big Bang model
beginning of time in, 197, 199
Big Bang vs., 50–51
described, 51
microwave background radiation
in, 50
superstring theory for, 175–176
time's arrow and, 43–44, 54, 58
bilayers, 255, 256
Bill & Ted's Excellent Adventure
(film), 402
"bird's nest" breakfast, 432–433
Black Cloud, The, 305
black holes, 52, 129, 175–176
Bohr, Niels, 167
Boltzmann Brain scenario, 89–90,
91–92, 272
Boltzmann, Ludwig, 56–57, 88, 227,
271, 376
Book of Revelation, 428
Born, Max, 155, 168
Born rule, 168
Boscovich, Roger, 32
bosons, 174

Bostrom, Nick, 92
Boy Who Came Back from Heaven, The,
219–220
brain. See also consciousness; mind-body
dualism
author's scan, 327, 328–329
Capgras delusion, 334
changes over time, 331
connectome of neurons in, 329–333, 343
damage and mental illness, 334–335
memory and, 324, 331
neurons and glial cells in, 327–328
of octopuses, 325–326
quantum processes in, 368–371
as radio receiver, 334
replacing neurons with machines, 342
simulating with China's population,
342–343
Bridgeman, Bruce, 322
Brief History of Time, A, 199
Brosnan, Sarah, 409
Brown, Jonathon, 426
Brown, Robert, 248
Brownian forces, 248
Bruno, Giordano, 48
Buridan, Jean, 27

C. elegans nematode connectome, 330
caloric, 219
Calvin, John, 135
Camus, Albert, 432
Capgras delusion, 334
carbon-dioxide, hydrogenation of, 260,
261–262
Carhart-Harris, Robert, 137
caring. See also morality and ethics
changes over time, 392–393
desire as an aspect of, 392
by God, 204
as source of mattering, 422
Carroll, Sean
"bird's nest" cooked by, 432–433
brain scan of, 327, 328–329
brush with death by, 1
churchgoing experience of,
428–429, 432
curiosity about psychic powers,
153–154, 155

Carroll, Sean (*cont.*)
    questions about the universe asked
        by, 195
    as token atheist at conference, 417–418
    transcendent experience of, 130
    transition to atheism by, 428–430
catalysis, 262
Catechism of the Catholic Church, 128
Catholic Catechism, 128
causality
    Aristotle's four causes, 25–26
    concerns about lack of, 372
    de Berk trial and, 39
    downward causation, 375–377
    as emergent, 44, 375
    not in fundamental level of reality,
        29, 54
    not in laws of physics, 63
    Principle of Sufficient Reason, 40, 41, 42
    "reasons why" as "causes," 42–43
    responsibility and, 63–64
    scientific shift away from, 28–29
    time's arrow and, 43–44, 59
    universe's existence and, 197, 199,
        201–202
    as useful invention, 64–66
    "Why?" questions, 42–46
cells
    as basic units of life, 247
    chaos harnessed by, 247–248
    compartmentalization of, 251, 252–259
    human beings as configurations of,
        341–342
    membranes as Markov blankets,
        257–259
Chalmers, David, 350, 355, 357, 363, 364, 366
chaos theory, 35
chemiosmotic process, 261
Chialvo, Dante, 333
Chinese Room thought experiment,
    338–341
choice. *See* free will
chronesthesia, 323–324
coarse-graining, 97–99
Cochrane, Kent, 324–325
coffee-and-cream system, 228–230, 231–234
cognitive biases, 120, 149
cognitive dissonance, 119–120

coherentism, 117
color
    experience of, 349–350, 358, 360–361
    Mary the Color Scientist, 351–354
compartmentalization, 252–259
    cell membranes as Markov blankets,
        257–259
    Schelling segregation model, 252–254
    spontaneous, in lipids, 254–257
    as ubiquitous feature of life, 251
compatibilism, 379
complexity. *See also* simplicity
    algorithmic or Kolmogorov, 232
    cream-and-coffee example, 228–230,
        231–234
    derivation of emergence and, 100
    of emergent levels of reality, 236
    entropy and the development of,
        227–235
    fractals and, 234
    of human beings, 32
    irreducible, 288–290
    phase transitions and, 102
    purpose and, 235
Comte, Auguste, 104, 105
confirmation bias, 120
connectome
    C. elegans nematode, 330
    changes over time, 331
    consciousness and the mapping of, 343
    critical point of, 333
    described, 329–330
    as hierarchical network, 331–333
    human, 329–330
    as small-world network, 332–333
consciousness. *See also* brain; memory;
        mind-body dualism
    argument from, 338
    artificial intelligence, 337–338
    awareness as key to, 322
    behavior over time and, 344
    brain structure and, 336–337
    burden of self-consciousness, 319
    Cartesian theater of, 320
    Chinese Room thought experiment,
        338–341
    as collection of phenomena, 349
    connectome mapping and, 343

dictionary definition of, 322
doubt of naturalism and, 319–320
dual-process theory, 321–322, 406
Easy Problems of, 350
Hard Problem of, 350, 355
inner experience of, 344–345, 349–350, 358–362
as invented concept, 343
Mary the Color Scientist, 351–354
MEG detection of thinking, 329
move from sea to land and, 319
multiple processes in, 320–321
multiple realizability of, 343–344
not a factor in quantum mechanics, 166, 366–367
"now" of, 323
pervading the universe, 363–366
as phase transition, 348, 363
phenomenal, 349–350
philosophical zombies and, 355–358
in photons, 364–365
physicalism and, 5
planning capacity of, 322–323
reality and illusion and, 110, 111–112
replacing neurons with machines and, 342
strong emergence and, 110
understanding and, 340–341
consequentialism, 406, 407, 414
conservation of energy, 200
conservation of information, 33–34, 61
conservation of momentum, 25, 27, 93
constructivism, 410–411, 415–417
Copernicus, Nicolaus, 19
Core Theory, 172–177
completeness of, 178–183
described, 176
energy of life and, 249
future progress and, 193–194
life after death and, 218–221
longterm accuracy of, 157–158
mind-body dualism and, 212, 215–216
missing answers in, 189
other branches of physics and, 189–190
rigidity of, 212, 215
scientific confidence in, 4, 177
correlated information, 297
cosmological argument for God, 199

cosmology, 44, 48–49, 199
cosmos. *See* universe
Craig, William Lane, 199
cream-and-coffee system, 228–230, 231–234
creationism, 118, 133
Crick, Francis, 268
Crick, James, 265
crossing symmetry, 181–182
crystals, 264, 265

Dagg, Joachim, 289
Dalton, John, 12
dark matter, 183
Darwin, Charles, 12, 226, 250–251, 278
de Berk, Lucia, 38–39
de Waal, Frans, 409
death
    author's brush with, 1
    impossibility of living forever, 2
    life after, 218–221, 387–388
    near-death experiences, 219–220
    Sagan's, 387–388
decoherence, 169
deduction, 40
Defense of Marriage Act, 139
dendrites, 328
Dennet, Daniel, 320, 345
deontology, 406, 407, 410–411, 413–414
deoxyribonucleic acid (DNA), 265–268, 280
Descartes, René, 27, 84–87, 90–91, 116, 126, 205, 207–208, 209, 210, 211, 213–214, 319, 333
desire, 392, 421
destiny, determinism vs., 36
determinism, 32–33, 34–36
Devinsky, Orrin, 383
diachronic meaning, 426
Diagnostic and Statistical Manual, 141
Dialogues Concerning Natural Religion, 226
Discourse on Method, 86
disequilibrium, 262
Dobzhansky, Theodosius, 226
domains of applicability, 21, 96–97, 103–104
Doors of Perception, The, 136
down quarks, 174, 175

downward causation, 375–377
Dragon's Egg, 305
drug addiction, 384
Druyan, Ann, 387–388
dual-process theory, 321–322, 406

E. *coli* evolution experiment, 273–274,
　　276–277
Eagleman, David, 322
Earth
　in ancient vs. modern cosmology, 48–49
　center-of-mass motion of, 187
Easy Problems of consciousness, 350
effective field theory, 186–192
Einstein, Albert, 12, 49, 79, 111,
　　124–126, 180
ekinology, 54
élan vital (life force), 219
electromagnetism, 173, 174
electron transport chain, 246
electrons, 160–162, 177
eliminativism, 19, 20, 110
Elisabeth of Bohemia, 205–207, 208–211,
　　212, 213–214, 333
emergent levels of reality. *See also*
　　fundamental level of reality
　as autonomous, 97, 106–107
　causality as, 44
　coarse-graining for, 97–99
　complexity and derivation of, 100
　complexity of, 236
　defined, 94
　as derived, 17, 99–100
　domain of applicability for, 21, 96–97
　effective field theory in, 186–192
　entanglement and, 100
　hierarchy of sciences and, 104
　molecules/fluid example, 95–96, 97, 98,
　　99–101
　as multiply realizable, 99
　newness and difference of, 105–110
　in poetic naturalism, 112
　questions raised by, 107–108
　reality and illusion and, 110–111
　strong emergence, 104, 108–110, 375
　strong reductionism and, 110
　ubiquitousness of, 95
　vocabularies different for, 96, 113–114

Emerson, Keith, 429
empiricism, 133–134
energy
　ATP as life's battery, 244–247, 248–249
　conservation of, 200
　free, 240–242, 245, 274, 275
　negative, 201
　of photons, 242–243
　of space (vacuum energy), 303–304,
　　305, 311
　zero amount in universe, 200–201
*Enquiry Concerning Human
　　Understanding*, 157
entanglement, 100, 168, 169
entropy
　Boltzmann's definition of, 57–58
　complexity's development and,
　　227–235
　cream-and-coffee example, 228–230,
　　231–234
　described, 227–228
　free energy as low entropy, 240
　human experience of time and, 60–61
　information and, 297–298
　second law of thermodynamics and,
　　57, 59
　time's arrow and, 43–44, 57–59, 60–61
Epicurus, 415
epigenetic phenomena, 277
episodic memory, 323–325
epistemology, 10
equilibrium, 58
ethics. *See* morality and ethics
Everett, Hugh, III, 167–168, 169, 170
evil, 146–147
evolution. *See also* natural selection;
　　replication with variation
　definition of life and, 238
　as design without a designer, 226
　E. coli experiment, 273–274, 276–277
　of eyes, 289–290
　of giraffes, 291–292
　information and, 293, 295–298
　irreducible complexity and, 288–290
　modern synthesis theory, 276
　of mousetraps, 288–289
　move from sea to land, 317–319
　new kinds of things and, 293–294

Omphalos hypothesis about, 87–88
  purpose and, 293, 294–295
  spandrels in, 277
  of *Tiktaalik roseae*, 317–318
  as unsupervised and impersonal,
    298–299
existence, questioning, 84–87, 195–204
existential therapy, 3, 428–433
eyes, evolution of, 289–290

fairness, 409
faith, 127–128
Faraday, Michael, 33
fate, determinism vs., 36
fatty acids, 255, 256–257
Fear and Trembling, 403
fermions, 174
Feynman diagrams, 181–182
Feynman, Richard, 159, 398
Fidelibus, Alex, 289
field theory. *See* quantum field theory
fine-tuning argument for God, 302–313
  anthropic principle and, 306
  arrogance of, 302
  general statement of, 302
  multiverse and, 306–309
  neutron mass and, 304–305
  "old evidence" in, 303
  probabilities unclear for, 303–305
  theism and current life, 309–310
  theism and current universe, 310–313
  vacuum energy and, 303–304, 305, 311
Fisher, Matthew, 368
fitness landscape, 281–283, 286–287
Fodor, Jerry, 372, 373
footbridge problem, 407–408
forces, 172, 174, 184–185
Forward, Robert, 305
foundationalism, 115–116
fractals, 234
free energy, 240–242, 245, 274, 275
free will, 378–384
  ability to have acted differently,
    380–381
  compatibilism and, 379, 382–383
  drug addiction and, 384
  Klüver-Bucy syndrome and, 383–384
  libertarian freedom, 381–382

mixing vocabularies and, 378–379,
    381–382
  responsibility and, 383–384
  time's arrow and, 380
frequentists, 70
Friston, Karl, 257, 258, 259
Früh-Green Gretchen, 263
fundamental level of reality. *See also*
    emergent levels of reality
  causes absent from, 29, 54
  Core Theory and, 189–190
  sparse ontology of, 19
  stories about, 20–21

Galileo Galilei, 12, 23–24, 26, 196
gender and sexuality, 139–142
general theory of relativity, 49, 50,
    124–126, 175
genetic algorithms, 283–286
genetic drift, 276, 282
genetic heredity, 276–277. *See also* natural
    selection
giraffe evolution, 291–292
glial cells, 327–328
gluons, 174, 175, 188
goals of this book, 3
God
  ability to sense directly, 135–136
  Abraham and Isaac tale, 403–404,
    405, 415
  attempted proofs of, 28, 86, 196, 199
  Bayesian reasoning about, 80–82,
    145–149
  caring, 204
  challenges for thinking about, 145
  definitions of, 148
  existence of universe and, 196, 202,
    203–204
  fine-tuning argument for, 302–313
  of the gaps, 212
  Intelligent Design and, 133, 225–226
  Laplace's lack of need for, 31, 197
  as necessary being, 196, 203
  Newton's belief in, 30
  Nietzsche on, 144
  problem of evil and, 146–147
  quantum indeterminacy and, 299–301
  varying definitions of, 145

Gödel, Kurt, 369
Gödel's Incompleteness Theorem, 369, 370
Gosse, Philip Henry, 87–88
Gould, Stephen Jay, 277
gravitons, 174, 183–184
gravity, 31, 173, 183–184, 201
Greene, Joshua, 406, 407
Guth, Alan, 308

Haeckel, Ernst, 32
Haidt, Jonathan, 414
happiness, overemphasis on, 425–426
Hard Problem of consciousness, 350, 355
Hartle, James, 199
Hawking, Stephen, 52, 111, 155, 199
heartbeats per lifetime, 389
Heinlein, Robert A., 336, 338
hierarchy of sciences, 104
Higgs boson, 176
Higgs field, 176
History of the World in 10 1/2 Chapters, 420
horizontal gene transfer, 277
Hoyle, Fred, 271–272, 305
Hubble, Edwin, 50
human beings
    complexity of, 32
    as configurations of cells, 341–342
    entropy and experience of time by, 60–61
    free will in, 378–384
    heartbeats per lifetime, 389
    impossibility of living forever, 2
    made of particles and fields, 186
    multiple ways of describing,
        372–373
    physics and understanding of, 179
    as source of mattering, 421–422
    Star Trek transporter and, 15–16
    strong emergence and, 104, 109–110
    vastness of universe and, 1–2, 53
humanism, 11
Hume, David, 41, 157, 226, 394–395, 397,
    410, 411
Huxley, Aldous, 136–137
Huxley, Thomas Henry, 275
Huygens, Christiaan, 27
hydrogenation of carbon-dioxide, 260,
    261–262
hydrothermal vents, 262–263

Ibn Sina (Avicenna), 26, 27
identity
    gender, 139–140
    thought experiments about, 15–16, 17
illusion. *See* reality and illusion
induction, 41
inflation, 308–309
information
    conservation of, 33–34, 61
    definitions of, 34
    entropy and, 297–298
    evolution and, 293, 295–298
    genetic, DNA copying mechanism for,
        266–267
    having vs. understanding, 340–341
    microscopic vs. macroscopic, 297
    replication of, origin of life and,
        264–265, 274–275
    inheritance, 276–277. *See also* natural
        selection
Inside Out (film), 321
instrumental rationality, 401–402
Intelligent Design, 133, 225–226, 288, 289
intentional stance, 345, 372–373
irreducible complexity, 288–290
Isaac and Abraham tale, 403–404, 405, 415
It's a Wonderful Life (film), 393

Jackson, Frank, 351, 354
Johnson, Chris, 393
Joyce, Gerald, 269
Just World Fallacy, 39

Kahneman, Daniel, 321
Kandel, Eric, 331
Kant, Immanuel, 226, 271, 404, 410–411
Kelley, Deborah, 263
Kepler, Johannes, 19
Kierkegaard, Søren, 403, 404
Klüver-Bucy syndrome, 383–384
knowledge
    Boltzmann Brain scenario and, 89–90
    brain changes due to learning, 331
    cosmology connected to, 44
    degrees of reliability of, 69–70
    empirical vs. logical, 84
    faith and, 127–128
    foundationalism, 115–116

by persons, 353
planets-of-belief metaphor for, 116–117
proof and, 123–126
uncertainty about, 126–129
knowledge argument, 351–354
Koch, Christof, 364
Kolmogorov complexity, 232
Kuhn, Thomas, 102–103

Laplace, Pierre-Simon, 30–33, 69, 173, 191,
     197
Laplace's Demon, 33, 34, 35, 380
Large Hadron Collider, 73, 106, 124, 129, 176
Last Thursdayism, 88
laws of physics. *See also* fundamental level
     of reality
  causes absent from, 63
  as completely known for everyday life,
     178–183, 191–192
  life and, 311–312
  loopholes in completeness of, 192–193
  psychic powers ruled out by, 154–155,
     156, 157, 216–218
  as "reasons why," 42–43
  time's arrow not in, 44, 55–56, 60
Leibniz, Gottfried Wilhelm, 32, 40, 41,
     135, 195
Lemaître, Georges, 50
Lenski, Richard, 273, 275
Leonardo (robot puppet), 346–347
Lerner, Melvin, 39
Lewontin, Richard, 277
libertarian freedom, 381–382
Libet, Benjamin, 382
life. *See also* abiogenesis (origin of life)
  ATP as battery of, 244–247, 248–249
  biological systems vs. machines, 247
  cell as basic unit of, 247
  centrality of, 312
  compartmentalization in, 251, 252–259
  after death, 218–221, 387–388
  definition elusive for, 238
  extraterrestrial, 237–238
  fine-tuning argument for God, 302–313
  as finite, 236
  free energy needed for, 241, 242
  heartbeats per lifetime, 389
  metabolism in, 251, 260–264

NASA definition of, 238
not forever, 420–421
phase transitions of, 348
as process, 2, 218–219
purpose of, 260
replication with variation in, 252
Schrödinger's definition of, 239, 241, 249
second law of thermodynamics and,
     239–240, 241
vitalism theory of, 219, 250
lifespan, scaling laws of, 388–389
Lincoln, Tracey, 269
lipids, self-organization in, 254–257
listening, importance of, 423
Locke, John, 379
logical syllogisms, 396–398, 400
Lost City hydrothermal vent, 263
love, Bayesian reasoning and, 391
Lowell, Percival, 237
Lucretius, 88, 201
Luther, Martin, 403

Malarkey, Alex, 219–220
Malenbranche, Nicolas, 211–212
manifest image, 20
Many-Worlds Interpretation, 167–168,
     169–170
Markov blankets, 257–259
marriage, 139–140
Mars, 237–238
Mary the Color Scientist, 351–354
mathematical proof, 123–124, 131–132
Maxwell, James Clerk, 12, 33, 173
McDonald, John, 289
meaning
  concerns about creating, 390–393
  diachronic and synchronic, 426
  erosion of confidence in, 145
  not a scientific endeavor, 389
  not in Core Theory, 389
  in poetic naturalism, 389–390
  reconciling with naturalism, 5
  of right and wrong (meta-ethics),
     405–406
  starting place for, 391–392
measurement or observation
  in field theory, 173–174
  in Many-Worlds Interpretation, 169–170

measurement or observation (*cont.*)
  measurement problem, 35–36, 166–167
  predicted values and, 165
  in quantum mechanics, 162–163, 165,
    173, 370
  superposition and, 163
Meditations on First Philosophy, 85,
  207, 208
Meletus, 87
memory
  brain and, 324, 331
  development in children, 325
  memories as useful inventions, 65–66
  reliability of, 88
  required to pass the Turing test, 345
  semantic vs. episodic, 323–325
  state of the universe and, 62
  time's arrow and, 61–62
Mendel, Gregor, 276
messenger RNA, 267
metabolism, 251, 260–264, 270
meta-ethics, 405–406. *See also* morality
  and ethics
methodological empiricism, 133–134
methodological naturalism, 133
micelles, 255
Michelson, Albert, 155
Milgram, Stanley, 332
Milky Way, 49–50
Miller, Stanley, 251
mind. *See* consciousness
mind-body dualism. *See also* psychic
    powers
  appearance of consciousness and,
    348–349
  brain findings casting doubt on,
    333–335
  causal closure of the physical and, 374
  Core Theory and, 212, 215–216
  Descartes's defense of, 210, 211
  Descartes's formulation of, 207–208
  Elisabeth of Bohemia's objections to,
    205, 208–211, 212, 213–214, 333
  pineal gland as seat of soul, 211
  property dualism, 213, 355–356
  Ryle on, 213
  as substance dualism, 208, 355
Mitchell, Melanie, 283, 285

Mitchell, Peter, 247, 261
mitochondria, 246
Mlodinow, Leonard, 111
modal reasoning, 65
Mohan, Varun, 231, 232
Moon Is a Harsh Mistress, The, 336, 338
morality and ethics
  consequentialism, 406, 407, 414
  constructivism, 410–411, 415–417
  deontology, 406, 407, 410–411,
    413–414
  ethics vs. meta-ethics, 405–406
  fairness, 409
  footbridge problem, 407–408
  of most people, 417
  nascent sensibility and, 409–410
  not a scientific endeavor, 400–402
  "ought" in systems of, 394–402
  relativism, 410
  Ten Commandments, 419–420
  Ten Considerations, 420–427
  trolley problem, 404–405, 406
  utilitarianism, 412–413
  virtue ethics, 408
Morgenbesser, Sidney, 195
motion
  Aristotle's conception of, 26, 27–28
  conservation of momentum, 25, 27, 93
  contributors to understanding of, 27
  Galileo's experiments on, 23–24, 26
  God as unmoved mover, 28, 196
  Newton's science of, 27
mousetrap evolution, 288–289
Moyle, Jennifer, 247, 261
multiple realizability, 99, 188, 343–344
multiplicative weight updates
    algorithm, 280
multiverse idea, 306–309
Myth of Sisyphus, The, 432

Nagel, Thomas, 348
National Association of Biology
    Teachers, 298
natural selection. *See also* evolution;
    replication with variation
  fitness landscape in, 281–283, 286–287
  genetic drift in, 276, 282
  genetic heredity and, 277

horizontal gene transfer in, 277
in modern synthesis theory, 276
replication with variation and, 275–277
as search algorithm, 279–286
sexual, 291, 292
Natural Theology, 289
naturalism. *See also* poetic naturalism
as atheistic ontology, 11
author's transition to, 428–429
consciousness and doubt of, 319–320
defined, 3–4
historical pedigree of, 11
methodological, 133
"ought" in systems of morality and,
395–396
philosophical difficulties of, 14
physicalism vs., 355
reconciling meaning and values with, 5
scientific acceptance of, 133–134
skepticism about, 13–14
three principles of, 20
near-death experiences, 219–220
nematode connectome, 330
neurons
connectome of, 329–333, 343
overview, 327–328
replacing with machines, 342
neutrinos, 176–177
new physics, 189–190, 193
Newcomb, Simon, 154
Newton, Isaac, 12, 19, 27, 30, 191, 212
Nietzsche, Friedrich, 144
"now" of conscious perception, 323
Nozick, Robert, 413
nucleotides, 265
Nutt, David, 137

observation. *See* measurement or
observation
octopuses, 325–326
Omphalos, 87–88
On the Origin of Species…, 226, 251, 275, 278
"On the Ultimate Origin of Things," 195
"Only Way, The" (song), 429
ontology
approaches to, 10–11
defined, 10
epistemology compared to, 10

importance of categories in, 16–17
naturalism as, 11
planets of belief compared to, 117
pluralistic, 11–12
rich vs. sparse, 17–19
simplification of, 12–13
Orgel, Leslie, 268
Ouellette, Lauren, 231
"ought" in systems of morality, 394–402
out-of-body experiences, 220

pain, inner experience of, 361–362
Paley, William, 225, 289
panpsychism, 363–366
paradigm shift, 102–103
paranormal abilities. *See* psychic powers
Parfit, Derek, 195, 412
Parmenides, 201
particles. *See also specific kinds*
arising out of fields, 172, 173
as completely known for everyday
life, 182
crossing symmetry and, 181–182
in quantum fields, 174
virtual, 180
Passions of the Soul, The, 214
Past Hypothesis, 58, 62, 64, 227, 235, 380
Payne, Gaposchkin, Cecilia, 12
Pearl, Judea, 257
Penrose, Roger, 369
Penzias, Arno, 50
persons
concept of, 352–353
knowledge held by, 353
ways of talking about, 220–221
Phaedrus, 322
phase transitions
appearance of consciousness as, 348, 363
complexity and, 102
critical point of, 333
defined, 101
in the history of the cosmos, 102
in ideas (paradigm shift), 102–103
of life on Earth, 348
origin of life as, 274
water example, 101–102
phenomenal consciousness, 349–350
Philoponus, John, 26

philosophical zombies, 338, 355–358
phlogiston, 219
phospholipids, 255, 256
photons, 174, 242–243, 364–365
photosynthesis, 245–246, 368
physicalism, 5, 355
physics. *See also* laws of physics
    fitness landscape in, 282
    simplicity of, 235
    simplification in, 25
pineal gland as seat of soul, 211
planets of belief
    defense mechanisms for, 119–121
    described, 116–117
    testing, 117–119
Plantinga, Alvin, 136, 298
Plato, 322
Poeppel, David, 327, 328
poetic naturalism. *See also* naturalism
    defined, 3–4
    experience of redness in, 360–361
    fortitude required by, 431–432
    identity concepts and, 142–143
    levels of reality in, 112
    meaning in, 389–390
    "poetic" part of, 94
    purpose in, 389–390
    values in, 14, 21
    ways of talking in, 19, 20–21
positive illusions, 426–427
Priestley, Joseph, 358
prime numbers, 124
Principle of Sufficient Reason, 40, 41, 42
Principle of the Best, 41–42
Principles of Philosophy, 86
probabilities
    of abiogenesis, 271–272
    for fine-tuning argument for God,
        303–305
    frequentists vs. Bayesians on, 70
    in Many-Worlds Interpretation, 169
    in quantum mechanics, 163
proof, 123–126, 129, 131–132
property dualism, 213, 355–358, 364–365
proton-motive force, 247, 261
psychic powers
    author's childhood curiosity, 153–154, 155
    defined, 154

dismissal of, 158
    ruled out by laws of physics, 154–155, 156,
        157, 216–218
    spoon bending, 155, 156, 192, 216–218
psychoactive drugs, 131, 136–137
purpose
    biological systems vs. machines and, 247
    complexity and, 235
    erosion of confidence in, 9–10, 145
    evolution and, 293, 294–295
    of life, 260
    not in Core Theory, 389
    not inherent in universe, 220
    in poetic naturalism, 389–390
Purpose-Driven Life, The, 390
Putnam, Hilary, 361

qualia
    desire to account for, 350
    as illusions, 361
    inner experience of consciousness,
        344–345, 349–350, 358–362
    lacking in philosophical zombies, 338,
        355–357
    as stories we tell, 359–361
quantum field theory
    bosons and fermions in, 174
    completeness of, 178–183
    crossing symmetry in, 181–182
    effective field theory, 186–192
    field, defined, 33, 172–173
    of gravity, 31, 173
    importance of, 33
    loopholes in, 192–193
    measurement or observation in, 173–174
    as multiply realizable, 188
    particles and, 172, 173
    power of, 178
    restrictive nature of, 178, 186
quantum mechanics. *See also* quantum
        field theory
    in brain processes, 368–371
    consciousness not a factor in, 166,
        366–367
    indeterminacy and God, 299–301
    location uncertainty in, 164
    measurement or observation in, 162–163,
        165, 173, 370

measurement problem of, 35–36, 166–167

mysteriousness of, 159–160

predictive accuracy of, 165

probabilities in, 163

restrictive nature of, 186

Schrödinger equation, 164 165, 198

time in, 198

variety of models of, 172, 193

wave function collapse in, 35 36, 165

quarks, 174, 175, 188

Randi, James, 154

rationalism, 134–135

reality and illusion. *See also* emergent levels
of reality; fundamental level of reality

consciousness and, 111–112

emergence and, 110–111

gender and, 141–142

positive illusions, 426–427

qualia and, 361

species dysphoria and, 142–143

red. *See* color

regulatory RNA, 268

relativism, 410

religion. *See also* atheism; God

gender and sexuality and, 141

scope of, 10–11

transcendent experience in, 131,
135–136

replication with variation. *See also*
evolution

in abiogenesis, 264–270, 274–275

in crystals, 264, 265

DNA and RNA for, 265–270

natural selection and, 275–278

as ubiquitous feature of life, 252

responsibility, 63–64, 383

Rhine, J. B., 154

ribosomes, 267, 268

ribozymes, 268

Rich, Alexander, 268

rich ontologies, 17–19

RNA, 266–267, 268

RNA world, 267 270

Robby the Robot, 283–286

robot puppet (Leonardo), 346–347

Rukeyser, Muriel, 19

Russell, Bertrand, 63, 84, 88

Russell, Michael, 260, 262, 263, 270

Rutherford, Ernest, 105

Ryle, Gilbert, 213

Sagan, Carl, 387–388, 393

scaling laws, 388–389

Schelling, Thomas, 252, 254

Schiaparelli, Giovanni, 237

Schrödinger equation, 164–165, 198

Schrödinger, Erwin, 164, 239, 265

scientific image, 20

scientific revolutions, 159

Scott, David, 23

search algorithms

defined, 279

fitness landscape for, 281–283, 286–287

genetic, 283–286

natural selection as, 279–286

Searle, John, 338, 339, 340, 341, 342,
398, 399

second law of thermodynamics, 57, 59,
240, 241

segregation model, 252–254

self-consciousness, burden of, 319

self-organization

in lipids, 254–257

Schelling segregation model, 252–254

ubiquitousness of, 252

self-serving bias, 120

Sellars, Wilfrid, 20

semantic memory, 323, 324

sexual selection, 291, 292

sexuality and gender, 139–142

Ship of Theseus thought experiment, 16,
17, 20

Simmons, Carolyn, 39

simplicity. *See also* complexity

as gauge of truth, 79–80

of physics, 235

simplification in physics, 25

ultimate, 12–13

simulation argument, 92

Sisyphus myth, 432

skepticism

about naturalism, 13–14

about laws of physics' completeness,
178–179, 191–192

Descartes's practice of, 84–87, 90–91

skepticism (*cont.*)
  Greek school of, 84
  radical, not useful, 91
  simulation argument, 92
small-world network, 332–333
Smith, Huston, 298
social robotics, 346–347
Socrates, 87
spacetime, 49, 197
sparse ontologies, 17–19
special relativity theory, 180
species dysphoria, 142–143
Spinoza, Baruch, 40
spontaneous generation of life, 250
spoon bending, 155, 156, 192, 216–218
standard model of particle physics, 175
Stanford Encyclopedia of Philosophy, 202
Star Trek transporter, 15–16, 20
Starry Night, The, 94, 95
state of the universe
  defined, 33
  memory and, 62
  next moment determined by, 32–33,
    36, 93
    predicting human behavior and, 36–37
    in quantum mechanics, 197
"Statement on Teaching Evolution," 298–299
stillness, impossibility for the
  living, 421
Street, Sharon, 410, 416
string theory, 307–308, 309
strong emergence, 104, 108–110, 375
strong nuclear force, 174
strong reductionism, 110
substance dualism. *See* mind-body dualism
substrate independence, 342
sun, free energy from, 240, 242, 245
superposition, 163
superstring theory, 175–176
supervention, 190
syllogisms, 396–398, 400
synapses, 328, 331
synchronic meaning, 426
synthesis theory, 276
System 1 and 2 modes of thought,
  321–322, 406
Szent-Györgyi, Albert, 263
Szostak, Jack, 269

Tavris, Carol, 120, 121
Taylor, Shelley, 426
teleology, 32, 54
Ten Commandments, 419–420
Ten Considerations, 420–427
  Desire Is Built into Life, 421
  It Pays to Listen, 423
  It Takes All Kinds, 424–425
  Life Isn't Forever, 420–421
  Reality Guides Us, 426–427
  There Is No Natural Way to Be,
    423–424
  The Universe Is in Our Hands, 425
  We Can Always Do Better, 422–423
  We Can Do Better Than Happiness,
    425–426
  What Matters Is What Matters to
    People, 421–422
Tertullian, 415–416
theorem, defined, 132
Theory of Everything, 189
theory of mind, 346
thinking. *See* consciousness
Thompson, Judith Jarvis, 407
thought experiments
  Chinese Room, 338–341
  footbridge problem, 407–408
  Laplace's Demon, 33, 34, 35
  Mary the Color Scientist, 351–354
  philosophical zombies, 355–358
  Ship of Theseus, 16, 17, 20
  Star Trek transporter, 15–16, 20
  trolley problem, 404–405, 406
Tiktaalik roseae, 317–318
time
  beginning of, 197–199, 200
  chronesthesia, 323–324
  consciousness and behavior over, 344
  direction not intrinsic to, 55–56
  humans' ability to tell, 226
  "now" of conscious perception, 323
  quantum, 198
time's arrow, 54–62
  Big Bang model and, 43–44, 54, 58
  free will and, 380
  increasing entropy and, 43–44, 57–59,
    60–61
  lacking in equilibrium, 58

memory and, 61–62
not in laws of physics, 44, 55–56, 60
pendulum motion and, 55
"reasons why" and, 43–44
Tononi, Giulio, 364
transcendent experiences, 130–131, 136–138
transfer RNA, 268
Treatise of Human Nature, A, 394–395
trolley problem, 404–405, 406
Tulving, Endel, 323, 325
Turing, Alan, 337, 338, 341, 345, 370
Turing test
    Chinese Room and, 338–341
    memory required to pass, 345
    overview, 337–338
    replacing neurons with machines
        and, 342
    response to stimuli in, 345

understanding, 3, 338–341
universe
    in ancient vs. modern cosmology, 48–49
    author's childhood questions
        about, 195
    beginning of, 195–201
    consciousness pervading, 363–366
    expansion and future fate of, 51–53
    fine-tuning argument for God, 302–313
    human significance and, 1–2, 53
    multiple ways of talking about, 374
    questioning existence of, 195–204
    theistic and atheistic stories of, 430–431
    vast size of, 50, 53
up quarks, 174, 175
Urey, Harold, 251
utilitarianism, 412–413

vacuum energy, 303–304, 305, 311
values, 14, 21. *See also* morality
    and ethics
van Gogh, Vincent, 94, 95
van Helmont, Jan Baptist, 250
virtual particles, 180
virtue ethics, 408
vitalism, 219, 250
von Neumann, John, 264–265
von Neumann Universal Constructor,
    265, 274
Voynich manuscript, 296–297

Warren, Rick, 390
Watson, James, 265
wave function in quantum mechanics,
    35–36, 164–165, 166–168,
    169–170
weak nuclear force, 174, 177
Weinberg, Steven, 180
Werness, Brent, 231, 232
West, Geoffrey, 388
What Is Life? 239, 265
Wheatley, Thalia, 413–414
"Why?" questions, 42–46
Wilczek, Frank, 176
Wilson, Kenneth, 187, 188
Wilson, Robert, 50
Wittgenstein, Ludwig, 84
Woese, Carl, 268
wonder, 430

zero (conserved quantities), 200–201
Zhou Enlai, 159
zombie photons, 365
zombies, philosophical, 338, 355–358

TURN THE PAGE FOR AN EXCERPT

Already hailed as a masterpiece, *Something Deeply Hidden* shows for the first time that facing up to the essential puzzle of quantum mechanics utterly transforms how we think about space and time. Sean Carroll's reconciling of quantum mechanics with Einstein's theory of relativity changes, well, everything. Rarely does a book so fully reorganize how we think about our place in the universe. We are on the threshold of a new understanding—of where we are in the cosmos, and what we are made of.

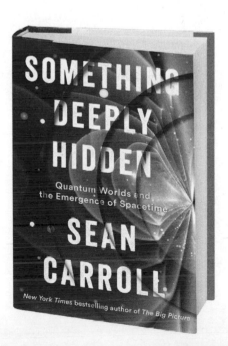

1

What's Going On:

Looking at the Quantum World

Albert Einstein, who had a way with words as well as with equations, was the one who stuck quantum mechanics with the label it has been unable to shake ever since: *spukhafte*, usually translated from German to English as "spooky." If nothing else, that's the impression we get from most public discussions of quantum mechanics. We're told that it's a part of physics that is unavoidably mystifying, weird, bizarre, unknowable, strange, baffling. Spooky.

Inscrutability can be alluring. Like a mysterious, sexy stranger, quantum mechanics tempts us into projecting all sorts of qualities and capacities onto it, whether they are there or not. A brief search for books with "quantum" in the title reveals the following list of purported applications:

Quantum Success
Quantum Leadership
Quantum Consciousness
Quantum Touch
Quantum Yoga
Quantum Eating
Quantum Psychology
Quantum Mind
Quantum Glory
Quantum Forgiveness
Quantum Theology

Quantum Happiness
Quantum Poetry
Quantum Teaching
Quantum Faith
Quantum Love

For a branch of physics that is often described as only being relevant to microscopic processes involving subatomic particles, that's a pretty impressive résumé.

To be fair, quantum mechanics—or "quantum physics," or "quantum theory," the labels are all interchangeable—is not only relevant to microscopic processes. It describes the whole world, from you and me to stars and galaxies, from the centers of black holes to the beginning of the universe. But it is only when we look at the world in extreme close-up that the apparent weirdness of quantum phenomena becomes unavoidable.

One of the themes in this book is that quantum mechanics doesn't deserve the connotations of spookiness, in the sense of some ineffable mystery that it is beyond the human mind to comprehend. Quantum mechanics is *amazing*; it is novel, profound, mind-stretching, and a very different view of reality from what we're used to. Science is like that sometimes. But if the subject seems difficult or puzzling, the scientific response is to solve the puzzle, not to pretend it's not there. There's every reason to think we can do that for quantum mechanics just like any other physical theory.

Many presentations of quantum mechanics follow a typical pattern. First, they point to some counterintuitive quantum phenomenon. Next, they express bafflement that the world can possibly be that way, and despair of it making sense. Finally (if you're lucky), they attempt some sort of explanation.

Our theme is prizing clarity over mystery, so I don't want to adopt that strategy. I want to present quantum mechanics in a way that will make it maximally understandable right from the start. It will still seem strange, but that's the nature of the beast. What it won't seem, hopefully, is inexplicable or unintelligible.

We will make no effort to follow historical order. In this chapter we'll look at the basic experimental facts that force quantum mechanics upon us, and in the next we'll quickly sketch the Many-Worlds approach to making sense of those observations. Only in the chapter after that will we offer a semi-historical account of the discoveries that led people to contemplate such a dramatically new kind of physics in the first place. Then we'll hammer home exactly how dramatic some of the implications of quantum mechanics really are.

With all that in place, over the rest of the book we can set about the fun task of seeing where all this leads, demystifying the most striking features of quantum reality.

Physics is one of the most basic sciences, indeed one of the most basic human endeavors. We look around the world, we see it is full of stuff. What is that stuff, and how does it behave?

These are questions that have been asked ever since people started asking questions. In ancient Greece, physics was thought of as the general study of change and motion, of both living and nonliving matter. Aristotle spoke a vocabulary of tendencies, purposes, and causes. How an entity moves and changes can be explained by reference to its inner nature and to external powers acting upon it. Typical objects, for example, might by nature be at rest; in order for them to move, it is necessary that something be causing that motion.

All of this changed thanks to a clever chap named Isaac Newton. In 1687 he published *Principia Mathematica*, the most important work in the history of physics. It was there that he laid the groundwork for what we now call "classical" or simply "Newtonian" mechanics. Newton blew away any dusty talk of natures and purposes, revealing what lay underneath: a crisp, rigorous mathematical formalism with which teachers continue to torment students to this very day.

Whatever memory you may have of high-school or college homework assignments dealing with pendulums and inclined planes, the basic ideas of classical mechanics are pretty simple. Consider an object such as a rock. Ignore everything about the rock that a geologist might consider interesting, such as its color and composition. Put aside the possibility that the basic structure of the rock might change, for example if you smashed it to pieces with a hammer. Reduce your mental image of the rock down to its most abstract form: the rock is an object, and that object has a *location in space*, and that location *changes with time*.

Classical mechanics tells us precisely how the position of the rock changes with time. We're very used to that by now, so it's worth reflecting on how impressive this is. Newton doesn't hand us some vague platitudes about the general tendency of rocks to move more in less in this or that fashion. He gives us exact, unbreakable rules for how everything in the universe moves in response to everything else—rules that can be used to catch baseballs or land rovers on Mars.

Here's how it works. At any one moment, the rock will have a position and also a velocity, a rate at which it's moving. According to Newton, if no forces act on the rock, it will continue to move in a straight line at constant velocity, for all time. (Already this is a major departure from Aristotle, who would have told you that objects need to be constantly pushed if they are to be kept in motion.) If a force does act on the rock, it will cause acceleration—some change in the velocity of the rock, which might make it go faster, or slower, or merely alter its direction—in direct proportion to how much force is applied.

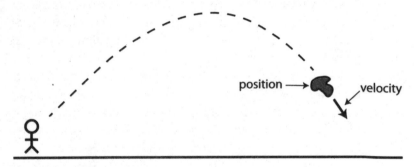

That's basically it. To figure out the entire trajectory of the rock, you need to tell me its position, its velocity, and what forces are acting on it. Newton's equations tell you the rest. Forces might include the force of gravity, or the force of your hand if you pick up the rock and throw it, or the force from the ground when the rock comes to land. The idea works just as well for billiard balls or rocket ships or planets. The project of physics, within this classical paradigm, consists essentially of figuring out what makes up the stuff of the universe (rocks and so forth) and what forces act on them.

Classical physics is, at heart, a straightforward picture of the world, but a number of crucial moves were made along the way to setting it up. Notice that we had to be very specific about what information we required to figure out what would happen to the rock: its position, its velocity, and the forces acting on it. We can think of those forces as being part of the outside world, and the important information about the rock itself as consisting of just its position and velocity. The acceleration of the rock at any moment in time, by contrast, is not something we need to specify; that's exactly what Newton's laws allow us to calculate from the position and the velocity.

Together, the position and velocity make up the *state* of any object in classical mechanics. If we have a system with multiple moving parts, the classical state of that entire system is just a list of the states of each of the individual parts. The air in a normal-sized room will have perhaps 10^{27} molecules of different types, and the state of that air would be a list of the position and velocity of every one of them. (Strictly speaking physicists like to use the momentum of each particle, rather than its velocity, but as far as Newtonian mechanics is concerned the momentum is simply the particle's mass times its velocity.) The set of all possible states that a system can have is known as the *phase space* of the system.

The French mathematician Pierre-Simon Laplace pointed out a profound implication of the classical-mechanics way of thinking. In principle, a vast intellect could know the state of literally every object in the universe, from which it could deduce everything that would happen in the future, as well as everything that had happened in the past. *Laplace's demon* is a thought experiment, not a realistic project for an ambitious computer scientist, but the implications of the thought experiment are profound. Newtonian mechanics describes a deterministic, clockwork universe.

The machinery of classical physics is so beautiful and compelling that it seems almost inescapable once you grasp it. Many great minds who came after Newton were convinced that the basic superstructure of physics had been solved, and future progress lay in figuring out exactly what realization of classical physics (which particles, which forces) was the right one to describe the universe as a whole. Even relativity, which was world-transforming in its own way, is a variety of classical mechanics rather than a replacement for it.

Then along came quantum mechanics, and everything changed.

✳

Alongside Newton's formulation of classical mechanics, the invention of quantum mechanics represents the other great revolution in the history of physics. Unlike anything that had come before, quantum theory didn't propose some particular physical model within the basic classical framework; it discarded that framework entirely, replacing it with something profoundly different.

The fundamental new element of quantum mechanics, the thing that makes it unequivocally distinct from its classical predecessor, centers on the question of what it means to *measure* something about a quantum system. What exactly a measurement is, and what happens when we measure something, and what this all tells us about what's really happening behind the scenes: together, these questions constitute what's called the *measurement problem* of quantum mechanics. There is absolutely no consensus within physics or philosophy on how to solve the measurement problem, although there are a number of promising ideas.

Attempts to address the measurement problem have led to the emergence of a field known as *the interpretation of quantum mechanics*, although the label isn't very accurate. "Interpretations" are things that we might apply to a work of literature or art, where people might have different ways of thinking about the same basic object. What's going on in quantum mechanics is something else: a competition between truly distinct scientific theories, incompatible ways of making sense of the physical world. For this reason, modern workers in this field prefer to call it "foundations of quantum mechanics." The subject of quantum foundations is part of science, not literary criticism.

Nobody ever felt the need to talk about "interpretations of classical mechanics"—classical mechanics is perfectly transparent. There is a mathematical formalism that speaks of positions and velocities and trajectories, and oh, look: there is a rock whose actual motion in the world obeys the predictions of that formalism. There is, in particular, no such thing as a measurement problem in classical mechanics. The state of the system is given by its position and its velocity, and if we want to measure those quantities, we simply do so. Of course, we can measure the system sloppily or crudely, thereby obtaining imprecise results or altering the system itself. But we don't have to; just by being careful, we can precisely measure everything there is to know about the system without altering it in any noticeable way. Classical mechanics offers a clear and unambiguous relationship between what we see and what the theory describes.

Quantum mechanics, for all of its successes, offers no such thing. The enigma at the heart of quantum reality can be summed up in a simple motto: what we *see* when we look at the world seems to be fundamentally different from what actually *is*.

Think about electrons, the elementary particles orbiting atomic nuclei, whose interactions are responsible for all of chemistry and hence almost everything interesting around you right now. As we did with the rock, we can ignore some of the electron's specific properties, like its spin and the fact that it has an electric field. (Really we could just stick with the rock as our example—rocks are quantum systems just as much as electrons are—but switching to a subatomic particle helps us remember that the features distinguishing quantum mechanics only really become evident when we consider very tiny objects indeed.)

Unlike in classical mechanics, where the state of a system is described by its position and velocity, the nature of a quantum system is something a bit less concrete. Consider an electron in its natural habitat, orbiting the nucleus of an atom. You might think, from the word "orbit" as well as from the numerous cartoon depictions of atoms you have doubtless been exposed to over the years, that the orbit of an electron is more or less like the orbit of a planet in the solar system. The electron (so you might think) has a location, and a velocity, and as time passes it zips around the central nucleus in a circle or maybe an ellipse.

Quantum mechanics suggests something different. We can *measure* values of the location or velocity (though not at the same time), and if we are sufficiently careful and talented experimenters we will obtain some answer. But what we're seeing through such a measurement is not the actual, complete, unvarnished state of the electron. Indeed, the particular measurement outcome we will obtain cannot be predicted with perfect confidence, in a profound departure from the ideas of classical mechanics. The best we can do is to predict the *probability* of seeing the electron in any particular location or with any particular velocity.

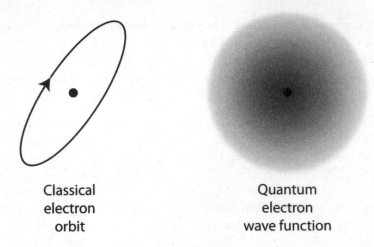

Classical
electron
orbit

Quantum
electron
wave function

The classical notion of the state of a particle, "its location and its velocity," is therefore replaced in quantum mechanics by something utterly alien to our everyday experience: a cloud of probability. For an electron in an atom, this cloud is more dense toward the center and thins out as we get farther away. Where the cloud is thickest, the probability of seeing the electron is highest; where it is diluted almost to imperceptibility, the probability of seeing the electron is vanishingly small.

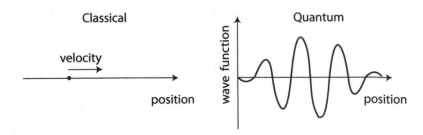

This cloud is often called a *wave function*, because it can oscillate like a wave, as the most probable measurement outcome changes over time. We usually denote a wave function by Ψ, the Greek letter Psi. For every possible measurement outcome, such as the position of the particle, the wave function assigns a specific number, called the *amplitude* associated with that outcome. The amplitude that a particle is at some position x_o, for example, would be written $\Psi(x_o)$.

The probability of getting that outcome when we perform a measurement is given by the amplitude squared.

$$\text{Probability of a particular outcome} =$$
$$|\text{Amplitude for that outcome}|^2$$

This simple relation is called the *Born rule*, after physicist Max Born.* Part of our task will be to figure out where in the world such a rule came from.

We're most definitely *not* saying that there is an electron with some position and velocity, and we just don't know what those are, so the wave function encapsulates our ignorance about those quantities. In this chapter we're not saying anything at all about what "is," only what we observe. In chapters to come, I will pound the table and insist that the wave function is the sum total of reality, and ideas such as the position or the velocity of the electron are merely things we can measure. But not everyone sees things that way, and for the moment we are choosing to don a mask of impartiality.

Let's place the rules of classical and quantum mechanics side by side to compare them. The state of a classical system is given by the position and velocity of each of its moving parts. To follow its evolution, we imagine something like the following procedure:

RULES OF CLASSICAL MECHANICS

1. Set up the system by fixing a specific position and velocity for each part.
2. Evolve the system using Newton's laws of motion.

---

* There's a slight technicality, which we'll mention here and then pretty much forget about: the amplitude for any given outcome is actually a complex number, not a real number. Real numbers are the ones that appear on the number line, any number between minus infinity to plus infinity. Anytime you take the square of a real number, you get another real number that is great than or equal to zero, so as far as real numbers are concerned there's no such thing as the square root of a negative number. Mathematicians long ago realized that square roots of negative numbers would be really useful things to have, so they defined the "imaginary unit" *i* as the square root of -1. An imaginary number is just a real number, called "the imaginary part," times *i*; whenever we square an imaginary number, we get a negative real number. Then a complex number is just a combination of a real number and an imaginary one. The little bars in the notation |Amplitude|² in the Born rule mean that we actually add the squares of the real and the imaginary parts. All that is just for the sticklers out there; henceforth we'll be happy to say "the probability is the amplitude squared" and be done with it.

That's it. The devil is in the details, of course. Some classical systems can have a lot of moving pieces.

In contrast, the rules of standard textbook quantum mechanics come in two parts. In the first part, we have a structure that exactly parallels that of the classical case. Quantum systems are described by wave functions rather than by positions and velocities. Just as Newton's laws of motion govern the evolution of the state of a system in classical mechanics, there is an equation that governs how wave functions evolve, called *Schrödinger's equation*. We can express Schrödinger's equation in words as: "The rate of change of a wave function is proportional to the energy of the quantum system." Slightly more specifically, a wave function can represent a number of different possible energies, and the Schrödinger equation says that high-energy parts of the wave function evolve rapidly, while low-energy parts evolve very slowly. Which makes sense, when we think about it.

What matters for our purposes is simply that there is such an equation, one that predicts how wave functions evolve smoothly through time. That evolution is as predictable and inevitable as the way objects move according to Newton's laws in classical mechanics. Nothing weird is happening yet.

The beginning of the quantum recipe reads something like this:

Rules of Quantum Mechanics (Part One)

1. Set up the system by fixing a specific wave function Ψ.
2. Evolve the system using Schrödinger's equation.

So far, so good—these parts of quantum mechanics exactly parallel their classical predecessors. But whereas the rules of classical mechanics stop there, the rules of quantum mechanics keep going.

All the extra rules deal with measurement. When you perform a measurement, such as the position or spin of a particle, quantum mechanics says there are only certain possible results you will ever get. You can't predict which of the results it will be, but you can calculate the probability for each allowed outcome. And after your measurement is done, the wave function *collapses* to a completely different function, with all of the new probability concentrated on whatever result you just got. So if you measure a quantum system, in general the best you can do is predict probabilities for various outcomes, but if you were to immediately measure the same quantity again, you will always get the same answer—the wave function has collapsed onto that outcome.

Let's write this out in gory detail.

Rules of Quantum Mechanics (Part Two)

3. There are certain observable quantities we can choose to measure, such as position, and when we do measure them, we obtain definite results.

4. The probability of getting any one particular result can be calculated from the wave function. The wave function associates an amplitude with every possible measurement outcome; the probability for any outcome is the square of that amplitude.

5. Upon measurement, the wave function collapses. However spread out it may have been pre-measurement, afterward it is concentrated on the result we obtained.

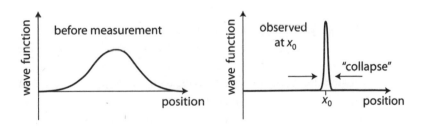

In a modern university curriculum, when physics students are first exposed to quantum mechanics, they are taught some version of these five rules. The ideology associated with this presentation—treat measurements as fundamental, wave functions collapse when they are observed, don't ask questions about what's going on behind the scenes—is sometimes called the *Copenhagen interpretation* of quantum mechanics. But people, including the physicists from Copenhagen who purportedly invented this interpretation, disagree on precisely what that label should be taken to describe. So instead we can just refer to it as "standard textbook quantum mechanics."

The idea that these rules represent how reality actually works is, needless to say, outrageous. .

What precisely do you mean by a "measurement"? How quickly does it happen? What exactly constitutes a measuring apparatus? Does it need to be human, or have some amount of consciousness, or perhaps the ability to encode information? Or maybe it just has to be macroscopic, and if so how macroscopic does it have to be? When exactly does the measurement occur, and how quickly? How in the world does the wave function collapse so dramatically? If the wave function were very spread out, does the collapse happen faster than the speed of light? And what happens to all of the possibilities that were seemingly allowed by the wave function but which we didn't observe? Were they never really there? Do they just vanish into nothingness?

To put things most pointedly: Why do quantum systems evolve smoothly and deterministically according to the Schrödinger equation *as long as we aren't looking at them*, but then dramatically collapse when we do look? How do they know, and why do they care? (Don't worry, we're going to answer all these questions.)

Science, most people think, seeks to understand the natural world. We observe things happening, and science hopes to provide an explanation for what is going on.

In its current textbook formulation, quantum mechanics has failed in this ambition. We don't know what's really going on, or at least the community of professional physicists cannot agree on what it is. What we have instead is a *recipe* that we enshrine in textbooks and teach to our students. Isaac Newton could tell you, starting with the position and velocity of a rock that you have thrown into the air in the Earth's gravitational field, just what the subsequent trajectory of that rock was going to be. Analogously, starting with a quantum system prepared in some particular way, the rules of quantum mechanics can tell you how the wave function will change over time, and what the probability of various possible measurement outcomes will be should you choose to observe it.

The fact that the quantum recipe provides us with probabilities rather that certainties might be annoying, but we could learn to live with it. What bugs us, or should, is our lack of understanding about what is actually happening.

Imagine that some devious genius figured out all the laws of physics, but rather than revealing them to the rest of the world, they programmed a computer to answer questions concerning specific physics problems, and put an interface to the program on a web page. Anyone who was interested could just surf over to that site, type in a well-posed physics question, and get the correct answer.

Such a program would obviously be of great use to scientists and engineers. But having access to the site wouldn't really qualify as understanding the laws of physics. We would have an oracle that was in the business of providing answers to specific questions, but we ourselves would be completely lacking in any intuitive idea of the underlying rules of the game. The rest of the world's scientists, presented with such an oracle, wouldn't be moved to declare victory; they would continue with their work of figuring out what the laws of nature actually were.

Quantum mechanics, in the form in which it is currently presented in physics textbooks, represents an oracle, not a true understanding. We can set up specific problems and answer them, but we can't honestly explain what's happening behind the scenes. What we do have are a number of good ideas about what that could be, and it's past time that the physics community started taking these ideas seriously.

SEAN CARROLL

"Carroll is a sure-footed guide through some of the most perplexing and fascinating insights of modern physics."

—Brian Greene,
author of *The Elegant Universe*

For a complete list of titles,
please visit prh.com/SeanCarroll